Real Prospects for Energy Efficiency in the United States

America's Energy Future Panel on Energy Efficiency Technologies

NATIONAL ACADEMY OF SCIENCES
NATIONAL ACADEMY OF ENGINEERING
NATIONAL RESEARCH COUNCIL
OF THE NATIONAL ACADEMIES

THE NATIONAL ACADEMIES PRESS
Washington, D.C.
www.nap.edu

THE NATIONAL ACADEMIES PRESS 500 Fifth Street, N.W. Washington, DC 20001

NOTICE: The project that is the subject of this report was approved by the Governing Board of the National Research Council, whose members are drawn from the councils of the National Academy of Sciences, the National Academy of Engineering, and the Institute of Medicine. The members of the panel responsible for the report were chosen for their special competences and with regard for appropriate balance.

Support for this project was provided by the Department of Energy under Grant Number DEFG02-07-ER-15923 and by BP America, Dow Chemical Company Foundation, Fred Kavli and the Kavli Foundation, GE Energy, General Motors Corporation, Intel Corporation, and the W.M. Keck Foundation. Support was also provided by the Presidents' Circle Communications Initiative of the National Academies and by the National Academy of Sciences through the following endowed funds created to perpetually support the work of the National Research Council: Thomas Lincoln Casey Fund, Arthur L. Day Fund, W.K. Kellogg Foundation Fund, George and Cynthia Mitchell Endowment for Sustainability Science, and Frank Press Fund for Dissemination and Outreach. Any opinions, findings, conclusions, or recommendations expressed in this publication are those of the author(s) and do not necessarily reflect the views of the organizations that provided support for the project.

International Standard Book Number-13: 978-0-309-13716-4
International Standard Book Number-10: 0-309-13716-0
Library of Congress Control Number: 2010925716

Available in limited supply and free of charge from:

Board on Energy and Environmental Systems
National Research Council
500 Fifth Street, N.W.
Keck W917
Washington, DC 20001
202-334-3344

Additional copies of this report are available from the National Academies Press, 500 Fifth Street, N.W., Lockbox 285, Washington, DC 20055; (800) 624-6242 or (202) 334-3313 (in the Washington metropolitan area); Internet, http://www.nap.edu.

Printed on recycled stock

Printed in the United States of America

THE NATIONAL ACADEMIES

Advisers to the Nation on Science, Engineering, and Medicine

The **National Academy of Sciences** is a private, nonprofit, self-perpetuating society of distinguished scholars engaged in scientific and engineering research, dedicated to the furtherance of science and technology and to their use for the general welfare. Upon the authority of the charter granted to it by the Congress in 1863, the Academy has a mandate that requires it to advise the federal government on scientific and technical matters. Dr. Ralph J. Cicerone is president of the National Academy of Sciences.

The **National Academy of Engineering** was established in 1964, under the charter of the National Academy of Sciences, as a parallel organization of outstanding engineers. It is autonomous in its administration and in the selection of its members, sharing with the National Academy of Sciences the responsibility for advising the federal government. The National Academy of Engineering also sponsors engineering programs aimed at meeting national needs, encourages education and research, and recognizes the superior achievements of engineers. Dr. Charles M. Vest is president of the National Academy of Engineering.

The **Institute of Medicine** was established in 1970 by the National Academy of Sciences to secure the services of eminent members of appropriate professions in the examination of policy matters pertaining to the health of the public. The Institute acts under the responsibility given to the National Academy of Sciences by its congressional charter to be an adviser to the federal government and, upon its own initiative, to identify issues of medical care, research, and education. Dr. Harvey V. Fineberg is president of the Institute of Medicine.

The **National Research Council** was organized by the National Academy of Sciences in 1916 to associate the broad community of science and technology with the Academy's purposes of furthering knowledge and advising the federal government. Functioning in accordance with general policies determined by the Academy, the Council has become the principal operating agency of both the National Academy of Sciences and the National Academy of Engineering in providing services to the government, the public, and the scientific and engineering communities. The Council is administered jointly by both Academies and the Institute of Medicine. Dr. Ralph J. Cicerone and Dr. Charles M. Vest are chair and vice chair, respectively, of the National Research Council.

www.national-academies.org

PANEL ON ENERGY EFFICIENCY TECHNOLOGIES

LESTER B. LAVE, Carnegie Mellon University, *Chair*
MAXINE L. SAVITZ, Honeywell, Inc. (retired), *Vice Chair*
R. STEPHEN BERRY, University of Chicago
MARILYN A. BROWN, Georgia Institute of Technology
LINDA R. COHEN, University of California, Irvine
MAGNUS G. CRAFORD, Philips LumiLeds Lighting
PAUL A. DeCOTIS, Long Island Power Authority
JAMES H. DeGRAFFENREIDT, JR., WGL Holdings, Inc.
HOWARD GELLER, Southwest Energy Efficiency Project
DAVID B. GOLDSTEIN, Natural Resources Defense Council
ALEXANDER MacLACHLAN, E.I. du Pont de Nemours & Company (retired)
WILLIAM F. POWERS, Ford Motor Company (retired)
ARTHUR H. ROSENFELD, California Energy Commission
DANIEL SPERLING, University of California, Davis

Liaison from the Committee on America's Energy Future
JOHN B. HEYWOOD, Massachusetts Institute of Technology

America's Energy Future Project Director
PETER D. BLAIR, Executive Director, Division on Engineering and Physical Sciences

America's Energy Future Project Manager
JAMES ZUCCHETTO, Director, Board on Energy and Environmental Systems

Staff
MADELINE G. WOODRUFF, Senior Program Officer, Study Director
GREG EYRING, Senior Program Officer, Study Director
THOMAS R. MENZIES, Senior Program Officer
E. JONATHAN YANGER, Senior Program Assistant
KATHERINE BITTNER, Senior Program Assistant (until July 2008)

Foreword

Energy, which has always played a critical role in our country's national security, economic prosperity, and environmental quality, has over the last two years been pushed to the forefront of national attention as a result of several factors:

- World demand for energy has increased steadily, especially in developing nations. China, for example, saw an extended period (prior to the current worldwide economic recession) of double-digit annual increases in economic growth and energy consumption.
- About 56 percent of the U.S. demand for oil is now met by depending on imports supplied by foreign sources, up from 40 percent in 1990.
- The long-term reliability of traditional sources of energy, especially oil, remains uncertain in the face of political instability and limitations on resources.
- Concerns are mounting about global climate change—a result, in large measure, of the fossil-fuel combustion that currently provides most of the world's energy.
- The volatility of energy prices has been unprecedented, climbing in mid-2008 to record levels and then dropping precipitously—in only a matter of months—in late 2008.
- Today, investments in the energy infrastructure and its needed technologies are modest, many alternative energy sources are receiving insufficient attention, and the nation's energy supply and distribution systems are increasingly vulnerable to natural disasters and acts of terrorism.

All of these factors are affected to a great degree by the policies of government, both here and abroad, but even with the most enlightened policies the overall energy enterprise, like a massive ship, will be slow to change course. Its complex mix of scientific, technical, economic, social, and political elements means that the necessary transformational change in how we generate, supply, distribute, and use energy will be an immense undertaking, requiring decades to complete.

To stimulate and inform a constructive national dialogue about our energy future, the National Academy of Sciences and the National Academy of Engineering initiated a major study in 2007, "America's Energy Future: Technology Opportunities, Risks, and Tradeoffs." The America's Energy Future (AEF) project was initiated in anticipation of major legislative interest in energy policy in the U.S. Congress and, as the effort proceeded, it was endorsed by Senate Energy and Natural Resources Committee Chair Jeff Bingaman and former Ranking Member Pete Domenici.

The AEF project evaluates current contributions and the likely future impacts, including estimated costs, of existing and new energy technologies. It was planned to serve as a foundation for subsequent policy studies, at the Academies and elsewhere, that will focus on energy research and development priorities, strategic energy technology development, and policy analysis.

The AEF project has produced a series of five reports, including this one on energy efficiency technologies, designed to inform key decisions as the nation begins a comprehensive examination of energy policy issues this year. Numerous studies conducted by diverse organizations have benefited the project, but many of those studies disagree about the potential of specific technologies, particularly those involving alternative sources of energy such as biomass, renewable resources for generation of electric power, advanced processes for generation from coal, and nuclear power. A key objective of the AEF series of reports is thus to help resolve conflicting analyses and to facilitate the charting of a new direction in the nation's energy enterprise.

The AEF project, outlined in Appendix A, included a study committee and three panels that together have produced an extensive analysis of energy technology options for consideration in an ongoing national dialogue. A milestone in the project was the March 2008 "National Academies Summit on America's Energy Future" at which principals of related recent studies provided input to the AEF study committee and helped to inform the panels' deliberations. A report chronicling the event, *The National Academies Summit on America's Energy Future:*

Summary of a Meeting (Washington, D.C.: The National Academies Press), was published in October 2008.

The AEF project was generously supported by the W.M. Keck Foundation, Fred Kavli and the Kavli Foundation, Intel Corporation, Dow Chemical Company Foundation, General Motors Corporation, GE Energy, BP America, the U.S. Department of Energy, and our own Academies.

Ralph J. Cicerone, President
National Academy of Sciences
Chair, National Research Council

Charles M. Vest, President
National Academy of Engineering
Vice Chair, National Research Council

Preface

As part of the National Academies' America's Energy Future (AEF) project (see Appendix A), the Panel on Energy Efficiency Technologies (Appendix B) was appointed to assess the potential of technologies to save money as well as energy within the buildings, transportation, and industrial sectors during three time periods: 2009–2020, 2020–2035, and beyond 2035. Box P.1 contains the charge to the panel.

The focus of the panel's assessment was the potential of technology for improving energy efficiency, which the panel defined as accomplishing a given objective with less energy (see Appendix D for an extended technical definition). Conservation is generally understood to mean saving energy by changing behavior, such as by driving a smaller car or setting back the thermostat in winter. Given its task, the panel did not examine how much energy savings could be achieved by conservation. Instead, the panel identified energy savings that could be achieved through energy efficiency.

In fact, energy efficiency technologies have been available for decades, but unfortunately, few have been implemented. The panel identified myriad barriers to getting these technologies adopted. It noted that if society were to give a higher priority to efficiency, perhaps because of higher energy prices, energy shortages, or concern about greenhouse gas emissions, deployment would be faster and the savings would be greater.

As the panel discovered, energy efficiency occupies a unique place in the energy debate. Energy efficiency requires none of the environmental disruption seen in extracting coal, petroleum, natural gas, or uranium; depends on no wind turbines or hydroelectric dams or thermal power plants; emits no greenhouse

BOX P.1 Statement of Task for the AEF Panel on Energy Efficiency Technologies

This panel will examine the potential for reducing energy demand through improving efficiency in transportation, buildings, and industrial processes using (1) existing technologies, (2) technologies developed but not yet used widely, and (3) prospective technologies. In keeping with the charge to the overall scope of the America's Energy Future Study Committee, the panel will not recommend policy choices, but will assess the state of development of technologies. The energy efficiency panel will evaluate technologies based on their estimated times to readiness for deployment and will provide the following information for each:

- Initial deployment times of less than 10 years: costs, performance, impacts;
- Deployment times of 10 to 25 years: barriers, implications for costs, R&D challenges/needs;
- Deployment times greater than 25 years: barriers, R&D challenges/needs (especially basic research needs).

The primary focus of the study will be on the quantitative characterization of technologies likely to be available for deployment within the next 10 years. The panel will provide details on the technical potential of improving efficient use of energy in the United States using existing technologies as well as consider the applicability of existing technologies in other nations. It will also assess the potential for improving energy efficiency by using technologies developed but not yet used widely in the United States or abroad, and by using prospective technologies with substantial likelihood of commercial use during the three deployment timescales described above.

gases or other pollutants; and can mitigate energy security risks associated with imported oil. The obvious benefits of energy efficiency technologies in making America's energy supply more secure and environmentally sustainable, and the U.S. economy more competitive by reducing the prices of goods and services, deserve additional public attention.

The panel's chair and vice chair thank the panel members and John Heywood for their hard work and insights—and apologize again to their family members for taking them away from other activities. The panel appreciates inputs provided in presentations by experts at its meetings (see Appendix C) and in writing (Anup Bandivadekar, International Council on Clean Transportation; Peter Biermayer, Sam Borgeson, Rich Brown, Jon Koomey, and Alan Meier, Lawrence

Berkeley National Laboratory; Lynette Cheah, Massachusetts Institute of Technology; Steve Dunn, Southwest Energy Efficiency Project; Mark Frankel, New Buildings Institute; Mauricio Justiniano and Nancy Margolis, Energetics, Inc.; Mike Messenger, Itron, Inc.; and Christopher Weber, Carnegie Mellon University).

Madeline Woodruff, the study director, was indefatigable and cheerful throughout the writing of the report and responding to reviewer comments. Greg Eyring helped pull the report together at the end. Tom Menzies supplied valuable material, comments, and data. Jonathan Yanger provided staff assistance throughout the project. Peter Blair, Jim Zucchetto, and Kevin Crowley guided the panel through the Academies' processes and coordinated its work with that of the other panels and the AEF Committee.

Lester B. Lave, *Chair*
Maxine L. Savitz, *Vice Chair*
Panel on Energy Efficiency Technologies

Acknowledgment of Reviewers

This report has been reviewed in draft form by individuals chosen for their diverse perspectives and technical expertise, in accordance with procedures approved by the Report Review Committee of the National Research Council (NRC). The purpose of this independent review is to provide candid and critical comments that will assist the institution in making its published report as sound as possible and to ensure that the report meets institutional standards for objectivity, evidence, and responsiveness to the study charge. The review comments and draft manuscript remain confidential to protect the integrity of the deliberative process. We wish to thank the following individuals for their review of this report:

Philip Anderson, Princeton University,
William Brinkman, Princeton University,
Andrew Brown, Jr., Delphi Corporation,
Clark Bullard, University of Illinois at Urbana-Champaign,
Jonathan Creyts, McKinsey & Company, Inc.,
J. Michael Davis, Pacific Northwest National Laboratory,
John DeCicco, Environmental Defense Fund,
Theodore Geballe, Stanford University (professor emeritus),
Susan Hanson, Clark University,
Trevor Jones, ElectroSonics Medical, Inc.,
Mark Levine, Lawrence Berkeley National Laboratory,
Richard Morgenstern, Resources for the Future,
Peter H. Pfromm, Kansas State University,

Bernard I. Roberston, DaimlerChrysler (retired),
Marc Ross, University of Michigan (professor emeritus),
Jeffrey Siirola, Eastman Chemical Company,
Anne E. Smith, CRA International,
Robert Socolow, Princeton University,
Dale F. Stein, Michigan Technological University,
James E. Woods, Sain Engineering Associates, Inc., and
Ernst Worrell, Lawrence Berkeley National Laboratory and Ecofys.

Although the reviewers listed above have provided many constructive comments and suggestions, they were not asked to endorse the conclusions or recommendations, nor did they see the final draft of the report before its release. The review of this report was overseen by Elisabeth M. Drake, Massachusetts Institute of Technology, and Robert A. Frosch, Harvard University. Appointed by the NRC, they were responsible for making certain that an independent examination of this report was carried out in accordance with institutional procedures and that all review comments were carefully considered. Responsibility for the final content of this report rests entirely with the authoring panel and the institution.

Contents

Appendixes

Summary

Energy from fossil fuels and nuclear and renewable sources provides power for the myriad activities that take place in residences and commercial buildings, the transportation of people and goods, and both light and heavy industrial manufacturing. In 2008, U.S. primary energy use totaled 99.4 quadrillion Btu (Figure S.1), making the United States the world's largest consumer of energy. Yet although energy is essential to the U.S. economy, technologies exist today that can help make it possible to achieve significant energy savings and still maintain current lifestyles. The 1973–1974 oil embargo and each subsequent energy crisis prompted studies showing that the United States could save energy and money by investing in energy efficiency. But although U.S. energy use per dollar of GDP has declined over the past 30 years (EIA, 2008b), many of the energy efficiency technologies identified in those studies have not been implemented.

Today, efficiency in energy use has taken on special urgency. Price fluctuations, national security concerns over U.S. dependence on imported oil, and growing recognition of the need to reduce emissions of greenhouse gases have transformed energy efficiency from an option to a necessity.

The Panel on Energy Efficiency Technologies, convened as part of the National Academies' "America's Energy Future" project (see Appendix A), was asked to examine the potential for technologies available today or soon to enable Americans to use energy more efficiently; the costs of accomplishing this; and the hurdles and barriers that impede adoption of the technologies. Because saving energy and mitigating the environmental impacts of energy production and use—especially, emission of greenhouse gases—is a long-term challenge and technology

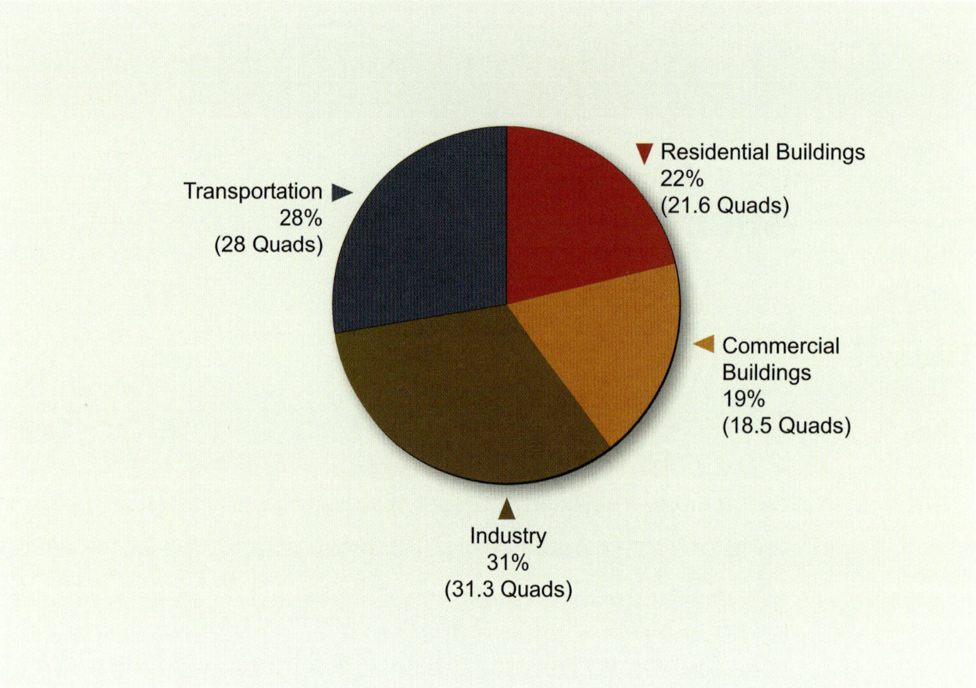

FIGURE S.1 *Total U.S. energy use by sector, 2008 (in quadrillion Btu, or quads). For each sector, "total energy use" is direct (primary) fuel use plus purchased electricity plus apportioned electricity-system losses. Economy-wide, total U.S. primary energy use in 2008 was 99.4 quads. Totals may not equal sum of components due to independent rounding.*
Source: EIA 2009a, as updated by EIA, 2009b.

continues to develop, the panel was also asked to look beyond 2035 in assessing the technological potential for increasing U.S. energy efficiency.

Although the terms "energy efficiency" and "energy conservation" are often used interchangeably, they refer to different concepts. Improving energy efficiency involves accomplishing an objective—such as heating a room to a certain temperature—while using less energy. Energy conservation can involve changing one's behavior so as to use less energy—e.g., driving a smaller car, or lowering the thermostat in winter. The panel's work focused on technology and energy efficiency, rather than energy conservation.

As a result of its broad look at energy use in other nations (Chapter 1); a detailed examination of the buildings (Chapter 2), transportation (Chapter 3), and industrial (Chapter 4) sectors and of numerous studies of energy use and potential savings in each; and its review of experience with and lessons learned from U.S. federal and state policies and programs (Chapter 5), the panel concluded that the potential for U.S. energy savings is large. Synthesizing the discussions and details presented across the chapters, the panel developed four overarching findings, which are presented below. The sector-specific findings given in Chapters 2 through 4 are also listed.

THE POTENTIAL FOR ENERGY SAVINGS

Table S.1 gives the panel's conservative and optimistic estimates of potential cost-effective annual energy savings in U.S. buildings, transportation, and industry for 2020 and 2030. These estimates represent technology assessments, not projections—that is, the estimates are assessments of the potential energy savings achievable with the use of energy efficiency technologies, assuming a rapid rate of deployment, but one nevertheless consistent with past deployment rates.

As indicated in Table S.1, the panel found that energy efficiency in buildings offers the greatest possibility for U.S. energy savings; by 2020, in the optimistic scenario, buildings would account for 53 percent of the total estimated savings,

TABLE S.1 Panel Estimate of the Potential for Cost-Effective Annual U.S. Energy Savings (in quads) Achievable with Energy Efficiency Technologies in 2020 and 2030

	Conservative Estimate		Optimistic Estimate	
	2020	2030	2020	2030
Buildings, primary (source) electricity	9.4	14.4	9.4	14.4
Residential	4.4	6.4	4.4	6.4
Commercial	5.0	8.0	5.0	8.0
Buildings, natural gas	2.4	3.0	2.4	3.0
Residential	1.5	1.5	1.5	1.5
Commercial	0.9	1.5	0.9	1.5
Transportation, light-duty vehicles	2.0	8.2	2.6	10.7
Industry, manufacturing	4.9	4.9	7.7	7.7
Total	18.6	30.5	22.1	35.8

Note: Savings are relative to the reference scenario of the EIA's *Annual Energy Outlook 2008* (EIA, 2008a) or, for transportation, a similar scenario developed by the panel. See Table 1.2 for more information on the baselines used in the panel's analysis of the buildings, transportation, and industry sectors.

industry for 35 percent, and transportation for 12 percent.[1] If all the potential energy savings the panel identified for residential and commercial buildings could be achieved, the effects on U.S. electricity generation needs could be dramatic.

Instead of increasing from 99 quadrillion Btu (99 quads) in 2008 (EIA, 2009a,b), to 111 quads in 2020, and then to 118 quads in 2030 (EIA, 2008a), U.S. energy use could, with full deployment of cost-effective, energy-efficient technologies, fall to 89–92 quads in 2020 and 82–88 quads in 2030.

The importance of the values in Table S.1, however, is not the specific numbers; rather, the point is that taking advantage of technologies that save money as well as energy to produce the same mix of goods and services could reduce U.S. energy use to 30 percent below the 2030 forecast level, and even significantly below 2008 energy use. The result would be lower costs and a more competitive economy that uses less fossil fuel, has lower emissions of greenhouse gases, and puts less pressure on environmental quality.

OVERARCHING FINDINGS

Overarching Finding 1

Energy-efficient technologies for residences and commercial buildings, transportation, and industry exist today, or are expected to be developed in the normal course of business, that could potentially save 30 percent of the energy used in the U.S. economy while also saving money. If energy prices are high enough to motivate investment in energy efficiency, or if public policies are put in place that have the same effect, U.S. energy use could be lower than business-as-usual projections by 19–22 quadrillion Btu (17–20 percent) in 2020 and by 30–36 quadrillion Btu (25–31 percent) in 2030.[2,3]

---

[1]The transportation fraction would be higher if heavy-duty vehicles and aviation had been included in the panel's analysis.

[2]The basis for comparison for the buildings and industry sectors is the reference scenario of the U.S. Department of Energy's Annual Energy Outlook 2008 (EIA, 2008a) and the panel's similar but slightly modified baseline for the transportation sector.

[3]The AEF Committee's report (NAS-NAE-NRC, 2009) estimated the amount of possible savings as 15–17 quads (about 15 percent) by 2020 and 32–35 quads (about 30 percent) by 2030. Since the release of that report, further analysis by the panel refined the amount of possible savings in 2020 to 17–20 percent.

Saving this amount of energy by using energy-efficient technologies would reverse the growth in energy use forecasted by the U.S. Department of Energy's (DOE's) Energy Information Administration (EIA, 2008a) and thus have a positive impact on needed U.S. electricity generation capacity.

As the report *America's Energy Future: Technology and Transformation* (NAS-NAE-NRC, 2009) points out, energy efficiency costs less than building new energy production facilities, which typically take years longer to start up than do energy efficiency measures and also have substantial environmental impacts (e.g., increased CO_2 emissions).

Overarching Finding 2

The full deployment of cost-effective, energy-efficient technologies in buildings alone could eliminate the need to add to U.S. electricity generation capacity. Since the estimated electricity savings in buildings from Table S.1 exceeds the EIA forecast for new net electricity generation in 2030, implementing these efficiency measures would mean that no new generation would be required except to address regional supply imbalances, replace obsolete generation assets, or substitute more environmentally benign generation sources.

As indicated by the differences between the conservative and optimistic estimates presented in Table S.1, there are considerable uncertainties in projections of both the timing and the quantity of potential energy savings. Formidable barriers impede the deployment of energy-efficient technologies, even if their adoption is projected to save money over time. These barriers include potentially high up-front costs; alternative uses for investment capital that are deemed more attractive; volatility of energy prices, leading to uncertainty in the payback time; and the lack of information available to consumers about the relative performance and costs of technology alternatives.

Although the panel was not able to review all the barriers to implementing energy-efficient technologies, it did review some of the experience gained at the national level with policies and programs aimed at overcoming the barriers. Many policy initiatives have been effective, including efficiency standards (for vehicles and appliances) combined with DOE research and development; promotion of combined heat and power (largely through the Public Utilities Regulatory and Policy Act of 1978); the ENERGY STAR® product-labeling program; building energy codes; and utility- and state-sponsored end-use efficiency programs (see

Chapter 5). Large states such as California and New York have also succeeded in overcoming many of the barriers to use of energy-efficient technologies, achieving high levels of energy savings (Chapter 5).

Overarching Finding 3

The barriers to improving energy efficiency are formidable. Overcoming these barriers will require significant public and private support, as well as sustained initiative. The experience of leading states provides valuable lessons for national, state, and local policy makers in the leadership skills required and the policies that are most effective.

The long lifetimes of buildings and some capital equipment present a particularly important barrier to implementing energy-efficient technologies. These investments—particularly buildings—can last for decades or even centuries, blocking the implementation of more efficient substitutes. Hence, actions now and over the next decade to use (or not use) energy-efficient technologies and design practices that are available today have long-term consequences for energy use.

Overarching Finding 4

Long-lived capital stock and infrastructure can lock in patterns of energy use for decades. Thus, it is important to take advantage of opportunities (during the design and construction of new buildings or major subsystems, for example) to insert energy-efficient technologies into these long-lived capital goods.

SECTOR ANALYSIS AND FINDINGS

Buildings

As shown in Figures S.1 and S.2, the myriad activities associated with residential and commercial buildings consumed about 40 quads, or 41 percent of the primary energy used in 2008 in the United States, including three-quarters of the electricity and half of the natural gas. Space heating, cooling, and ventilation are the largest consumers of energy in buildings, followed by lighting.

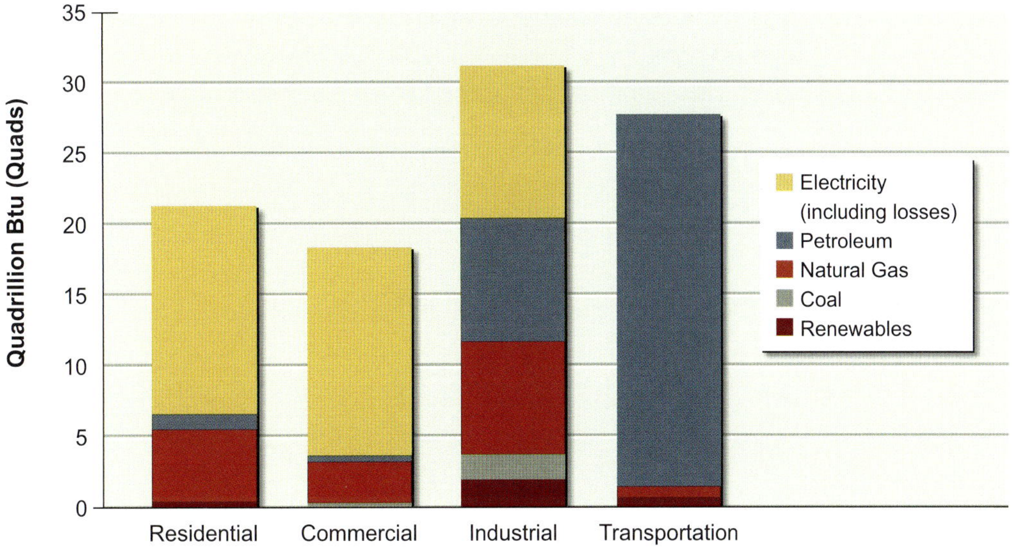

FIGURE S.2 *Total energy consumption in the United States in 2008, by sector and fuel. Shown are electricity consumption—with the losses in generation, transmission and distribution allocated to the end-use sectors—and the fuels used on-site in each sector. Electricity is generated off-site using fossil, renewable, and nuclear energy sources. Source: EIA 2009a, as updated by EIA, 2009c.*

Many studies cited in Chapter 2 that have evaluated the quantity of realistically achievable savings as a function of the cost of saved energy show consistent results, despite differences in assumptions and approaches. As determined by the panel from its review of such studies, median predictions of achievable and cost-effective energy savings are 1.2 percent per year for electricity and 0.5 percent per year for natural gas, amounting to a 25–30 percent energy savings for the U.S. buildings sector as a whole over 20 years. If this level of savings were to be achieved, it would offset the EIA (2008a) projected increase in energy use in the buildings sector.

Conservation supply curves are a tool for displaying the results of detailed assessments of the energy savings that could be achieved with specific technologies as a function of cost. The curves developed for buildings in this report (see Chapter 2) indicate that the projected baseline energy use in 2030 (EIA, 2008a) can be reduced by about 30 percent at a cost less than current average retail energy prices.

TABLE S.2 Estimated Average Cost of Conserved (Saved) Energy in Residential and Commercial Buildings Compared with National Average Retail Energy Prices, 2007

	Average Cost of Conserved (Saved) Energy	National Average Retail Energy Price
Residential		
Electricity	2.7¢/kWh	10.6¢/kWh
Natural gas	$6.9/million Btu	$12.7/million Btu
Commercial		
Electricity	2.7¢/kWh	9.7¢/kWh
Natural gas	$2.5/million Btu	$11.0/million Btu

Note: For the specific savings, see Figures 2.8 and 2.9.
Source: Brown et al., 2008. Energy prices are from EIA, 2008c.

As shown in Table S.2, the estimated average costs of the energy saved (usually termed the "cost of conserved energy," or CCE) in residential and commercial buildings for electricity and natural gas use as a result of energy efficiency measures were dramatically lower than the corresponding average retail prices for electricity and natural gas in 2007, indicating that large savings in energy costs were available.

Not reflected in Table S.2 are the results of integrated approaches designed to yield system-wide or building-wide savings. These approaches can involve integrating the design of the heating, ventilation, and air-conditioning system with that of the envelope system and the lighting system and its controls, all at the beginning of the design process. A small but growing number of new commercial buildings incorporate these design approaches to reach a 50 percent savings in energy use—mainly in heating, cooling, air-conditioning, water heating, and lighting—compared with prevailing building codes. With appropriate policies and programs in place, such energy-efficient buildings could become the norm in new construction.

Beyond the savings that could be realized through wider use of existing, energy-efficient technologies, advanced technologies, including light-emitting diode (LED) lamps, innovative window systems, new types of cooling systems, and power-saving electronic devices, are under development and are likely to become commercially available within the next decade.

Findings for Buildings

B.1 Studies assessing the potential for energy savings in buildings take several different approaches, looking at whole-building results as well as results by end-use and technology. Nevertheless, their results tend to be consistent.

B.2 The potential for large, cost-effective energy savings in buildings is well documented. Median predictions of achievable, cost-effective savings are 1.2 percent per year for electricity and 0.5 percent per year for natural gas, amounting to a 25–30 percent energy savings for the buildings sector as a whole over the next 20–25 years. If this level of savings were to be achieved, it would offset the EIA (2008a) projected increase in energy use in this sector over the same period.

B.3 Studies of energy efficiency potential are subject to a number of limitations and biases. On the one hand, factors such as not accounting for new and emerging energy efficiency technologies can lead such studies to underestimate energy-savings potential, particularly in the midterm and long term. On the other hand, some previous studies were overly optimistic about the cost and performance of certain efficiency measures, thereby overestimating energy-savings potential, particularly in the short term. Although these limitations must be acknowledged, they do not affect the panel's overall finding that the potential for energy savings in buildings is large.

B.4 Many advanced technologies under development and likely to become commercially available within the next decade—including LED lamps, innovative window systems, new types of cooling systems, and power-saving electronic devices—will further increase the energy-savings potential in buildings. In addition, new homes and commercial buildings with relatively low overall energy use have been demonstrated throughout the country. With appropriate policies and programs, they could become the norm in new construction.

B.5 Despite substantial barriers to widespread energy efficiency improvements in buildings, a number of countervailing factors could drive increased energy efficiency, including rising energy prices, growing concern about global climate change and the resulting willingness of consumers and businesses to take action to reduce emissions, a movement toward "green buildings," and growing recognition of the significant nonenergy benefits offered by energy efficiency measures.

Transportation

The 28 percent of the nation's primary energy used for transportation comes almost entirely (97 percent) from petroleum (see Figure S.2). Transportation consumes 14 million of the 20 million barrels per day of petroleum used in the United States. Since 12 million barrels per day of petroleum are imported, the energy used for transportation is a major factor in national security. Moreover, transportation accounts for 30 percent of all U.S. carbon emissions arising from energy use, as well as for significant fractions of other air pollutants.

The potential for displacing petroleum in U.S. transportation resides both in increasing the efficiency with which liquid fuels (especially petroleum) are used and in shifting some of the vehicle fleet to alternative fuels such as electricity (including that generated using hydrogen fuel cells) and biofuels.

Most of the energy used in transportation—some 75 percent—is consumed in moving passengers and goods on highways, leading to a focus on highway vehicles. An extensive menu of technologies is available today (see Chapter 3)—and additional technologies will likely be available in the future—that could reduce highway fuel consumption by cars and light trucks (light-duty vehicles; LDVs).[4] Improvements in internal-combustion engines (ICEs) and transmissions, reductions in rolling resistance and aerodynamic drag, and reductions in vehicle weight and size are all achievable with technologies that are available now but are used only at a low level, or with technologies that will be available soon.

If the energy savings from improvements in passenger vehicles are to be large, Americans' penchant for increasing vehicle size and performance will have to give way to the goal of reducing fuel consumption—that is, improvements in fuel efficiency must have priority over increases in vehicle size and performance.

The panel found that evolutionary improvements in gasoline vehicles using ICEs are likely to prove the most cost-effective technology for improving fuel efficiency and reducing petroleum consumption, at least through 2020. Because changing the manufacturing, servicing, and fuel infrastructure to serve electric or fuel cell vehicles would be expensive and time-consuming, the new technology would have to offer major advantages. For the medium term, plug-in hybrid-electric vehicles (PHEVs) and the associated electricity fueling infrastructure could be deployed more rapidly and more cheaply than hydrogen fuel-cell vehicles and the associated hydrogen fuel production and distribution infrastructure. Thus, if

---

[4]Light-duty vehicles include passenger cars and trucks less than 8500 lb.

high-energy-storage battery technology progresses sufficiently, PHEVs would be a promising mid- to long-term option. In contrast, it would take decades—perhaps until 2050—for hydrogen fuel-cell vehicles (HFCVs) to have a major impact on U.S. oil use.

Table S.3 shows plausible reductions in fuel consumption and CO_2 emissions stemming from evolutionary improvements in LDVs as well as the use of new vehicle types, assuming that most of the gain does not go to increases in vehicle size and performance. As shown, evolutionary improvements could reduce the fuel

TABLE S.3 Potential Relative Vehicle Petroleum Use and Greenhouse Gas Emissions from Vehicle Efficiency Improvements Through 2035

	Petroleum Consumption (gasoline equivalent)		Greenhouse Gas Emissions[a]	
Propulsion System	Relative to Current Gasoline ICE	Relative to 2035 Gasoline ICE	Relative to Current Gasoline ICE	Relative to 2035 Gasoline ICE
Current gasoline	1.00		1.00	
Current turbocharged gasoline	0.90		0.90	
Current diesel	0.80		0.80	
Current hybrid	0.75		0.75	
2035 gasoline	0.65	1.00	0.65	1
2035 turbocharged gasoline	0.60	0.90	0.60	0.90
2035 diesel	0.55	0.85	0.55	0.85
2035 HEV	0.40	0.60	0.40	0.60
2035 PHEV	0.20	0.30	0.35–0.45	0.55–0.70
2035 BEV	None		0.35–0.50	0.55–0.80
2035 HFCV	None		0.30–0.40	0.45–0.60

Note: These estimates assume that vehicle performance (maximum acceleration and power-to-weight ratio) and size remain the same as today's average new-vehicle values. That is, the improvements in propulsion efficiency are used solely to decrease fuel consumption rather than to offset increases in vehicle performance and size. Estimates have been rounded to the nearest 0.05. BEVs and HFCVs are expected to have shorter driving ranges than PHEVs between rechargings or refuelings.
BEV, battery-electric vehicle; HEV, hybrid-electric vehicle; HFCV, hydrogen fuel-cell vehicle; ICE, internal combustion engine; PHEV, plug-in hybrid vehicle.

[a]Greenhouse gas emissions from the electricity used in 2035 PHEVs, 2035 BEVs, and 2035 HFCVs are estimated from the projected U.S. average electricity grid mix in 2035. Greenhouse gas emissions from hydrogen production are estimated for hydrogen produced from natural gas.

Source: Bandivadekar et al., 2008. Estimates based on assessments by An and Santini, 2004; Wohlecker et al., 2007; Cheah et al., 2007; NPC, 2007; and NRC, 2004.

consumption of gasoline ICE vehicles by up to 35 percent over the next 25 years. Hybrid-electric vehicles—both HEVs and PHEVs—could deliver deeper reductions in fuel consumption, although they would still depend on gasoline or other liquid fuels. Vehicles powered by batteries and hydrogen fuel cells need not depend on hydrocarbon fuels; if they ran on electricity or hydrogen, they could have zero tailpipe emissions of CO_2 and other pollutants. If the electricity or hydrogen were generated without CO_2 emissions, they would have the potential to reduce total life-cycle CO_2 emissions dramatically.

To have a significant effect, advanced-technology vehicles must garner a sizable share of the market. Generally, a decade or more is required to develop a technology to the stage that it can be deployed, to introduce it on a commercial vehicle, and then to achieve significant sales. There are also technical constraints on the speed with which the market shares of advanced technologies can grow, such as the need for breakthroughs in battery performance and for a hydrogen-distribution infrastructure.

The panel examined the available literature to assess how the performance and costs of LDV technologies might change over time. It then developed both conservative and optimistic scenarios for technology penetration and examined their impacts on fuel consumption in the U.S. LDV fleet. Annual fuel savings in the conservative scenario could reach 16 billion gallons in 2020 and 66 billion gallons in 2035; in the optimistic scenario, the savings could be 21 billion gallons and 86 billion gallons, respectively.

The panel also examined other forms of highway transportation, as well as aircraft, railroad, and marine transport. Because ships and railroads are highly efficient, substantial efficiency improvements are unlikely. Jet aircraft efficiency improved 70 percent from 1960 to 2000 with promises of continuing improvement. For example, the new designs for the Boeing 747 and 787 are 20–25 percent more efficient. In addition, minimizing fuel costs has been a high priority for trucking companies, and reductions of 10–20 percent in the fuel economy of heavy- and medium-duty vehicles appear feasible over the next decade or so as a result of improved technology. Further opportunities to save fuel are presented by shifting some long-haul freight from truck to rails. A broad examination of the potential for improved freight system effectiveness is needed.

Findings for Transportation

T.1 In the transportation sector, the potential for energy savings and petroleum displacement resides both in increasing the efficiency with which liquid fuels (especially petroleum) are used and in shifting some of the vehicle fleet's energy demand to electricity (including hydrogen fuel-cell vehicles). The overall energy use and greenhouse gas emissions (and other environmental effects) associated with such a shift depend on how the electricity or hydrogen is generated.

T.2 An extensive menu of technologies exists today for increasing energy efficiency in transportation. Achieving the average new-vehicle fuel economy targets for 2020 set by the Energy Independence and Security Act of 2007 (EISA; P.L. 110-140), which represent a 40 percent increase over today's value (and a 30 percent reduction in average fuel consumption), is thus a feasible, although challenging, objective. Reaching the EISA targets, and continuing to decrease fuel consumption, will require a shift from the historic U.S. emphasis on ever-increasing vehicle power and size to an emphasis on using efficiency improvements to improve vehicle fuel consumption.

T.3 In the near term, fuel-consumption reductions will come predominantly from improved gasoline and diesel engines, improved transmissions, and reduced vehicle weight and drag. Through at least 2020, evolutionary improvements in vehicles with gasoline internal-combustion engines are likely to prove the most cost-effective approach to reducing petroleum consumption. Gasoline-electric hybrids will likely play an increasingly important role as their production volume increases and their cost, relative to that of conventional vehicles, decreases. Meeting the EISA standards is likely to require that, over the next decade or two, an ever-larger fraction of the new vehicle fleet be hybrids or plug-in hybrids.

T.4 Beyond 2020, continuing reductions in fuel consumption are possible. Plausible efficiency improvements in light-duty vehicles, alongside weight reduction and more extensive use of hybrid and plug-in hybrid (and possibly battery-electric) vehicles, could reduce transportation fuel consumption to below the levels implied by the higher 2020 fuel-economy standards mandated by the EISA. An especially important R&D focus is developing marketable vehicles that use electricity, which will require improving the performance and reducing the cost of high-energy-storage batteries.

T.5 A parallel longer-term prospect is fuel cells with hydrogen as the energy carrier. To be attractive, major improvements, especially in reducing costs, are needed. Widespread implementation requires significant investment in efficient, low-greenhouse-gas-emissions hydrogen supply and distribution systems. Onboard hydrogen storage is a key R&D issue. Establishing a new propulsion system technology and new fuel infrastructure on a large scale is a formidable task, and significant deployment of fuel-cell vehicles is unlikely before 2035.

T.6 There are opportunities to reduce energy use in freight transportation by improving both vehicle efficiency and freight system logistics and infrastructure. Reductions of 10–20 percent in the fuel economy of heavy- and medium-duty vehicles appear feasible over a decade or so. A broad examination is needed of the potential for improving the effectiveness of the freight system to reduce energy consumption further.

T.7 Air transport and waterborne shipping have become more energy-efficient in response to higher fuel prices. Jet engine and aircraft technology has the potential to improve the efficiency of new aircraft by up to 35 percent over the next two decades. However, improvements in aviation efficiency for passenger transport are unlikely to fully offset projected growth in air travel. Major additional issues are the full greenhouse gas and other environmental impacts of aviation fuel use at high altitude and of growing airline travel; the potential for using biomass-based fuels in jets; and whether the use of low-grade residual fuel in ocean-going vessels will continue.

T.8 Most transportation efficiency studies and proposals have focused on the considerable energy efficiency gains that could be achieved with improved vehicles rather than in the transportation system as a whole. This emphasis is appropriate given the potential for and impact of such gains. However, major insights and improvements can result from a broader and deeper understanding of transportation system issues. The potential overall impact of such broader, system-based changes, such as densifying and reorganizing land use and collective modes of travel, needs further exploration and quantification. Developing better data and tools that can be used to analyze and forecast how different policies and investments might affect vehicle use and travel is thus an important task.

Industry

Figure S.1 shows that U.S. industry consumes almost one-third of the energy used in the United States. Between 1985 and 2004, real GDP in U.S. industry increased by nearly 45 percent while total energy use was virtually unchanged, leading to a decrease in energy intensity by nearly one-third. However, much of this improvement in energy intensity was due to a change in the composition of manufacturing in the United States. The share of industrial GDP accounted for by energy-intensive industries such as petroleum refining and paper manufacturing declined and was replaced by less-energy-intensive sectors such as computers and electronics.

Independent studies (cited in Chapter 4) using various approaches show that the economic potential for improved energy efficiency in industry is large. On the basis of its assessment of those studies, the panel concluded that of the 34.3 quads of energy forecasted to be consumed by U.S. industry in 2020 (EIA, 2008a), 14–22 percent (4.9–7.7 quads) could be saved through cost-effective energy efficiency improvements (those with an internal rate of return of at least 10 percent or that exceed a company's cost of capital by a risk premium). A large part of this savings—2 quads at the upper end of the range—would be supplied by further use of combined heat and power systems. Table S.4 summarizes the potential for energy savings in industry as estimated by various studies.

Beyond 2020, a wide array of advanced industrial technologies could make significant contributions to reducing industrial energy consumption and CO_2 emissions. Possible revolutionary changes include novel heat and power sources, as well as innovative processes for new products that take advantage of developments in nanotechnology and micro-manufacturing. Examples include the microwave processing of materials and nano-ceramic coatings, which show great potential for boosting the efficiency of industrial processes. In addition, advances in resource recovery and utilization—e.g., aluminum recycling—could reduce the energy intensity of U.S. industries.

Findings for Industry

I.1 Independent studies using different approaches agree that the economic potential for improved energy efficiency in industry is large. Of the 34.3 quads of energy forecasted to be consumed by U.S. industry in 2020 (EIA, 2008a), 14–22 percent could be saved through cost-effective energy efficiency improvements (those with an internal rate of return of at least 10 percent or that exceed a company's cost of capital by a risk

TABLE S.4 Economic Potential for Energy Efficiency Improvements in Industry in the Year 2020: Sector-wide and by Selected Subsectors and Technologies

	Estimates for U.S. Industry			
	CEF Study (IWG, 2000) Scaled to *AEO 2008* (quads)	McKinsey and Company (2008) (quads)	Other U.S. Studies (quads)	Global Estimates from IEA (2007) (%)
Petroleum refining	n.a.	0.3	0.61–1.21 to 1.40–3.28[a]	13–16
Pulp and paper	0.14[b]	0.6	0.37 to 0.85[c]	15–18
Iron and steel	0.21[d]	0.3	0.79[e]	9–18
Cement	0.08[f]	0.1	0.29[g]	28–33
Chemical manufacturing	n.a.	0.3	0.19[h] to 1.1[i]	13–16
Combined heat and power	2.0	0.7	4.4–6.8[j]	
Total, industrial sector	7.7 (22.4%)	4.9 (14.3%)		18–26

Note: This table appeared in Lave (2009) before this report was completed. The data in Table S.4 have been updated since the Lave (2009) article was published. CEF study, *Scenarios for a Clean Energy Future* (IWG, 2000); *AEO 2008, Annual Energy Outlook 2008, with Projections to 2030* (EIA, 2008a); n.a., not available.

[a]Based on a range of 10–20 percent savings (LBNL, 2005) to 23–54 percent savings (DOE, 2006a) from a baseline forecast of 6.08 quads.

[b]6.1 percent of the 2.31 quads of energy consumption forecast for the paper industry in 2020 by the *Annual Energy Outlook 2008* (EIA, 2008a).

[c]Based on 16 percent savings (Martin et al., 2000a) and 37 percent savings (DOE, 2006b) from the baseline forecast of 2.31 quads.

[d]15.4 percent of the 1.36 quads of energy consumption forecast for the iron and steel industry in 2020 by the *Annual Energy Outlook 2008* (EIA, 2008a).

[e]Based on 58 percent savings (AISI, 2005) from the baseline forecast of 1.36 quads.

[f]19.1 percent of the 0.43 quads of energy consumption forecast for the cement industry in 2020 by the *Annual Energy Outlook 2008* (EIA, 2008a).

[g]Based on 67 percent savings (Worrell and Galitsky, 2004) from the baseline forecast of 0.43 quads.

[h]National Renewable Energy Laboratory, 2002.

[i]DOE, 2007.

[j]Bailey and Worrell, 2005.

premium). These innovations would save 4.9–7.7 quads annually by 2020.

I.2 Additional efficiency investments could become cost-competitive through energy RD&D. Enabling and crosscutting technologies, such as advanced sensors and controls, microwave processing of materials, nanoceramic coatings, and high-temperature membrane separation, can provide efficiency gains in many industries as well as throughout the

energy system—for example, in vehicles, feedstock conversion, and electricity transmission and distribution.

I.3 Industry has experienced a significant shift to offshore manufacturing of components and products. If the net energy embodied in imports and exports is taken into account, the energy consumption attributable to industry would be increased by 5 quads.

I.4 Energy-intensive industries such as aluminum, steel, and chemicals have devoted considerable resources to increasing their energy efficiency. For many other industries, energy represents 10 percent or less of costs and is not a priority. Energy efficiency investments compete for human and financial resources with other goals such as increased production, improved productivity, introduction of new products, and compliance with environment, safety, and health requirements. Outdated capital depreciation schedules, backup fees for combined heat and power systems, and other policies also hamper energy efficiency investment.

I.5 More detailed data, collected more frequently, are needed to better assess the status of and prospects for energy efficiency in industry. Proprietary concerns will have to be addressed to achieve this.

I.6 Drivers for energy efficiency in industry include rising and volatile energy prices, intense competitive pressure to lower costs, and an increased focus on corporate sustainability.

RELATED CONSIDERATIONS

Experience with Policies and Programs

As noted above, the most cost-effective energy efficiency policies and programs of the last three decades (see Chapter 5) were vehicle and appliance efficiency standards, regulatory reforms to promote the adoption of combined heat and power systems, ENERGY STAR® product labeling and promotion, building energy codes, and utility and state end-use efficiency programs. Common characteristics of the most effective policies include:

- Periodic analysis and revision to assess effectiveness and to account for new technologies and opportunities;

- Financial incentives (if used) structured so that they reward performance and stimulate further action by consumers and businesses, rather than simply subsidize "efficiency" indiscriminately; and
- Integration of policies into market transformation strategies that address the full range of barriers present in a particular situation.

Research and Development

Finally, the panel concluded that, based in part on a prior National Research Council study (NRC, 2001), U.S. DOE-funded R&D on energy-efficient technologies has been highly productive.

Energy efficiency is a dynamic resource. Basic and applied research can continue to develop technologies that deliver large energy savings. If the potential for energy efficiency technologies is to be realized beyond the next decade, the dynamic nature of the resource must be recognized and supported.

REFERENCES

American Iron and Steel Institute (AISI). 2005. Saving One Barrel of Oil per Ton. Washington, D.C. October.

An, F., and D. Santini. 2004. Mass impacts on fuel economies of conventional vs. hybrid electric vehicles. SAE Technical Paper 2004-01-0572. Warrendale, Pa.: SAE International.

Bailey, O., and E. Worrell. 2005. Clean Energy Technologies: A Preliminary Inventory of the Potential for Electricity Generation. Report LBNL-57451. Berkeley, Calif.: Lawrence Berkeley National Laboratory. September.

Bandivadekar, A., K. Bodek, L. Cheah, C. Evans, T. Groode, J. Heywood, E. Kasseris, K. Kromer, and M. Weiss. 2008. On the Road in 2035: Reducing Transportation's Petroleum Consumption and GHG Emissions. Laboratory for Energy and the Environment Report No. LFEE 2008-05 RP. Cambridge, Mass: MIT. July.

Brown, R., S. Borgeson, J. Koomey, and P. Biermayer. 2008. U.S. Building-Sector Energy Efficiency Potential. Berkeley, Calif.: Lawrence Berkeley National Laboratory.

Cheah L., C. Evans, A. Bandivadekar, and J. Heywood. 2007. Factor of Two: Halving the Fuel Consumption of New U.S. Automobiles by 2035. LFEE Report 2007-04 RP. MIT Laboratory for Energy and the Environment. Cambridge, Mass: Massachusetts Institute of Technology. Available at http://web.mit.edu/sloan-auto-lab/research/beforeh2/files/cheah_factorTwo.pdf.

DOE (Department of Energy). 2006a. Energy Bandwidth for Petroleum Refining Processes. Prepared by Energetics Inc. Washington, D.C. October. Available at http://www1.eere.energy.gov/industry/ petroleum_refining/bandwidth.html.

DOE. 2006b. Pulp and Paper Industry Energy Bandwidth Study. Prepared by Jacobs Engineering Group and the Institute of Paper Science and Technology for the American Institute of Chemical Engineers and DOE. Washington, D.C. August.

DOE. 2007. Energy and Environmental Profile of the U.S. Petroleum Refining Industry. Prepared by Energetics, Inc., for the U.S. Department of Energy, Industrial Technologies Program. Washington, D.C. November. Available at http://www1.eere.energy.gov/ industry/petroleum_refining/analysis.html.

EIA (Energy Information Administration). 2008a. Annual Energy Outlook 2008. DOE/EIA-0383(2008). Washington, D.C.: Department of Energy, Energy Information Administration.

EIA. 2008b. Annual Energy Review 2007. DOE/EIA-0384(2007). Washington, D.C.: Department of Energy, Energy Information Administration.

EIA. 2008c. Monthly Energy Review (April). DOE/EIA-0035(2008/04). Washington, D.C.: Department of Energy, Energy Information Administration. Available at http://tonto.eia.doe.gov/FTPROOT/multifuel/mer/00350804.pdf.

EIA. 2009a. Annual Energy Review 2008. DOE/EIA-0384(2008). Washington, D.C.: Department of Energy, Energy Information Administration.

EIA. 2009b. Monthly Energy Review (June). Washington, D.C.: Department of Energy, Energy Information Administration.

EIA. 2009c. Monthly Energy Review (October). Washington, D.C.: Department of Energy, Energy Information Administration.

IEA (International Energy Agency). 2007. Tracking Industrial Energy Efficiency and CO_2 Emissions. Paris, France: IEA.

Lave, L.B. 2009. The potential of energy efficiency: An overview. The Bridge 39(2):5-14.

LBNL (Lawrence Berkeley National Laboratory). 2005. Energy Efficiency Improvement and Cost Saving Opportunities for Petroleum Refineries, An ENERGY STAR® Guide for Energy and Plant Managers. LBNL-57260-Revision. Prepared by C. Galitsky, S. Chang, E. Worrell, and E. Masanet. Berkeley, Calif.: LBNL. February.

Martin, N., N. Anglani, D. Einstein, M. Khrushch, E. Worrell, and L.K. Price. 2000. Opportunities to Improve Energy Efficiency and Reduce Greenhouse Gas Emissions in the U.S. Pulp and Paper Industry. LBNL-46141. Berkeley, Calif.: Lawrence Berkeley National Laboratory.

McKinsey and Company. 2008. The Untapped Energy Efficiency Opportunity of the U.S. Industrial Sector: Details of Research, 2008. New York: McKinsey and Company.

NAS-NAE-NRC (National Academy of Sciences-National Academy of Engineering-National Research Council). 2009. America's Energy Future: Technology and Transformation. Washington, D.C.: The National Academies Press.

NPC (National Petroleum Council). 2007. Technologies for Transportation Efficiency. Topic paper of the Transportation Efficiency Subgroup. Council Committee on Global Oil and Gas. Washington, D.C.: NPC. July.

NRC (National Research Council). 2001. Energy Research at DOE: Was It Worth It? Washington, D.C.: National Academy Press.

NRC. 2004. The Hydrogen Economy: Opportunities, Costs, and R&D Needs. Washington, D.C.: The National Academies Press.

NREL (National Renewable Energy Laboratory). 2002. Chemical Industry of the Future: Resources and Tools for Energy Efficiency and Cost Reduction Now. DOE/GO-102002-1529; NREL/CD-840-30969. October. Available at http://www.nrel.gov/docs/fy03osti/30969.pdf.

Wohlecker, R., M. Johannaber, and M. Espig. 2007. Determination of weight elasticity of fuel economy for ICE, hybrid and fuel cell vehicles. SAE Technical Paper 2007-01-0343. Warrendale, Pa.: SAE International.

Worrell, E., and C. Galitsky. 2004. Energy Efficiency Improvement and Cost Saving Opportunities for Cement Making—An ENERGY STAR (R) Guide for Energy and Plant Managers. LBNL-54036. Berkeley, Calif.: Lawrence Berkeley National Laboratory.

1 Energy Use in Context

This report of the Panel on Energy Efficiency Technologies reviews the technologies that could increase energy efficiency[1] in U.S. buildings, transportation, and industry over the next few decades, especially during the period 2010–2020. It describes the technologies' state of development; the potential for their use to achieve energy savings in buildings, transportation, and industry; and their performance, costs, and environmental impacts, most notably emission of greenhouse gases.[2] The panel was convened as part of the National Academies' America's Energy Future (AEF) project (described in Appendix A). The panel's charge is given in the preface in Box P.1.

Because continued technological advances make energy efficiency a dynamic resource, this report also reviews advanced technologies—some of which could become available and cost-effective in the 2020–2035 timeframe and beyond—and the research and development (R&D) needed to support their development.

To make a difference, energy efficiency technologies will have to be adopted widely, and so this report also addresses the sometimes formidable barriers to

---

[1]The terms "energy efficiency" and "energy conservation" are often used interchangeably, but although both can save energy, they refer to different concepts (see Box 1.4 and Appendix D). Improving energy efficiency involves accomplishing an objective, such as heating a room to a certain temperature, while using less energy. Energy conservation involves behavior expressed in actions taken to reduce energy use and can involve lifestyle changes—e.g., lowering the thermostat in winter. This report focuses on energy efficiency.

[2]Although greenhouse gas emissions are the primary environmental impact considered in this report, the full evaluation of a specific application of a technology or measure should consider many other effects, including local effects, on the environment and natural resources.

achieving high market penetration, and it outlines some of the experience gained with key policies and programs aimed at overcoming these barriers.

1.1 ENERGY USE IN THE UNITED STATES

The United States is the world's largest consumer of energy. In 2008 it used 99.4 quadrillion Btu (99 quads) of primary energy (Figure 1.1), 20 percent of world consumption. The next largest user, China, accounted for 15 percent of world consumption, but its per capita use was less than one-fifth that of the United

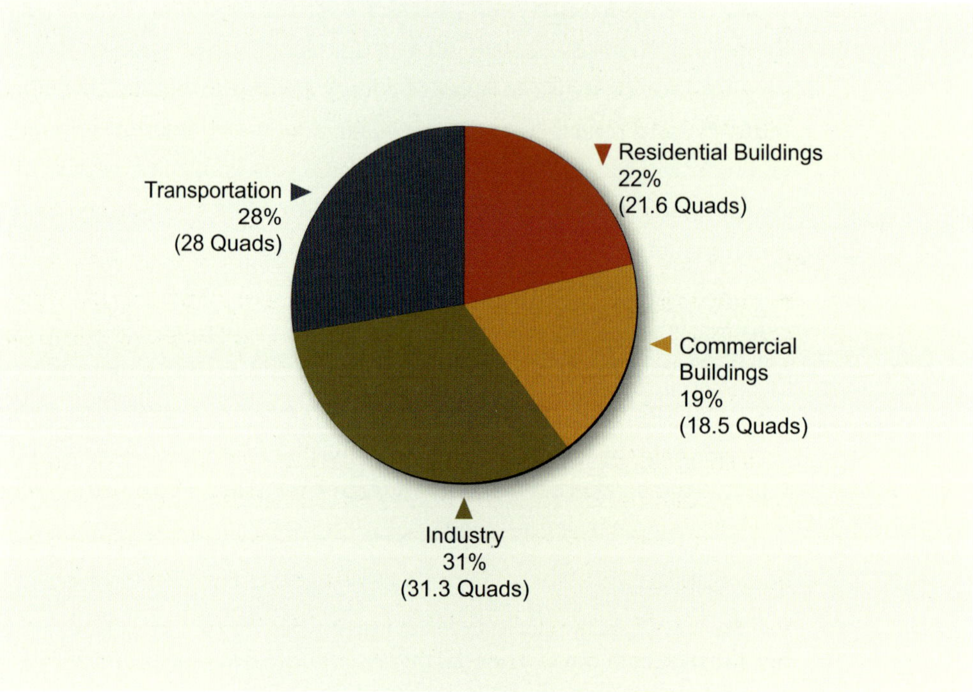

FIGURE 1.1 *Total U.S. energy use by sector, 2008 (in quadrillion Btu, or quads). For each sector, "total energy use" is direct (primary) fuel use plus purchased electricity plus apportioned electricity-system losses. Economy-wide, total U.S. primary energy use in 2008 was 99.4 quads. Totals may not equal sum of components due to independent rounding.*
Source: EIA 2009a, as updated by EIA, 2009b.

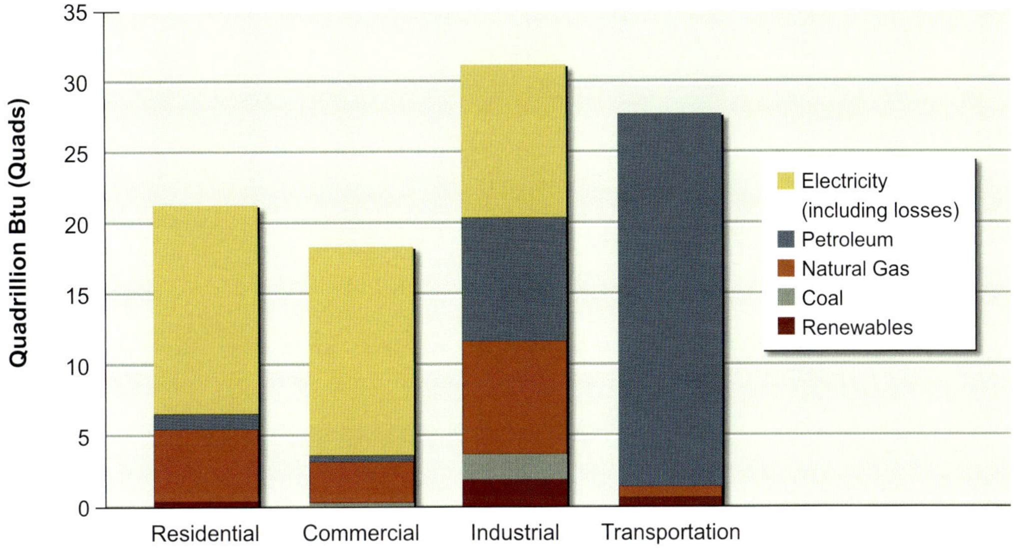

FIGURE 1.2 *Total energy consumption in the United States in 2008, by sector and fuel. Shown are electricity consumption—with the losses in generation, transmission and distribution allocated to the end-use sectors—and the fuels used on-site in each sector. Electricity is generated off-site using fossil, renewable, and nuclear energy sources. Source: EIA 2009a, as updated by EIA, 2009c.*

States.[3] In 2008, about 40 percent of the energy consumed in the United States was used in the myriad activities and services associated with residential and commercial buildings; 28 percent was used in transportation; and 31 percent was used in industry. U.S. energy consumption in 2008 by sector and by fuel type is shown in Figure 1.2.

Additional details on the sources and sectoral uses of energy in the United States are shown in Figure 1.3, which indicates for 2008 the amount of primary energy used for electricity generation (40.0 quads) and how much generated electricity and other energy was used in residential and commercial buildings (20.1 quads), transportation (27.9 quads), and industry (23.9 quads). In 2008, 73 percent of the generated electricity was used in the residential and commercial buildings sector and almost all of the rest by industry, with only a small amount used for transportation. Figure 1.3 also shows on the far right how much of the

[3]Energy Information Administration (EIA) data on the energy consumption of various countries are available from EIA at http://www.eia.doe.gov/emeu/aer/txt/ptb1103.html.

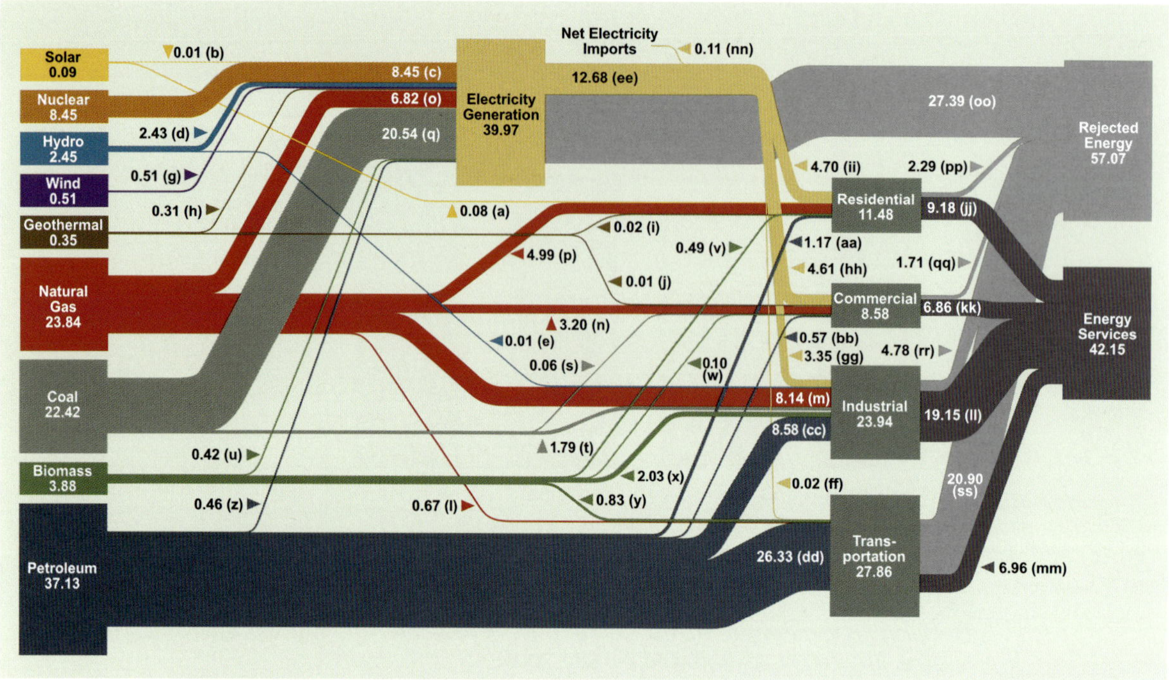

FIGURE 1.3 *Energy flows in the United States in 2008, quadrillion Btu (quads). The figure illustrates the delivery of energy from primary fuel sources, shown in the boxes on the left, to the residential, commercial, industrial, and transportation sectors, which are shown in the boxes at the center-right. Energy is delivered to these sectors primarily in three forms: (1) electricity, which is produced principally from coal, natural gas, and nuclear sources, and to a much lesser extent from renewable sources (hydro, solar, wind, and biomass); (2) liquid fuels, principally petroleum, with a small contribution from biomass-derived fuels (such as corn ethanol); and (3) natural gas for heating and as an industrial feedstock. Small quantities of coal and biomass are also used as industrial feedstocks. The width of each bar indicates the relative contribution of that energy source; the absolute contribution (in quads) is indicated by the numerical labels next to each bar. The bar for electricity represents retail electricity sales only and does not include self-generated electricity. The boxes on the right side of the figure show that a total of about 99 quads of energy were consumed in the United States in 2008, but only 42 percent (42 quads) was used to provide energy services. The remaining 58 percent (57 quads) was rejected—i.e., not used to provide energy services—because of inefficiencies in energy production, distribution, and use.*
Source: Lawrence Livermore National Laboratory and the U.S. Department of Energy, based on data in the Annual Energy Review 2008 (EIA, 2009a). Available at https://publicaffairs.llnl.gov/news/energy/energy.html.

primary energy input resulted in actual energy services and how much was lost because of inefficiencies in energy production, distribution, and use.

Energy use in the United States has grown steadily since 1949, although with a dip in the 1970s during the oil crisis; today energy consumption is double what it was in 1963 and 40 percent higher than it was in 1975 (the low point following the oil crisis). At the same time, U.S. energy intensity—the amount of energy used per dollar of gross domestic product (GDP)—has fallen steadily over many decades, with the exception of 1890 to 1920 and a few years after 1945 (Figure 1.4). Energy intensity decreased even as the United States became an industrial giant and built its railroads, highways, and other infrastructure. From 1973 to 2006, energy intensity fell by half, a rate of reduction of 2.1 percent per year.

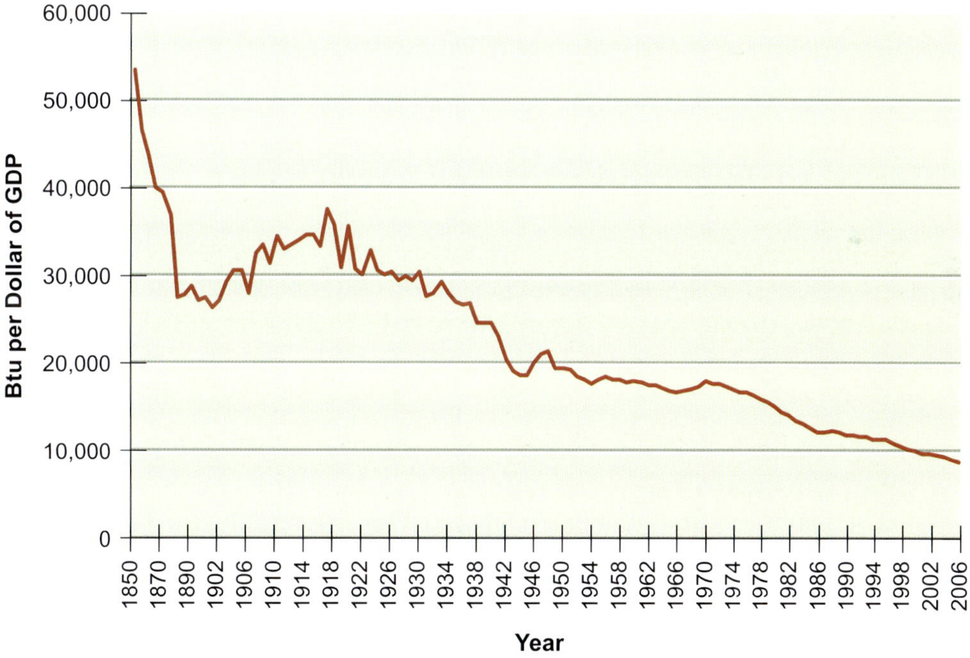

FIGURE 1.4 *U.S. energy intensity (energy use per dollar of GDP), 1850–2006. Note the scale change on the x-axis between 1890 and 1902. Values for energy use before 1900 are inconclusive, because they depend on an estimate of the amount of wood used, which was the predominate energy source in 1850.*
Source: Lave (2009), based on data in Schurr and Netschert (1960) and EIA (2008b).

About 70 percent of the decline in energy intensity since 1973 is estimated to have resulted from improvements in energy efficiency (IEA, 2004). If the trend toward lower energy intensity seen from 1973 to 2006 were to continue as a result of greater energy efficiency, by 2030 U.S. energy intensity would have dropped by 36 percent.

U.S. GDP is forecast to grow by 2.5 percent per year—slightly more than the 2.1 percent per year drop seen in U.S. energy intensity—and thus by 68 percent over the period from 2009 to 2030. With no change in U.S. energy intensity, U.S. energy use would grow by 68 percent. But if energy intensity continued to drop by 2.1 percent per year, by 2030 the energy intensity of the U.S. economy could be 30 percent lower than it is today, and total energy use in 2030 would increase by only 8.7 percent. And if the United States could accelerate the reduction in energy intensity to 2.5 percent per year, it could enjoy the projected 68 percent increase in GDP by 2030 without using any more energy than in 2009.

Can the United States achieve greater energy efficiency so that its energy intensity slows in relation to growth in U.S. GDP? What are the prospects for using energy more efficiently so as to reduce overall U.S. energy use as well?

1.2 THE POTENTIAL FOR IMPROVED ENERGY EFFICIENCY

Using energy more efficiently became a national concern during the first half of 2008 as energy prices hit record highs. It has also been gaining in importance as a result of growing concern about how to reduce emissions of greenhouse gases. Greater energy efficiency would reduce the need for fossil fuels, which provide 86 percent of the U.S. energy supply, and would thereby enhance not only environmental quality but also national security. Fortunately, the potential for greater U.S. energy efficiency is high.

It is the case that, despite the impressive gains made by the United States over the last 30 years, almost all other developed nations use less energy per capita and less energy per dollar of GDP (see Table 1.1 for examples). Denmark's levels of usage, for example, are about half those of the United States by both measures. While there are structural variations that account for part of this gap, studies have consistently shown that some 50 percent of the observed differences in energy intensity result from differences in energy efficiency (Darmstadter et al., 1977; Schipper, 2004; Weber, 2008). Box 1.1 provides some information on comparisons of energy intensity across nine nations.

TABLE 1.1 Energy Use in 2006 in Selected Nations, Per Capita and Per Dollar of Gross Domestic Product, Using Purchasing Power Parities (2000 dollars)

	Million Btu per Person	Btu per Dollar of GDP
United States	335	8841
France	181	6596
Japan	179	6492
Germany	178	6428
Denmark	161	5267

Note: Purchasing power parity exchange rates take into account the relative cost of living in the countries being compared. A similar table in Lave (2009) for energy use 2005 used data that reflected market exchange rates and that were taken from the EIA's interactive website (http://tonto.eia.doe.gov/cfapps/ipdbproject/IEDIndex3.cfm), in which data are continuously updated.
Source: EIA, 2006.

Moreover, energy-efficient and cost-effective technologies are available today to supply services (such as lighting, heating, cooling, refrigeration, transport, industrial motor drive, and computing) that are integral to modern life and that constitute the underlying drivers of the demand for energy. Hundreds of technologies, some already available commercially and others just beginning to enter the market, can provide these services more efficiently than is the case today, and they can collectively save large amounts of energy. Box 1.2 provides an example. Others are discussed in detail in Chapters 2 through 4.

Nevertheless, achieving greater energy efficiency in the United States will take considerable time and effort because, among other impediments, long-lived infrastructure, plants, and equipment—such as buildings, automobile assembly lines, and industrial and residential boilers—will have to be replaced or retrofitted. The range and number of other barriers to fuller use of energy efficiency technologies are suggested in Box 1.3 and discussed more fully in Chapters 2 through 4.

1.3 APPROACH TO AND SCOPE OF THIS STUDY

The panel's assessment looks at energy use in U.S. buildings (both residential and commercial), transportation, and industry over three timeframes—the present to 2020, 2020–2035, and 2035–2050. The first period receives major attention because so many cost-effective technologies are ready for implementation today or will be ready within a few years. The panel examined the literature on energy

BOX 1.1 Comparing Energy Intensity Across Nations

Early work by Darmstadter et al. (1977) and Schipper (2004) examined energy efficiency across nine developed nations. Adjusting for the mix of goods and services produced, heating- and cooling-degree days, and distance traveled, Schipper (2004) showed the results of the comparison (Figure 1.1.1).

In comparison with Denmark, for example, the United States uses 53 percent more energy per dollar of GDP. If the two nations had the average structure of the other nations but their own level of energy efficiency (comparing the second bars), the United States would still use 19 percent more energy per dollar of GDP than Denmark. In other words, the U.S. GDP mix is inherently more energy intensive than that of the average nation. If both nations had their current structures but had the same level of energy efficiency as the average among the nine nations shown in Figure 1.1.1 (comparing the third bars), the United States would use 28 percent more energy per dollar of GDP than Denmark. Another way of interpreting these estimates is that the U.S. economy would use about 19 fewer quads each year (about 19 percent of its total energy use) if it had the average energy efficiency of the other nine nations shown in Figure 1.1.1. The work by Darmstadter et al. (1977) suggested that the United States would use 25 fewer quads, remarkably similar results for comparisons made two decades apart.

Correcting for structural differences, the International Energy Agency has shown that between 1973 and 1998 the energy intensity of the United States and the eight major European economies fell 34 percent, and the energy intensity of Japan fell 30 percent (IEA, 2004). In Japan, strong declines in the energy intensity of manufacturing and the services sectors led the way, while the energy intensity of passenger transport and the residential sector actually increased. In Europe, manufacturing had the strongest decline in energy intensity, followed by declines in the service and household sectors.

By 2030, Europe is likely to be less energy intensive than today, and so the United States is likely still to lag behind. Some technologies that are cost-effective today in Europe are not cost-effective in the United States because energy prices in Europe are much higher due to high taxes. Until U.S. energy prices are comparable to prices in Europe, or other factors have the same effect, Europe is likely to remain less energy intensive.

efficiency, was briefed on the results of recent energy efficiency studies, and in some cases performed its own analyses to fill the gaps.

For buildings, the panel developed energy efficiency supply curves for electricity and natural gas in the residential and commercial sectors that show the amount of energy that could be saved over a range of costs below the current

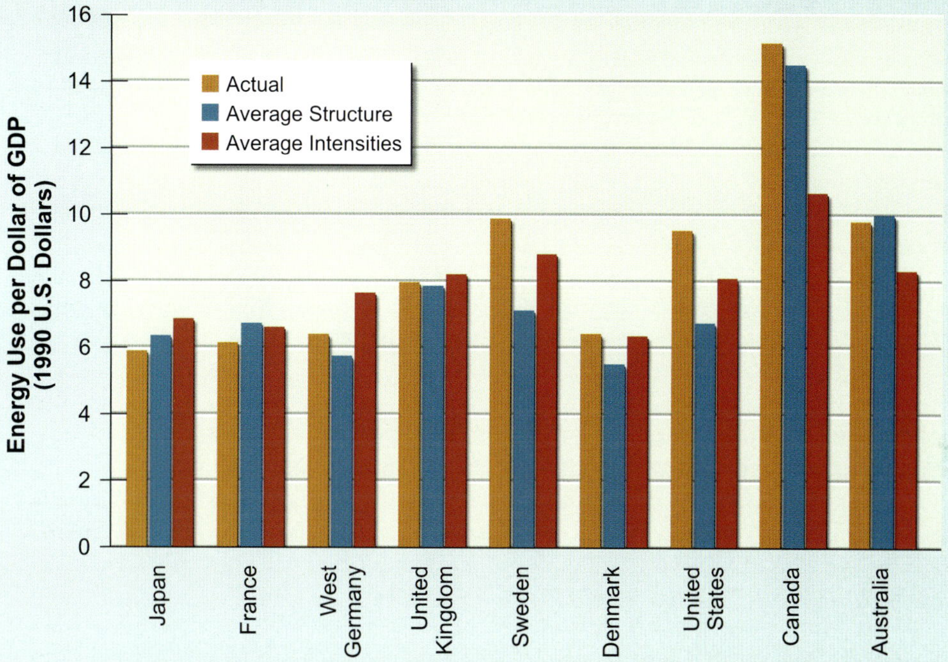

FIGURE 1.1.1 Summary and comparison of nine nations' energy intensity across all sectors and uses. The first bar for each nation indicates its overall energy intensity—expressed as energy use per dollar of GDP—in 1995. The value reflects such factors as each nation's economic structure (production of goods by heavy industry and light manufacturing, and provision of services), its weather, and distances traveled. The second bar indicates what each country's overall energy intensity would be if it had the average economic structure of the nations shown but its existing set of energy intensities. The third bar represents what the overall energy intensity would be if a country kept its own economic structure but had the energy intensities of the average among the nations shown. Source: Reprinted from Schipper (2004), copyright 2004, with permission from Elsevier.

price of energy. For transportation, the panel focused on the alternative technologies that could power the nation's cars and light trucks. By estimating the cost and energy savings associated with each technology and how R&D might improve the technology over time, as well as the timeframes in which specific technologies might be expected to penetrate the market, the panel was able to

BOX 1.2 The Benefits of More Efficient Products: An Illustration

Appendix E of this report provides information on how to calculate the net costs and benefits of energy savings. Figure 1.2.1 illustrates that the overall energy efficiency of providing light using incandescent lamps—starting from the burning of coal to produce electricity and continuing through to the production of visible light—is about 1.3 percent: about two-thirds of the energy in the coal is lost in generating electricity, about 9 percent is lost in transmitting and distributing the electricity, and an incandescent lightbulb's efficiency in transforming electricity to visible light is only 4 percent (Tsao et al., 2009).

In comparison, compact fluorescent lamps (CFLs) are about four times as efficient in transforming electricity to light as is an incandescent lamp (Azevedo et al., 2009; Tsao et al., 2009). Across the residential, commercial, and industrial sectors, a switch from incandescent lighting to CFLs today would save nearly 6 percent of the total amount of electricity generated in the United States today.[1] With further R&D, solid-state lamps (light-emitting diodes, LEDs) are expected to become 10 times as efficient as an incandescent lamp (Azevedo et al., 2009; Tsao et al., 2009).[2]

Across the residential, commercial, and industrial sectors, a switch from incandescent lighting to CFLs today would save approximately 228 terawatt-hours (TWh) of electricity per year relative to today's consumption, or nearly 6 percent of the total amount of electricity generated in the United States.

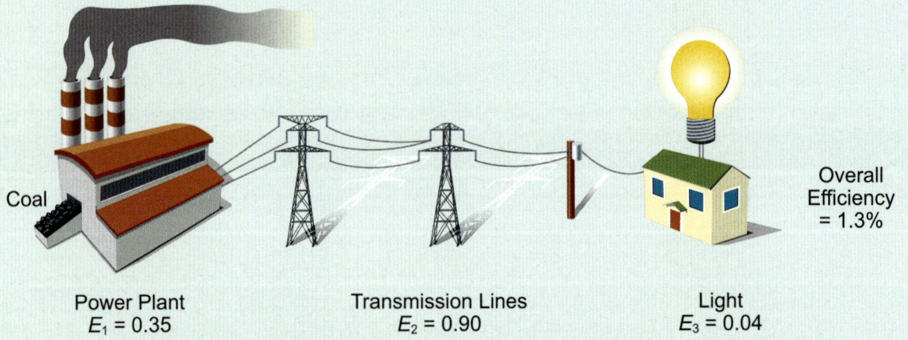

Overall Efficiency for Conversion of Chemical Energy to Light Energy $= E_1 \times E_2 \times E_3 = 0.35 \times 0.90 \times 0.04 = 0.013$

FIGURE 1.2.1 Example of how end-use efficiency influences overall fuel conversion efficiency. In this example, the efficiency of converting the chemical energy stored in coal to the electricity entering a building is about 32 percent (0.35 × 0.9). But after accounting for the low efficiency of the incandescent lightbulb, the efficiency of converting chemical energy to light energy is only 1.3 percent. (All values are approximate.)
Source: Adapted and updated from Hinrichs and Kleinbach, 2006.

TABLE 1.2.1 Annual Electricity Use for Lighting (TWh)

	Estimated 2008 Electricity Use for Lighting	Electricity Use for Inefficient Incandescent Lighting	Electricity Use for Fluorescent and Other Efficient Lighting	Electricity Use for Formerly Inefficient Lighting If Replaced by CFLs	Total 2008 Electricity Use for Lighting If All Lamps Were Efficient	Electricity Use for Lighting If All Lamps Are Later Replaced with LED Lamps
Residential	208	176.8 (85%)	31.2	44.2	75.4	37.7
Commercial	391	125.1 (32%)	265.9	31.3	297.2	148.6
Industrial	108	2.2 (2%)	105.8	0.6	106.4	53.2
Total	707	304.1 (43%)	402.9	76.1	479.0	239.5
Saved TWh					228.0	467.5

Assuming that the LEDs become twice as efficient (on average) as CFLs and other types of currently efficient lamps in 5 years, replacing all lamps with LEDs at that time would save an additional 230.5 TWh. Based on the current average carbon dioxide (CO_2) emissions rate for the U.S. electricity sector, about 650,000 metric tons of CO_2 per TWh consumed, the switch from incandescent lighting to CFLs would reduce U.S. CO_2 emissions by about 148 million metric tons per year (about 2 percent). If all fluorescent and other lighting were subsequently converted to LEDs at twice the energy efficiency of fluorescent and other high-intensity discharge lamps and fixtures on average, an additional 56 million metric tons of CO_2 emissions would be avoided annually.[3]

All of the above calculations use 2001 data from Navigant Consulting, Inc. (2002) (see also Table 1.2.1). The residential sector used 208 TWh of electricity in 2001, 10 percent of which went to fluorescent lights. Because of the growing adoption of CFLs, the panel's analysis assumes that the fluorescent share in 2008 was 15 percent and the incandescent share 85 percent. Thus, the amount of electricity consumed by incandescent lamps was about 176.8 TWh. Shifting to fluorescent lighting would reduce electricity use by 75 percent to 44.2 TWh. Together with the 15 percent of lighting that is already fluorescent, the 2008 usage would have been 75.4 TWh if all lights were efficient. If CFLs and other fluorescent lamps are replaced with LEDs in, say, 5 years, half as much electricity, or 37.7 TWh, would be used. Table 1.2.1 has the other calculations.

---

[1]The calculation is based on data in Navigant Consulting, Inc. (2002).

[2]The purchase price of a CFL is higher than that for an incandescent lamp, and the cost of an LED lamp is still higher. While both the CFL and LED save money and energy, compared with the incandescent lamp, whether the LED saves money compared with the CFL is sensitive to the number of hours each year the lamp is used, the purchase price, and the discount rate.

[3]The United States emitted 5.89 billion metric tons of CO_2 in 2006 as a result of all energy consumption.

BOX 1.3 Why Aren't Energy-Efficiency Opportunities More Attractive to Consumers and Businesses?

Why don't consumers and businesses take greater advantage of "cost-effective" energy efficiency opportunities? If so much energy can be saved, why doesn't everyone do it, especially when the cost savings over time tend to well outweigh the initial costs?

The answer is complex, as there is no one reason for this seeming "behavior gap." Each of this report's sector discussions, as well as the policy discussion at the end of the report, identify factors that impede the full use of energy efficiency technologies and measures. Such barriers fall into several categories; the following examples illustrate how some of them affect decisions:

- Cost savings may not be the only factor influencing a decision to invest in an energy efficiency measure. For example, consumers purchase vehicles based on many factors, such as size, performance, and interior space, in addition to fuel economy. Fuel economy may not come into the picture at all.
- Although energy and cost savings might be achievable with only a low first cost (investment), such savings may be a small-enough part of the family or company budget that they are not really relevant to decisions.
- The up-front financial investment might be small, but substantial investments in time and effort may be required to find and study information about the potential energy-saving technologies, measures, and actions.
- It is well established that purchasers tend to focus much more on first costs than on life-cycle costs when making investments. This behavior is no different when it comes to energy efficiency. There is also the phenomenon of risk aversion—new products may be unfamiliar or not work as expected. The default behavior is often simply the status quo. Knowing this, producers may never design and develop energy-efficient products.
- Some of the "behavior gap" can be attributed to structural issues. For example, landlords of rental residential buildings are not motivated to pay for more

develop illustrative scenarios of how total energy consumption in the light-duty vehicle (LDV) fleet could evolve.[4] For industry, the panel focused on the four most

[4]Technologies to increase the fuel efficiency of LDVs (automobiles and light trucks) have been studied extensively, so the panel focused on the energy savings that could be achieved with these technologies. There has been less study of efficiency in medium- and heavy-duty vehicles (MHDVs). An NRC committee is currently assessing the potential for efficiency improvements in these vehicles. The panel also did not treat aviation in depth. The potential energy savings in transportation would be higher if MHDVs and aviation had been included in the panel's analysis.

efficient technologies when their tenants pay the utility bills. And builders whose incentive is to minimize the cost of new homes may not install high-efficiency heating and cooling equipment and systems that increase purchase prices but save buyers money over time.

- Other factors may involve retailers of equipment and appliances. If there is low demand for efficient products, retailers may not stock them. Even purchasers who might be motivated to search elsewhere for an efficient product may have to deal with limited choices in the event of an emergency purchase, such as when a refrigerator fails.
- Other reasons for the behavior gap are the subject of much social science research. They involve factors such as habits in purchasing or use, which can be very difficult to change. Some apparent consumer preferences—typically learned from parents, neighbors, and friends—may change very slowly, if at all.
- Energy-savings investments by businesses and industries are not always seen as beneficial. If energy accounts for only a small part of total costs, or if the available capital is limited, other investments may be preferred—e.g., in reducing other costs, improving products, or developing new ones. If the consequences of a new-product or production-method failure are large, this in itself can maintain the status quo.
- Firms may not be aware of the potential savings achievable by replacing equipment such as motors with more efficient or variable-speed versions. When motors, large or small, are used throughout a facility, the savings from upgrading them can be substantial.
- Energy-efficiency investments by companies are made in the context of complex business cultures. "Champions," or commitment at the highest levels, may be required.

More details on how barriers such as these play out in the buildings, transportation, and industrial sectors are given in Chapters 2 through 4. Chapter 5 examines some of the policies and programs that have been aimed at overcoming these barriers.

energy-intensive U.S. industries and also reviewed studies that assessed the potential for energy efficiency across all of manufacturing.

1.3.1 Baselines and Key Assumptions

The panel began by identifying a "baseline" or "business-as-usual" case. For buildings and industry, this was the reference case scenario of the Energy Information Administration's (EIA's) *Annual Energy Outlook 2007* (EIA, 2007)[5] or *2008* (EIA, 2008a). For transportation, the panel developed its own baseline. For each of the three sectors, the panel estimated the level of energy-efficiency improvement beyond the baseline or reference case that could be attained with better technologies. Table 1.2 summarizes the key assumptions used in the panel's analyses of the three sectors; more details can be found in Chapters 2, 3, and 4.

1.3.2 Key Energy-Related Terms and Measures Used

Box 1.4 provides definitions of some of the key energy-related terms and measures used throughout this report.

In addition, the panel notes the importance of distinguishing between "primary energy" and "delivered energy," depending on the aspect of energy use being considered or analyzed.[6] "Delivered energy" is the energy value of the fuel or electricity that enters the point of use (e.g., a building). Primary energy accounts for the total amount of fuel needed to provide this delivered energy. Note that "primary energy" and "delivered energy" are often referred to as "source energy" and "site energy," respectively. The Department of Energy's (DOE's) EIA, in its statistical reports, uses the former terminology; this report does the same.

When discussing electricity or natural gas bills, or assessing the electricity demand that must be met by electricity production, delivered energy is the metric of interest. When determining the ultimate impact on total fuel consumption and energy resources of the energy demand by residences, commercial buildings, transportation, and industry, and related effects on the economy and the environment, primary energy is the appropriate metric. When quoting values for energy demand and production, it is important to specify which measure is being used.

The EIA defines primary energy as energy in the form that it is first accounted for in a statistical energy balance, before any transformation to secondary or tertiary forms of energy (see Box 1.4). The fuels counted by the EIA in U.S. primary energy consumption are given in Box 1.5. Distinguishing between primary and delivered energy is most useful for electricity, where the difference between

---

[5]With an assessment of whether the results might have differed if the EIA's *Annual Energy Outlook 2008* had been used.

[6]This section draws on information in U.S. Department of Energy, Energy Information Administration: "Energy Efficiency Measurement Discussion: Source Versus Site Energy," available at http://www.eia.doe.gov/emeu/efficiency/measure_discussion.htm#Site%20Energy%20Versus%20Primary%20Energy, and "Glossary," available at http://www.eia.doe.gov/glossary/index.html.

TABLE 1.2 Sources and Key Assumptions Used to Develop Energy Savings and Cost Estimates

	Buildings (Chapter 2)	Transportation (Chapter 3)	Industry[a] (Chapter 4)
Reference scenario	AEO 2007 reference scenario (EIA, 2007), but with an assessment of whether using the AEO 2008 (EIA, 2008a) reference scenario would have changed the results	Developed by the panel,[b] but similar to the AEO 2008 reference scenario (EIA, 2008a)	AEO 2008 reference scenario (EIA, 2008a)
Source of cost estimates	Critical assessment of the literature	Critical assessment of the literature	Critical assessment of the literature
Source of savings estimates	Critical assessment of the literature Panel-derived conservation supply curve analysis	Critical assessment of the literature on specific technologies For light-duty vehicles (LDVs), the panel derived illustrative scenarios of overall savings in fleet fuel consumption	Critical assessment of the literature on: • Industry-specific savings • Industry-wide savings • Savings from specific cross-cutting technologies
Key cost-effectiveness criteria	Levelized cost of energy savings is less than the national average electricity and natural gas prices	Recovery of discounted costs of energy savings over the life of an LDV[c]	Energy savings provide an internal rate of return on investment of at least 10 percent or exceed the company's cost of capital by a risk premium
Technology lifetimes	Technology specific	Average vehicle lifetime	Technology specific
Before-tax discount rate (%, annual)	7	7	15
Other considerations	Assessment accounts for stock turnover in buildings and equipment	For LDVs, assessment considers how the distribution of specific vehicle types in the new-vehicle fleet affects the on-the-road fleet	Assessment of savings in specific industries used to confirm industry-wide estimates

[a]Manufacturing only.
[b]This is a "no-change" baseline in which, beyond 2020 (when the original targets set by the Energy Independence and Security Act of 2007 are assumed to have been met), any fuel efficiency improvements are fully offset by increases in vehicle performance, size, and weight.
[c]A cost-effectiveness criterion was not applied in the illustrative scenario analysis for transportation. Rather, the panel estimated, using the criteria in this table, whether an initial investment in the specific technologies assessed was likely to be recouped over the life of the vehicle.
Source: Adapted from Table 3.A.2 in NAS-NAE-NRC, 2009.

BOX 1.4 **Definitions**

Energy, in its forms as work, heat, and electric power, provides such services as powering a car or providing light. It can come from renewable sources, such as the sun or wind; from chemical reactions, such as combustion of fossil or renewable fuels; or from nuclear reactions. Energy is usually measured in British thermal units (Btu) or kilowatt-hours (kWh) (or joules; 1 watt-hour is equivalent to 3600 joules). Delivered energy is a measure at the point-of-use (site) of the amount of energy delivered to a consumer without adjusting for the energy lost in transforming a fuel or other form of energy to electricity, transmitting the energy to the point where it will be used, and then distributing it to individual users. Primary energy is the amount of delivered energy adjusted upward to account for the energy that is lost in the transformation and delivery of that energy to an end user, such as a residential housing unit.

Energy intensity is a measure of the amount of energy used per unit of output for a company, industry, or the whole economy. For example, the energy intensity of steel production represents the amount of energy used to produce a ton of steel, or Btu per ton. Energy intensity can also represent the amount of energy used per dollar of output—or, for the whole economy, per dollar of gross domestic product (GDP). For example, the energy intensity of the U.S. economy is about 9000 Btu per dollar of GDP, equivalent to a bit less than one pound of coal or a bit more than one cup of gasoline per dollar.

Energy efficiency is a measure describing how much useful work can be obtained from a system from a given amount of input energy. (A more formal definition derived from thermodynamics is given in Appendix E.) This report deals primarily with technologies for realigning *improvements* in energy efficiency, which the panel defines as accomplishing a specified objective using less energy. The objective might be to heat a room to a certain temperature, or provide a certain amount of light. For example, a compact fluorescent lamp is more efficient than an incandescent lamp, in that it provides the same number of lumens and quality of light as an incandescent lamp for only one-quarter to one-third of the energy input. Energy efficiency can be expressed directly as a dimensionless ratio—in this case, the ratio of the energy fed into a lightbulb to that which is radiated as light.

Conservation is usually understood to mean action taken to reduce the amount of energy used by changing behavior, such as turning off personal computers when not in use or setting back the thermostat in winter. It can also involve using technologies, such as room occupancy sensors for lighting, which reduce energy use without someone having to remember to turn off the light.

BOX 1.5 **Forms of Primary Energy Consumption Included in EIA Statistics**

- Coal consumption; coal coke net imports
- Petroleum consumption (petroleum products supplied, including natural gas plant liquids and crude oil burned as fuel, but excluding ethanol blended into motor gasoline)
- Dry natural gas consumption—excluding supplemental gaseous fuels
- Nuclear electricity net generation, converted to British thermal units (Btu) using the heat rate for nuclear plants
- Conventional hydroelectricity net generation, converted to Btu using the heat rate for fossil-fueled plants
- Geothermal electricity net generation, converted to Btu using the heat rate for geothermal plants, and geothermal heat-pump energy and geothermal direct-use energy
- Solar thermal and photovoltaic electricity net generation, converted to Btu using the heat rate for fossil-fueled plants, and solar thermal direct-use energy
- Wind electricity net generation, converted to Btu using the heat rate for fossil-fueled plants
- Wood and wood-derived fuels
- Biomass waste (municipal solid waste from biogenic sources, landfill gas, sludge waste, agricultural byproducts, and other biomass)
- Fuel ethanol and biodiesel; losses and co-products from the production of fuel ethanol and biodiesel
- Electricity net imports, converted to Btu using the electricity heat content of 3412 Btu/kWh.

Source: EIA, 2007, p. 34.

the two can be large. (For example, about two-thirds of the fuel energy used in a thermal power plant is lost in generating electricity. About 9 percent of the generated electricity is lost during transmission and distribution of the electricity. So the amount of electricity entering a building or facility represents only 30 percent or less of the original fuel energy.) The EIA distinguishes primary energy from delivered energy only for electricity. For other fuels, delivered energy consumption is assumed to equal primary energy consumption. For these fuels, the energy used to transform fuels from one form to another, such as from crude oil to gasoline, is counted as energy consumed by the industry that performs the transformation—in this case, petroleum refining.

Analysts have debated the usefulness of measuring delivered (or site) energy. Many object to adding the energy value (as measured in British thermal units, or Btu) of delivered electricity to that of the fuels (such as natural gas) used directly at a site to arrive at a total value for "delivered energy" use at a site. This approach can lead to misleading conclusions—for example, that an all-electric building with resistance heating uses less energy than a comparable building with gas heated, because the delivered electricity, as measured in Btu, can be lower than the Btu value of the delivered natural gas. But the effects of electricity consumption on total energy use, and therefore the effects of consumer technology choices and upgrades, depend also on how the electricity is generated, which is not accounted for in delivered energy.

In this report, "delivered energy" refers to consumption of only a single fuel at a time for the buildings and industry sectors. In other words, energy delivered in different forms is not summed to produce a total value for delivered energy. For transportation, all petroleum-based fuels—gasoline, diesel oil, and others—are sometimes summed to obtain a total value for petroleum consumption.

This report identifies electricity consumption as primary (source) energy use whenever this is the metric being used. Otherwise, the value being quoted is in terms of delivered (site) energy.

1.3.3 Scope and Content

Chapters 2 through 4 provide the panel's detailed assessment of the technologies that could improve energy efficiency in the buildings, transportation, and industrial sectors of the U.S. economy. Each chapter estimates the cost of technologies that offer improved energy efficiency, the amount of energy each could save, and the timeframe in which the technologies are likely to be available. The panel's findings are presented at the end of each chapter. An important aspect of the panel's analysis was the factors that impede putting energy-efficient technologies to use. Past studies have identified many technologies that, despite their potential to lower costs and save energy, were never implemented widely. Why seemingly attractive technologies are not deployed is explored for each sector in Chapters 2 through 4 (see Box 1.3 for examples). Chapter 5 reviews key policies and programs that have been aimed at overcoming barriers to improving energy efficiency, including some that led to large energy savings, and it recounts some of the positive experiences of two large states in achieving increased energy savings. In addition, the panel presents in Chapter 5 four composite or overarching findings based on discussion presented throughout the chapters.

1.4 REFERENCES

Azevedo, I., M. Morgan, and F. Morgan. 2009. The transition to solid-state lighting. Proceedings of the IEEE 97(3):481-510. March.

Darmstadter, J., J. Dunkerley, and J. Alterman. 1977. How Industrial Societies Use Energy: A Comparative Analysis. Johns Hopkins University Press, p. 189.

EIA (Energy Information Administration). 2006. Annual Energy Outlook 2006. DOE/EIA-0383(2006). Washington, D.C.: Department of Energy, Energy Information Administration.

EIA. 2007. Annual Energy Outlook 2007. DOE/EIA-0383(2007). Washington, D.C.: Department of Energy, Energy Information Administration.

EIA. 2008a. Annual Energy Outlook 2008. DOE/EIA-0383(2008). Washington, D.C.: Department of Energy, Energy Information Administration.

EIA. 2008b. Annual Energy Review 2007. DOE/EIA-0384(2007); Washington, D.C.: Department of Energy, Energy Information Administration.

EIA. 2008c. Monthly Energy Review (April). Washington, D.C.: Department of Energy, Energy Information Administration.

EIA. 2009a. Annual Energy Review 2008. DOE/EIA-0384(2008); Washington, D.C.: Department of Energy, Energy Information Administration.

EIA. 2009b. Monthly Energy Review (June). Washington, D.C.: Department of Energy, Energy Information Administration.

EIA. 2009c. Monthly Energy Review (October). Washington, D.C.: Department of Energy, Energy Information Administration.

IEA (International Energy Agency). 2004. Oil Crises and Climate Challenges: 30 Years of Energy Use in IEA Countries. Paris, France: Organization for Economic Cooperation and Development.

Hinrichs, R.A., and M.H. Kleinbach. 2006. Energy: Its Use and the Environment, 4th Edition. Brooks/Cole Publishing Company.

Lave, L.B. 2009. The potential of energy efficiency: An overview. The Bridge 39(2):5-14.

NAS-NAE-NRC (National Academy of Sciences-National Academy of Engineering-National Research Council). 2009. America's Energy Future: Technology and Transformation. Washington, D.C.: The National Academies Press.

Navigant Consulting, Inc. 2002. U.S. Lighting Market Characterization, Volume 1, National Lighting Inventory and Energy Consumption Estimates. Washington, D.C.: Department of Energy, Office of Energy Efficiency and Renewable Energy.

Schipper, L. 2004. International comparisons of energy end use: Benefits and risks. In Encyclopedia of Energy, Vol. 1. Elsevier.

Schurr, S., and B. Netschert. 1960. Energy in the American Economy, 1850-1975: An Economic Study of Its History and Prospects. Johns Hopkins University Press.

Tsao, J., M. Coltrin, and M. Crawford. 2009. Solid-state lighting: An integrated human factors, technology and economic perspective. Quantifying past and potential-future trajectories. Proceedings of the IEEE, forthcoming.

Weber, C. 2008. An Examination of Energy Intensity and Energy Efficiency. Green Design Institute, Carnegie Mellon University.

2 Energy Efficiency in Residential and Commercial Buildings

The efficiency of the appliances and equipment used in homes and businesses has increased greatly over the past three decades. However, there is still much that can be done to reduce the amount and slow the growth of energy consumption in residential and commercial buildings.

This chapter describes how energy is used in buildings today and discusses the factors that have driven the growth of energy use. It then identifies opportunities for improving energy efficiency in the near term (through 2020) as well as the medium term (through 2030–2035). The chapter presents conservation supply curves that show the amount of energy that could be saved as a function of the cost of the saved energy and describes how whole-building approaches can produce new buildings with very low energy consumption. It reviews the market barriers to improving energy efficiency in buildings and presents some factors that are helping to overcome the barriers. Finally, the chapter presents the findings of the Panel on Energy Efficiency Technologies with regard to the potential for greater efficiency in residential and commercial buildings.

2.1 ENERGY USE IN BUILDINGS

In 2006, residential and commercial buildings accounted for 39 percent of the total primary energy used and 72 percent of the electricity used in the United States to supply power and fuel for heating, cooling, lighting, computing, and other needs. As Figures 2.1 (residential buildings) and 2.2 (commercial buildings) show, heating, ventilation, and air-conditioning (HVAC) consumed the most energy, followed by lighting.

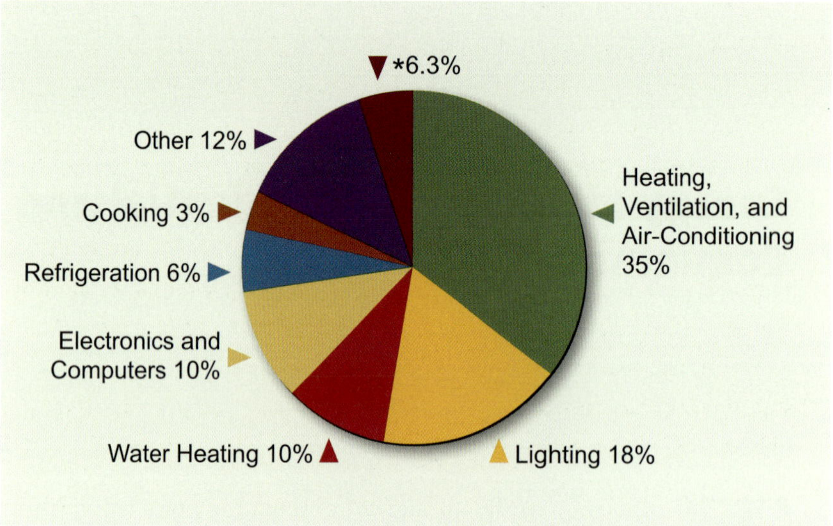

FIGURE 2.1 *Energy use in U.S. residential buildings by end-use, 2006.*
*Note: *, Energy Information Administration (EIA) adjustment factor that accounts for incomplete data in EIA's sampling and survey methodology.*
Source: Pew Center on Climate Change, based on data in DOE/EERE (2008), available at http://www.pewclimate.org/technology/overview/buildings.

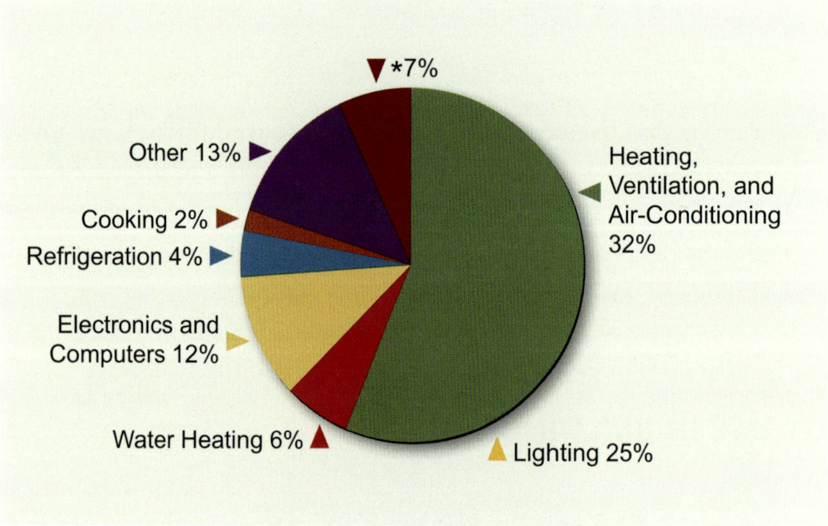

FIGURE 2.2 *Energy use in U.S. commercial buildings by end-use, 2006.*
*Note: *, Energy Information Administration (EIA) adjustment factor that accounts for incomplete data in EIA's sampling and survey methodology.*
Source: Pew Center on Climate Change, based on data in DOE/EERE (2008), available at http://www.pewclimate.org/technology/overview/buildings.

On the residential side, this energy was used in approximately 80.8 million single-family homes, 24.8 million multifamily housing units, and nearly 6.9 million mobile homes in the United States as of 2006 (EIA, 2008b). On the commercial side, there were approximately 75 billion square feet (7 billion square meters) of floor space in 5 million commercial buildings as of 2006 (EIA, 2008b). The building stock is long-lived: homes can last 100 years or more, commercial buildings often last 50 years or more, and appliances and equipment used in buildings can last 10–20 years (IWG, 1997). Nonetheless, there have been significant changes in energy use and energy efficiency in buildings over the past 30 years.

Energy use in buildings has increased over the past 30 years, but at a rate slower than the rate of increases in gross domestic product (GDP). As shown in Figure 2.3, in the residential sector over the period 1975–2005, delivered-energy

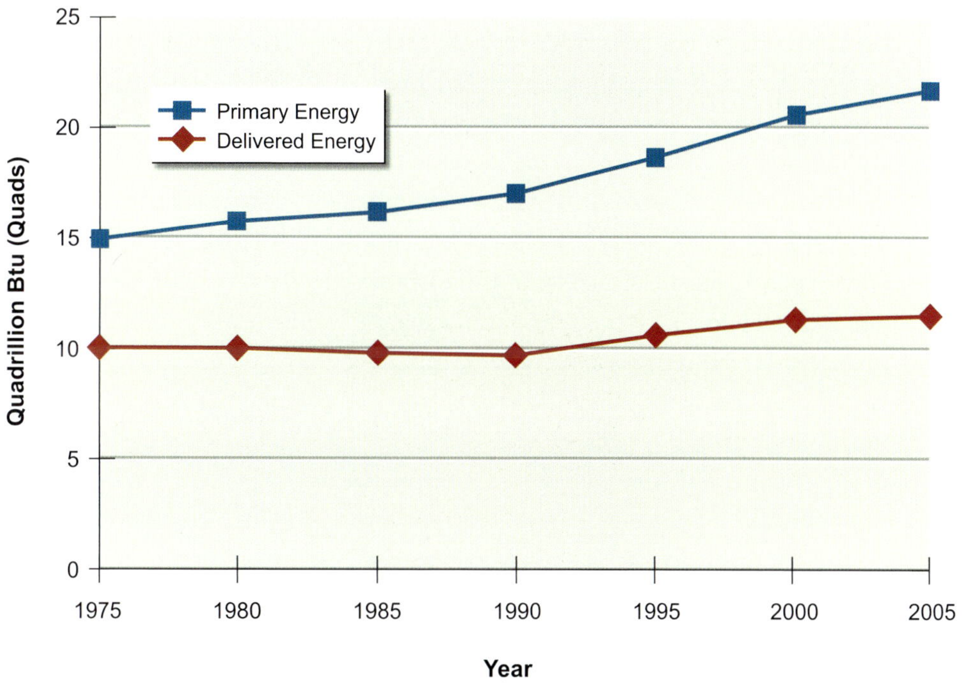

FIGURE 2.3 *U.S. residential energy use trends. Primary energy use (accounting for losses in electricity generation and transmission and distribution, and for fuels, such as natural gas, used on-site) has increased faster than delivered energy use (which does not account for such losses, but does include fuels used on-site) because use of electricity has increased faster than use of other fuels.*
Source: Data from EIA, 2007b.

use[1] increased about 15 percent whereas primary energy use increased 46 percent. This difference is due to the growing electrification of energy use in homes. In 1975, direct fuel use in homes was four times that of electricity use in terms of end-use energy content, but by 2005 this ratio had fallen to about 1.4 to 1.

Understanding the potential for improvements in building energy efficiency requires detailed energy-use data beyond those presented in Figures 2.1 and 2.2, because the sector potential is composed of a long list of appliance-specific and building-specific measures. Unfortunately, much of the available data on energy use in buildings is based on self-reporting or inferences rather than on direct measurement, and estimates of uncertainties around the data are seldom available. Expanded data gathering, particularly through direct measurement, would facilitate more rigorous evaluation of energy efficiency measures and would contribute to the accuracy and completeness of future studies.

Growth in the use of a variety of electrical appliances is one factor contributing to the growth of energy use in buildings in recent decades. Figure 2.4 shows the penetration (the percentage of U.S. households having an appliance) of selected appliances in U.S. households between 1980 and 2005. During this period the percentage of households having central air-conditioning more than doubled, and the penetration of microwave ovens increased by more than a factor of six and that of dishwashers by 57 percent. Personal computer use was essentially nonexistent in 1980, yet by 2005, 68 percent of all U.S. households had a personal computer. In addition, 56 percent of households had cable television service, nearly 22 percent had a satellite dish antenna, and more than 27 percent of households had at least one large-screen television as of 2005 (DOE, 2009).

Compared with the residential sector, the commercial sector experienced much faster growth in energy use over the period 1975–2005: delivered-energy use in the commercial sector increased approximately 50 percent, and primary energy use increased 90 percent (Figure 2.5). As in the residential sector, the growing electrification of energy use in the commercial sector led to a faster rise in primary energy use than in delivered-energy use (DOE, 2008a).

Residential energy intensity, defined as energy use per square foot of living space, declined over the past 30 years in spite of the growing penetration of

---

[1] "Delivered" energy refers to the electricity delivered to a site plus the fuels used directly on-site (e.g., natural gas for heating water). This measure does not account for the losses incurred in generating and transmitting and distributing the electricity. Delivered energy plus these losses is referred to as "primary" energy. See Box 1.4 in Chapter 1.

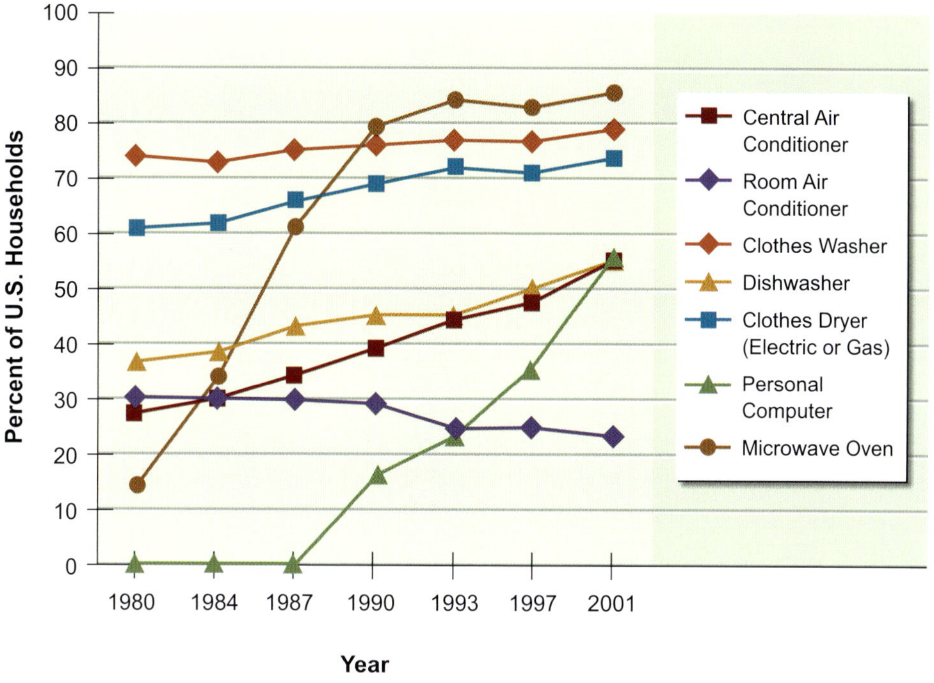

FIGURE 2.4 *Household appliance penetration trends. "Penetration" is the percentage of U.S. households having the appliance specified. Data for personal computers are unavailable before 1990.*
Source: U.S. Department of Energy, Energy Information Administration. Data through 2001 are from Regional Energy Profiles, Appliance Reports, Table 1: Appliances in U.S. Households, Selected Years, 1980–2001, available at http://www.eia.doe.gov/emeu/reps/appli/all_tables.html. Data for 2005 are from 2005 Residential Energy Consumption Survey—Detailed Tables, available at http://www.eia.doe.gov/emeu/recs/recs2005/hc2005_tables/detailed_tables2005.html.

appliances (see discussion below). However, the rate of decline depends on how energy intensity is measured. Total delivered-energy use per household fell 31 percent over the period 1978–2005, while primary energy use per household fell 16 percent (Table 2.1). Although household size in terms of square feet of floor area has been increasing, leading to a steeper decline in primary energy use per square foot of floor area (DOE, 2008a), the number of people living in a typical household declined from 2.8 in 1980 to 2.6 in 2001 (Battles and Hojjati, 2005). Thus primary energy use per household member remained relatively constant over the period 1980–2005. Smaller households use less absolute energy than larger households do, but more energy is used per person in the former. The 2005 residential

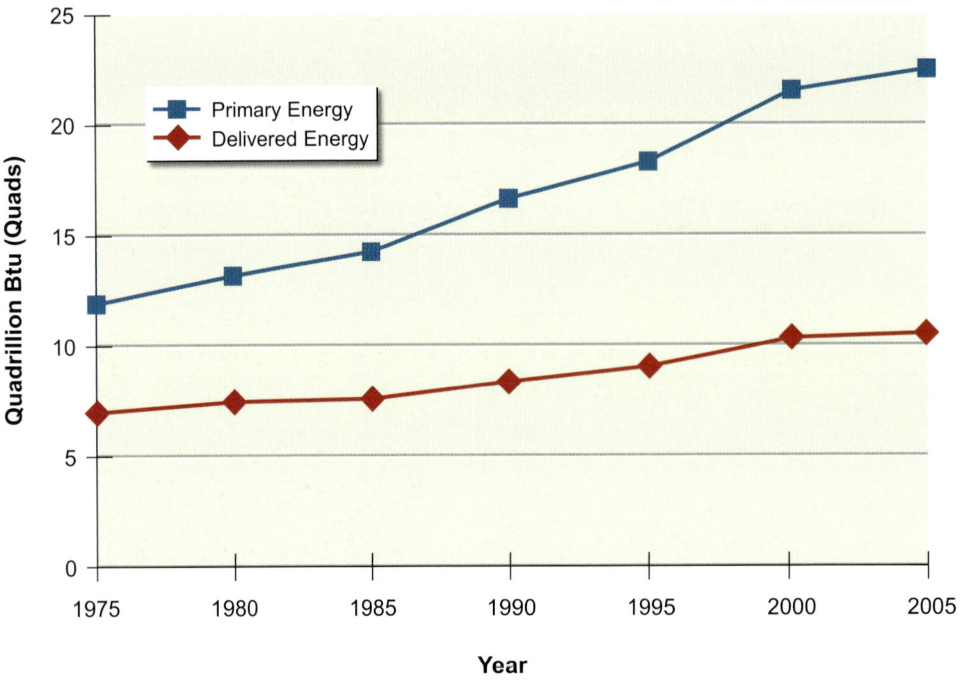

FIGURE 2.5 *U.S. commercial energy-use trends. Primary energy use (accounting for losses in electricity generation and transmission and distribution and for fuels, such as natural gas, used on-site) has increased faster than delivered energy use (which does not account for such losses but does include fuels used on-site) because use of electricity has increased faster than use of other fuels.*
Source: EIA, 2007b.

energy consumption survey showed that, on average, one-person households annually consumed 71 million Btu per capita; two-person households, 48 million Btu per capita; and three-person households, 35 million Btu per capita (DOE, 2009).

A geographic shift in population (e.g., that from the northeastern and midwestern regions of the United States to the more temperate southern and western regions of the country) was one of the factors leading to the decline in residential energy intensity. Energy intensity tends to be lower in the latter regions, especially on a delivered-energy basis. Improvements in energy efficiency resulting from the adoption of efficiency standards for appliances and the offering of utility-sponsored and government-sponsored demand-side management (DSM) programs also helped reduce residential energy intensity (Battles and Hojjati, 2005).

TABLE 2.1 Residential Sector Energy Intensity Trends

Year	Delivered (million Btu/household)	Primary (million Btu/ household)	Primary (1000 Btu/ft^2)	Primary (million Btu/household member)
1978	138	204		72
1980	114	176	101	63
1984	105	164	98	61
1987	101	163	94	63
1990	98	164	91	63
1993	104	172	92	66
1997	101	172		66
2001	92	164	79	64
2005	95	171	79	66

Note: Trend may look different depending on the metric used.
Source: DOE, 2009, available at www.eia.doe.gov/emeu/recs/recs2005/hc2005_tables/detailed_tables2005.html.

TABLE 2.2 Household Energy Expenditures by Income Level in 2001

Household Income[a]	Percentage of Households	Energy Expenditures (dollars)[a]	Percentage of Income Spent on Energy
Less than $9,999	10	1,039	16
$10,000 to $14,999	7	1,124	9
$15,000 to $19,999	8	1,290	7
$20,000 to $29,999	13	1,315	5
$30,000 to $39,999	13	1,398	4
$40,000 to $49,999	12	1,518	3
$50,000 to $74,999	20	1,683	3
$75,000 to $99,999	8	1,825	2
$100,000 or more	8	2,231	2

[a]2001 dollars.
Source: DOE/EERE, 2007.

Residential energy use varies by household income, as shown in Table 2.2. Upper-income households earning more than $100,000 annually in 2001 used about twice the energy used by lower-income households earning under $15,000 annually. But the energy burden (the fraction of income spent on energy) is much higher for lower-income households compared with middle- or upper-income households.

Commercial energy intensity measured in energy use per square foot of floor area declined over the 1979–1986 period but has fluctuated since 1986, as shown in Table 2.3. Commercial energy intensity has increased in particular types of buildings, such as health care and educational facilities. Efficiency improvements in lighting and air-conditioning have tended to reduce overall energy intensity, whereas greater use of amenities and devices such as computers and other plug loads have tended to increase it. Overall energy intensity in commercial buildings has declined in spite of a 45 percent increase in electricity use per square foot between 1983 and 2005 (Belzer, 2007). Energy use per square foot declined more on a delivered-energy basis than on a primary-energy basis during 1979–2003 owing to the increasing electrification of energy use.

There is great diversity in energy intensity in different commercial building types, as shown in Table 2.4. On the basis of delivered-energy and primary-energy use, food sales and food services facilities use more than two times as much energy per square foot of floor area as is used by office, retail, education, and lodging facilities. Likewise, health care facilities tend to have high energy use per square foot of floor area.

Table 2.5 presents a breakdown of energy end-use in residential and commercial buildings in 2005, as estimated by the Energy Information Administration (EIA). In housing, space heating represented about 48 percent of total energy

TABLE 2.3 Commercial Sector Energy Intensity Trends

Year	Delivered (1000 Btu/ft^2)	Primary (1000 Btu/ft^2)
1979	114.0	203.2
1983	97.5	187.1
1986	85.5	170.2
1989	91.6	180.4
1992	80.0	158.5
1995	90.5	180.1
1999	85.1	178.0
2003	91.0	191.0

Source: Energy Information Administration. Data through 1999 from http://www.eia.doe.gov/emeu/consumptionbriefs/cbecs/cbecs_trends/intensity.html. Data for 2003 from http://www.eia.doe.gov/emeu/cbecs/cbecs2003/detailed_tables_2003/detailed_tables_2003.html#consumexpen03.

TABLE 2.4 Commercial Sector Energy Intensity by Principal Building Activity, 2003

Principal Activity	Delivered (1000 Btu/ft²)	Primary (1000 Btu/ft²)
Education	83	159
Food sales	200	535
Food service	258	523
Health care	188	346
Lodging	100	193
Mercantile and service	87	204
Office	93	212
Public assembly	94	180
Public order and safety	116	221
Religious worship	43	77
Warehouse	45	94
Other	164	319

Source: Energy Information Administration. Data from http://www.eia.doe.gov/emeu/cbecs/cbecs2003/ detailed_tables_2003/detailed_tables_2003.html#consumexpen03.

use on a delivered basis and 31 percent on a primary basis. Water heating, space cooling, and lighting each represented 11–12 percent of total residential primary energy use. Electronic devices such as televisions, computers, and other types of office equipment represented about 8.5 percent of residential primary energy use in 2005, and this fraction increases as households acquire more and bigger electronic products.

Space heating in commercial buildings in 2005 accounted for 24 percent of delivered-energy use and 14 percent of primary energy use, on average. Lighting accounted for about 17 percent of delivered-energy use and more than 25 percent of primary energy use, on average. Likewise, the end-use of space cooling and ventilation accounted for nearly 13 percent of delivered-energy use and 19 percent of primary energy use, on average. The end-use data should be viewed as approximate owing to the lack of metered data by end-use. "Other" energy use in Table 2.5 includes laboratory, medical, and telecommunications equipment; pumps; and fuel use for combined heat and power production.

TABLE 2.5 Energy End-Uses in Buildings, 2005

	Residential Sector				Commercial Sector			
	Primary		Delivered		Primary		Delivered	
End-Use	(quads)	(%)	(quads)	(%)	(quads)	(%)	(quads)	(%)
Space heating	6.69	(30.7)	5.61	(48.2)	2.55)	(14.2	2.04	(24.0)
Space cooling and ventilation	2.67	(12.3)	0.84	(7.2)	3.42	(19.1)	1.09	(12.8)
Water heating	2.66	(12.2)	1.75	(15.0)	1.23	(6.8)	0.84	(9.9)
Lighting	2.40	(11.0)	0.75	(6.5)	4.57	(25.5)	1.44	(16.9)
Refrigeration	1.64	(7.5)	0.52	(4.4)	0.74	(4.1)	0.23	(2.7)
Electronics[a]	1.86	(8.5)	0.58	(5.0)	1.70	(9.5)	0.53	(6.2)
Laundry and dishwashers	1.05	(4.8)	0.38	(3.2)	NA[b]		NA[b]	
Cooking	0.98	(4.5)	0.48	(4.1)	0.35	(2.0)	0.27	(3.2)
Other	0.83	(3.8)	0.41	(3.5)	2.37	(18.2)	1.12	(13.2)
Adjustment[c]	1.02	(4.7)	0.32	(2.8)	0.98	(5.5)	0.92	(10.9)
Total	21.78	(100)	11.63	(100)	17.91	(100)	8.49	(100)

[a]Electronics include TVs, computers, and other office equipment.
[b]NA, not available.
[c]Adjustment to reconcile discrepancies between sources.
Source: DOE/EERE, 2007.

2.2 ENERGY EFFICIENCY TRENDS

Improvements in energy efficiency are a key factor in the decline in energy intensity in buildings over the past 30 years. Driven largely by research, development, and demonstration (RD&D), building energy codes, ENERGY STAR® labeling, and state and federal efficiency standards (see Chapter 5), the efficiency of new appliances has improved dramatically since the 1970s. For example, the average electricity use of new refrigerators sold in 2007 was about 498 kWh per year, 71 percent less than the average electricity use of new refrigerators sold 30 years earlier (AHAM, 2008). This is in spite of the fact that refrigerators have become larger and offer more features, such as automatic defrosting, ice makers, and through-the-door water and ice dispensers. Likewise, the average efficiency of other products, including air conditioners, gas furnaces, clothes washers, and dishwashers, has improved significantly over the past 30 years. Yet progress has been minimal for other products, such as water heaters. Less policy attention has been paid to the energy use of these other appliances and equipment, accounting in part for this divergence of trends in energy efficiency improvements.

Significant energy efficiency gains have also been made in lighting. The sales and use of compact fluorescent lamps (CFLs), which use about 75 percent less electricity per unit of light output relative to incandescent lamps, have increased greatly in the past decade. As shown in Figure 2.6, CFL shipments (based on data on imports, since all CFLs are imported into the United States) increased from about 21 million units in 2000 to 185 million units by 2006. But as a result of various factors—growing state, regional, and utility energy efficiency programs, along with a federal procurement program aimed at reducing the size and improving the quality of CFLs; stepped-up marketing efforts by some large retailers; and national promotion campaigns led by the federal ENERGY STAR® program—CFL shipments jumped to about 400 million units in 2007. This means that CFLs represented about 20–25 percent of all screw-in lightbulbs (incandescent and fluorescent) sold in 2007. Given that CFLs last 5 to 10 times longer than incandescent lamps, CFLs actually accounted for the majority of the total "light service" (i.e., lumen-hours) sold in 2007. CFLs do have some drawbacks, such as their use of mercury and difficulty with dimming. However, the small amount of mercury released to the environment if a CFL is disposed of in a landfill is much less than

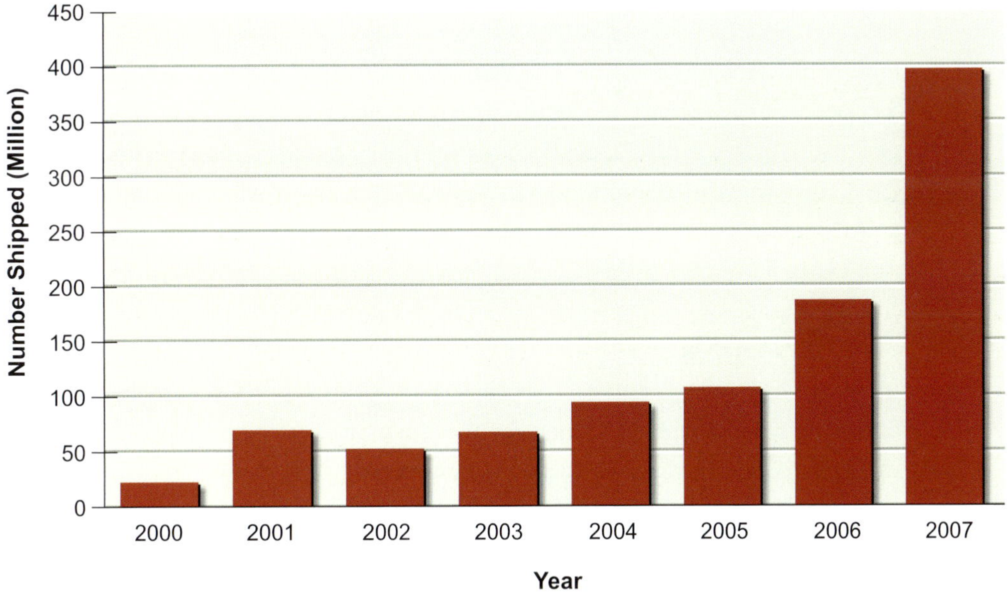

FIGURE 2.6 *Shipments of compact fluorescent lamps.*
Source: U.S. Department of Commerce data obtained from USA Trade Online, available at https://orders.stat-usa.gov/on_sam.nsf/fsetOrder/UTO.

the mercury avoided through the reduction in electricity generation, given average mercury emissions associated with electricity generation in the United States (ENERGY STAR®, 2008). Solid-state lighting, which addresses these shortcomings of CFLs, is now emerging in the marketplace (see Section 2.6.1).

Energy-efficient fluorescent lighting fixtures containing T8 fluorescent lamps (these have a diameter of 1 inch) and high-frequency electronic lamp ballasts used in commercial buildings use 30–40 percent less power per unit of light output compared with older fixtures containing T12 lamps (which have a diameter of 1.5 inches) and electromagnetic ballasts (Suozzo et al., 2000). The U.S. Department of Energy (DOE) played a major role in the development of electronic ballasts during the early 1980s (NRC, 2001; Geller and McGaraghan, 1998). Utility and state DSM programs and the Environmental Protection Agency's (EPA's) "Green Lights" program (which developed into the ENERGY STAR® program), helped move the product from the laboratory into the marketplace. As shown in Table 2.6, the market share for electronic ballasts increased from about 1 percent in the late 1980s to 47 percent by 2000 and then to 73 percent in 2005 (DOE/EERE, 2007). Minimum efficiency standards promulgated by the U.S. DOE in 2000 that took effect in 2005 are facilitating the transition from magnetic to more efficient electronic ballasts.

Periodic EIA surveys of commercial buildings show growth in the use of energy efficiency and conservation measures from 1992 to 2003, as indicated in Table 2.7. The increase in the use of energy-efficient lighting devices such as CFLs, electronic ballasts, and specular light reflectors is most noteworthy. At the same time, a significant fraction, and in some cases a majority of commercial buildings, still do not use common energy efficiency measures such as energy management and control systems or HVAC economizer cycles (which make use of outdoor air for cooling when temperature and humidity levels permit).

The adoption of ENERGY STAR®-labeled products and new homes has also increased substantially in recent years. For example, the construction and certification of ENERGY STAR® new homes—which must be at least 15 percent more efficient for heating, cooling, and water heating than homes built to meet the International Energy Conservation Code (IECC) or 15 percent more efficient than the prevailing state energy code, whichever is more rigorous—grew from about 57,000 new homes in 2001 to 189,000 new homes in 2006. That is, 11.4 percent of all new homes built in 2006 were certified as ENERGY STAR®-compliant. The market share for ENERGY STAR® new homes exceeded 25 percent in 10 states and 50 percent in 2 states—Nevada and Iowa, in 2006 (EPA, 2007a).

TABLE 2.6 Shipments of Fluorescent Lamp Ballasts

Year	Number of Magnetic Type Shipped (million)	Number of Electronic Type Shipped (million)	Electronic Market Share (%)
1986	69.4	0.4	1
1988	74.6	1.1	1
1990	78.4	3	4
1992	83.7	13.3	14
1994	83.5	24.6	23
1996	67	30.3	31
1998	63.9	39.8	38
2000	55.4	49.3	47
2001	46.9	52.5	53
2002	40.7	53.8	57
2003	35.2	54.4	61
2004	30.5	59.2	66
2005	22.2	61.3	73

Source: DOE/EERE, 2007.

TABLE 2.7 Growth in the Use of Energy Efficiency Measures in Commercial Buildings

	Percentage of Floorspace with Measure	
Efficiency Measure	1992	2003
HVAC economizer cycle	27	33
HVAC variable air volume system	21	30
Energy management and control system	21	24
Compact fluorescent lamps	12	43
Electronic lamp ballasts	NA[a]	72
Specular light reflectors	22	40
Multipaned windows	44	60
Tinted or reflective window glass	37	59
Daylighting sensors	NA[a]	4

[a]NA, not available.

Source: Energy Information Administration, *2003 Commercial Buildings Energy Consumption Survey: Building Characteristics Tables* and *Commercial Buildings Characteristics 1992*, both available on the EIA website.

2.3 THE POTENTIAL FOR ENERGY EFFICIENCY IN BUILDINGS

2.3.1 Review of Studies

Energy use in buildings embraces dozens of end-uses, and for each there are a variety of efficiency-improvement technologies (and levels of intensity of application—e.g., insulation thickness) available. The more rigorous and analytically valid studies of efficiency potential[2] aggregate similar measures—often hundreds or thousands of them (especially if the research disaggregates the measures by building type, climate, and so on)—into supply curves of energy efficiency potential. The supply curves depict graphically the energy savings available from a given measure (or aggregation of measures) as a function of the cost of saved energy. Section 2.5 presents conservation supply curves for both residential and commercial buildings.

This section presents a review of what the Panel on Energy Efficiency Technologies believes is a representative sample of the most credible such studies. Most of these studies concentrate only on the United States, but the one study reviewed by the panel that looked at worldwide savings potential (IEA, 2006) comes to conclusions very similar to those of the U.S. studies regarding percentage savings.

All of these studies rely on similar methodologies. They look at end-use data on a level of disaggregation and detail far higher than that available in most energy-demand forecasts. Efficiency measures are compared to the efficiency inherent in one or several average base cases. Capital stock turnover is explicitly considered: new appliances and buildings are added to the stock while old buildings and products are slowly retired, and retrofits are considered for items such as building envelopes. None of the studies assumes early retirement as an efficiency measure. Because some degree of "natural" improvement in energy efficiency is assumed in the "business as usual" case, some of the initially projected savings are in the end subtracted.

The panel reviewed and synthesized several of the most important (though not all) relevant studies carried out at the national, regional, state, and utility levels. Some of the major national or regional energy-savings potential studies performed over the past 12 years include those by Optimal Energy, Inc. (2003); IWG

---

[2]Analysts of building energy use often use the term "potentials" studies to refer to studies of the potential for energy savings. This report uses the latter terminology.

(1997, 2000); Energy Innovations (1997); Nadel and Geller (2001); NPCC (2005); EETF (2006); and Creyts et al. (2007).

Studies of the potential for energy efficiency improvement in buildings typically assess this potential in terms of three categories: *technical potential,* which is the broadest and includes technologies with improved performance but not necessarily lower costs; *economic potential,* which includes those technologies that are judged to be economically attractive; and *achievable potential,* which is a subset of economic potential that takes account of various market failures and barriers. This section looks most closely at the economic potential: that is, how much potential there is for energy savings at prices of energy up to or moderately above current or projected electricity market prices.

The economic potential as assessed in any study depends on the following: how many end-uses are examined in detail (since it is hard to posit a supply curve for saved energy from the "miscellaneous" or "other" categories of energy use); the timeframe of the study; the policy authority of the agency that commissioned the study (e.g., a typical utility-sponsored study will not look at the technical and economic potential of codes and standards, and a state-funded study might not consider measures that require federal action); and how the study will be used (studies that lead to mandatory goals typically show less potential than studies with broader and more flexible uses).

The results of the studies also depend in part on the policies that the authors assume will be used to achieve the potential. If the authors assume that the use of a technology can be boosted through standards or through generous financial incentives that cause close to 100 percent adoption, they will have a larger efficiency resource potential than the studies that do not make this assumption. For example, a study carried out by five national laboratories (IWG, 1997) postulates penetration rates of 35 percent and 65 percent in its two scenarios; an assumption that cost-effective technologies could be implemented at near-100 percent levels would have produced substantially different results. The subsequent study *Scenarios for a Clean Energy Future* included explicit assumptions about policies, programs, and their impacts (IWG, 2000).

If a study assumes that the adoption of strong standards or incentives that achieve near-100 percent market acceptance induces manufacturers or designers to invest in new product development to introduce a next-generation product (which studies generally do not assume), then the results will show more potential savings than if next-generation products are not included. Thus, for the limited number of products for which these policies were adopted, most studies understate the efficiency advances that actually occurred (Goldstein and Hoffman, 2004). This issue is discussed further in the succeeding paragraphs.

Nadel et al. (2004) reviewed 11 studies of energy savings potential in buildings, covering the period 2000 through 2004. This meta-analysis indicated a potential in the United States for substantial technical, economic, and achievable energy savings (Table 2.8). Across all sectors (residential, commercial, and industrial), the studies reviewed by Nadel et al. (2004) showed a median technical savings potential of 33 percent for electricity and 41 percent for natural gas (see Table 2.8). The median achievable savings potential was 24 percent for electricity (an average of 1.2 percent per year) and 9 percent for natural gas (an average of 0.5 percent per year). The review compared the findings on achievable potential to recent-year actual savings from portfolios of electricity and natural gas efficiency programs in leading states and found substantial consistency. (Note that the natural-gas savings potential suggested by these studies is less than that indicated

TABLE 2.8 Summary of Results from the ACEEE Meta-Analysis: Studies of the Potential for Energy Savings in Buildings, 2000–2004

			Potential (%)		
Region	Year	No. of Years	Technical	Economic	Achievable
Electricity					
California	2003	10	18	13	10
Massachusetts	2001	5		24	
New York	2003	20	36	27	
Oregon	2003	10	31		
Puget	2003	20	35	19	11
Southwest	2002	17			33
Vermont	2003	10			31
United States	2000	20			24
Median			33	21.5	24
Natural Gas					
California	2003	10		21	10
Oregon	2003	10	47	35	
Puget	2003	20	40	13	9
Utah	2004	10	41	22	
United States	2000	20			8
Median			41	22	9

Source: Nadel et al., 2004.

in the conservation supply curves presented in Section 2.5, because the studies assumed only limited policy and program interventions for the purpose of estimating achievable natural-gas savings potential.)

Nadel et al. (2004) also reported on savings potential by sector. The median technical potential for saving electricity across the studies was 32 percent for the residential sector, 36 percent for the commercial sector, and only 21 percent for the industrial sector. The median achievable potentials were 26 percent, 22 percent, and 14 percent, respectively. For natural gas, savings potentials appear to be higher in the residential sector than in the commercial sector. The median technical potential for gas savings across the studies is 48 percent in the residential sector and 20 percent in the commercial sector. The median achievable gas-savings potential drops to 9 percent in the residential sector and 8 percent in the commercial sector.

These savings percentages are based on the business-as-usual cases specific to each study. But because constructing a business-as-usual forecast is problematic, there is some uncertainty about what is the appropriate basis for calculating savings. However, such calculations are still useful for the purpose of rough estimates; they would have to be refined for use in program planning.[3]

The overall median achievable electricity savings potential across the studies is 1.2 percent per year, with similar medians for each of the sectors. However, the annual achievable potential is often lower for studies extending further in time (e.g., 20 years) than for shorter-term studies. Nadel et al. (2004) suggest that this is primarily because existing technologies can be heavily adopted over the first decade, and the new technologies and practices that would emerge during the second decade are not included in most studies.

A detailed comparison of the results of various studies is desirable, but it is problematic because many studies examine hundreds or even thousands of discrete efficiency measures, making such a comparison difficult and costly to perform.

---

[3]Each study of the potential for energy savings attempts to address the issue of reconciling the base case in the study (which involves much more detailed data than the base case in the energy forecast) with the overall results of the forecast. While such a process introduces some levels of uncertainty into the calculation, and greater levels of uncertainty to the casual reader who is trying to interpret the results without the help of the large spreadsheets used in the savings analysis, the studies reviewed seem to have done a good job of avoiding double counting or missed potentials. The errors that remain have little practical consequence because they do not affect supply resource planning, nor do they affect efficiency program planning or evaluation.

There is to the panel's knowledge no literature offering such comparison and contrast or even comprehensively reviewing individual studies.

Studies of technical and economic energy-savings potential generally capture energy efficiency potential at a single point in time based on technologies that are available at the time a study is conducted. But new efficiency measures continue to be developed and to add to the long-term efficiency potential. This trend is illustrated by comparing two studies on available electricity-savings opportunities that were prepared for New York State in 1989 and 2003 (see Figure 2.7).

In the first of the two studies, Miller et al. (1989) examined more than 70 efficiency measures and found an economic potential of 27 percent electricity savings, based on a 5 percent real discount rate. This study included such measures

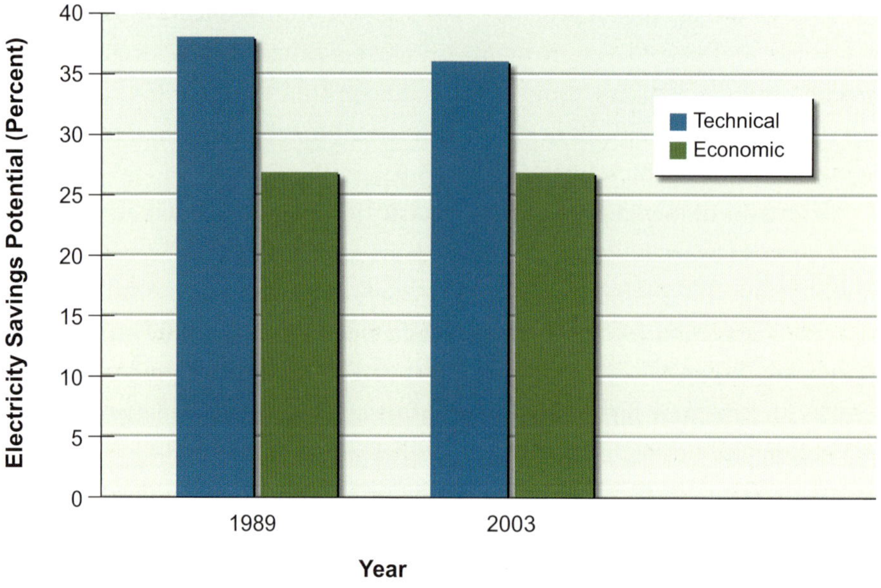

FIGURE 2.7 *Comparison of the electricity-savings potential identified in two studies for New York State (percent of annual electricity consumption). Despite wide use by 2003 of many technologies considered in Miller (1989), which thus became part of the "baseline" for 2003, the economic potential for savings in 2003 remained large, primarily because new technologies had become available, including many that were still under development in 1989. "Technical potential" is the maximum amount of energy use that could be displaced by energy efficiency measures, disregarding factors such as cost-effectiveness and the willingness of end-users to adopt the measures. "Economic potential" is the subset of technical potential that is cost-effective compared with conventional energy supplies.*
Source: ACEEE (2007), based on data in Miller et al. (1989) and Optimal Energy (2003).

as energy-saving fluorescent lamps (e.g., 34 W tubes replacing 40 W tubes) and efficient magnetic ballasts, as well as what at the time were called "very high efficiency" lamps and ballasts (e.g., T8 lamps with electronic ballasts).

In the second study, Optimal Energy, Inc. (2003) examined more than 100 efficiency measures, including many that had been added or updated after the 1989 study. The 2003 study also found that, 4 years later, the economic savings potential remained around 27 percent because of promising new technologies. Many measures included in the 1989 study were dropped because they were already widely adopted by 2003 and were therefore included in the base-case forecast (e.g., 34 W fluorescent lamps and efficient magnetic ballasts). Instead, much of the savings in the 2003 study came from measures that were still under development in 1989 (e.g., new "super T8" lamps and "pulse start" metal halide lamps) or were otherwise not included in the 1989 study.

As revealed by these studies, the potential for cost-effective energy efficiency improvements is very large. The exact potential is uncertain, but even if it is 30 percent less or 30 percent greater than the median case presented in this report, it still represents the largest, least expensive, and shortest-lead-time resource for balancing energy supply and demand.

It should be noted that the plausible uncertainty around the median savings figures reported here is not symmetric. The risk of overestimating efficiency potential is minimal, owing to the methodologies that are used in the studies. Instead, the studies openly and intentionally make assumptions that lead to "conservatively" low estimates of the efficiency resource, as discussed in Section 2.3.2.

Some states and utilities have achieved a significant share of the energy efficiency potential indicated by the studies. For example, a review was conducted recently of the degree to which the efficiency potential identified for California (Rufo and Coito, 2002) was realized through electric utility efficiency programs. Messenger (2008) found that utility DSM programs achieved about 25 percent of the projected 10-year energy-savings potential in the commercial sector and nearly 27 percent of the projected energy-savings potential in the residential sector from just 3 years (2004–2006) of utility DSM program activity. The amount of savings potential achieved varied among end-uses and building types. The utilities spent about 2 percent of their sales revenue on DSM programs in order to capture this savings. Further information on utility DSM programs is provided in Chapter 5. A study of the energy savings to date by sector in New York State relative to the potential savings that were identified reveals that the greatest energy efficiency potential for buildings remains in the residential sector, despite recent achievements, because energy savings are more costly and more difficult to obtain in the residential sector than in the commercial sector (DeCotis et al., 2004).

2.3.2 Limitations of Studies of Energy Efficiency Potential

Studies of the potential for energy efficiency are intended to provide specific answers to well-framed policy questions. But the question of how much efficiency is available at what price is not well framed, because the meaning of "available" is ambiguous with respect to several critical issues:

- *The timeframe over which the potential is available.* The efficiency potential within 3 years from retrofitting homes is a very different policy question from the potential within 30 years from retrofitting homes, both in terms of the number and the type of efficiency measures that can be implemented.
- *The level of incentive required for realizing the potential.* The greater the incentives paid to achieve the savings, all else being the same, the greater the savings achieved. The limiting case is a 100 percent payment, that is, free installation (no cost to the building occupant), in which case a very high level of implementation is possible.
- *The motivation of society in pursuing the energy savings.* The amount of savings available when a nation or region is facing a crisis—for example, the need to relieve the California electricity crisis in 2001, or the need in New York City to achieve reductions in electricity demand quickly in order to maintain reliability—is much larger than in a situation in which energy efficiency merely reduces normal utility bills.

A number of biases can lead studies to understate energy efficiency potential (Goldstein, 2008), including (1) sponsoring agencies' motivation to underestimate savings potential in order to avoid challenges to their energy-savings goals; (2) the exclusion of new and emerging technologies; (3) failure to consider that energy efficiency technologies are likely to improve in performance and decline in cost over time; (4) failure to consider the potential to adopt energy efficiency measures in an integrated manner with synergistic effects (such as the whole-building approach to improving efficiency); (5) failure to consider efficiency measures' non-energy benefits, which in some cases are substantial and can be valued more than the energy benefits (Romm, 1999); and (6) the development of studies of potential in contexts in which the risk to the researcher of making an error is asymmetric (that is, a data point that overestimates savings or underestimates cost can be much more problematic than one that does the reverse).

There are, however, cases in which studies of energy efficiency potential have been overly optimistic. For example, although CFLs had been available in the marketplace and identified as a viable energy efficiency measure since the late 1970s, they only started to gain widespread acceptance by consumers in about 2000. This slow market penetration is attributed to a variety of early CFL deficiencies, including large size relative to incandescent lamps, performance issues, poor light quality, and high first cost of CFLs produced in the 1980s and 1990s (Sandahl et al., 2006). Once these deficiencies were addressed, CFLs became more appealing to consumers. However, studies that projected large energy-savings potential from CFLs during the 1980s and 1990s (Geller et al., 1986; Rosenfeld, 1985) overestimated their potential, at least prior to 2000.

Likewise, heat-pump water heaters have been produced on a limited basis since the early 1980s. These devices use one-third to one-half as much electricity as that used by electric-resistance water heaters, with the energy savings paying back the incremental first cost in 5 years or less (Ashdown et al., 2004). However, heat-pump water heaters are not being produced on a large scale and have had little market penetration in the United States. This is due to performance problems with early heat-pump water heaters (e.g., poor reliability), lack of a supply infrastructure (e.g., no production by major water heater manufacturers), and the nature of the water heater market (e.g., split incentives and many purchases made in a rush). Once again, studies which assumed that these problems would be overcome during the past two decades (Geller et al., 1986) overestimated achievable energy-savings potential.

These examples do not reflect errors in estimates of the energy efficiency potential ultimately available. Rather, they reflect overly optimistic assumptions regarding technological maturity and/or underestimates of the difficulty of overcoming the market barriers and failures that prevent broad commercialization and market acceptance.

2.4 APPROACHES TO UNDERSTANDING ENERGY EFFICIENCY POTENTIAL

There are several approaches to reviewing the technologies and design principles available today to make buildings more energy-efficient. Each illuminates a subset of important engineering and physics issues but obscures other subsets. Each approach has its advantages and disadvantages. Therefore, this report does not adopt one preferred style of presentation but instead looks at three different approaches:

- An integrated whole-building or system-wide approach,
- An approach by end-use and technology description, and
- An approach by individual "widgets" or detailed energy efficiency technologies and measures.

2.4.1 Integrated Whole-Building or System-Wide Approach

The first approach looks at integrated whole-building or system-wide energy use and describes the types of technological improvements that could create savings of a given percentage for whole buildings or whole systems. For example, a small but growing subset of new commercial buildings achieve a savings of 50 percent (relative to prevailing model Energy Code ASHRAE 90.1) in regulated energy use (heating, cooling, air-conditioning, water heating, and lighting).[4] Reviews of highly efficient commercial buildings (NBI, 2008; ASHRAE, 2008; Torcellini et al., 2006) show that such buildings incorporate the following measures:

- *High-efficiency electrical lighting systems* that not only incorporate state-of-the-art lamps, ballasts, and luminaires (lighting fixtures), but also use luminaires to provide the desired lighting in the right places (e.g., as task lighting) and use controls that limit electrical lighting when daylighting is available;
- *Fenestration systems and designs* that reduce heat gain in climates with high cooling requirements;
- *HVAC controls* that provide for the effective operation of the HVAC system during part-load conditions;[5] and

---

[4]"Regulated energy use" refers to energy use covered by building energy codes. Such codes do not apply to plug-in office equipment, for example.

[5]There are many reasons why efficiency at part load can be lower. Examples range from systems that do not modulate but simply turn on or off, to chiller designs that are optimized for efficiency at full load and work poorly at part load (perhaps because they are not tested or marketed on the basis of such performance), to overall systems controls that continue to operate one part of the system at full-power use even though other parts are at partial power and do not require that support.

- *On-site power generation* such as combined heat and power systems or solar photovoltaic generation to reduce purchased energy.

Low-energy buildings do not always operate as they were designed to do. Experience shows that in order to maximize real-world energy savings, it is critical to properly commission and monitor the performance of low-energy buildings and to ensure that control systems are working properly and are adjusted to account for occupancy conditions (Torcellini et al., 2006; Mills, 2009).

The net incremental first cost of achieving a 50 percent reduction in energy use through an integrated approach[6] can be at or near zero; the savings from downsizing and simplifying HVAC systems generally pay fully for the additional costs of measures such as additional insulation, better windows, and daylighting (Goldstein, 2008). But the next increment of savings, up to 60 percent, has very few exemplars.

For residential buildings, a whole-house approach can result in a 50 percent or greater savings in heating and cooling and a 30–40 percent savings in total-home energy use, and can do so cost-effectively (DOE, 2004a; Dunn, 2007).[7] This conclusion is also supported by the fact that more than 8,000 applications were submitted for the federal tax credit for 50 percent savings for new, single-family homes during the first year of its availability—calendar year 2006—despite substantial delays in the availability of guidance from the Internal Revenue Service on how to perform the savings calculations and computer software for doing so. Although comprehensive evaluations of the tax credit are not yet available, the market share of new homes qualifying for the credit grew from below 1 percent before the credit to 1 percent of eligible homes in 2006 and 3 percent in 2007. For 2008, the number of qualifying homes grew to more than 23,000 (about 4.6 percent of all homes built), according to a survey of home energy raters (S. Baden, RESNET, personal communication, May 1, 2008).

---

[6]An "integrated approach" involves integrating the design of the HVAC system with that of the envelope system and the lighting system and its controls. Current design practice involves designing the envelope of the building independent of such integrative consideration, then passing the design onto HVAC engineers, who design the HVAC system without looking back at what could be done differently at the envelope or without looking forward to how lighting designs could enable improved HVAC designs.

[7]The savings are relative to a new home built to just meet the prevailing model energy code, namely, the 2006 International Energy Conservation Code.

2.4.2 Approach by End-Use and Technology Description

Because whole-building studies focus on the level of savings achieved and the cost of getting there, they often do not specify the kinds of energy-saving measures used and how relevant they would be to broad-scale application across the economy. Instead, the savings estimates sometimes are based on measured results from demonstration buildings or multibuilding projects, or on case studies; they sometimes consider simulated energy savings based on integrated designs of new buildings or retrofits; and they sometimes are based on more than one approach.

Other studies, however, rely on the end-use and technology description approach to identifying energy efficiency potential. This approach assigns energy use to major end-use categories and reviews the specific technologies and measures available for reducing energy use in each category (often ordered by cost-effectiveness). The end-use approach is based on text and explanation of technologies and measures. Most of these technologies and measures could be incorporated into existing buildings.

As an example, space heating is the largest user of energy in residential buildings, and cooling is the second-largest or close to second-largest user. Similar energy-saving measures and strategies can be applied to both. These efficiency measures and strategies include the following (Scheckel, 2007; Amann et al., 2007):

- *Increasing insulation in all components compared with what is done according to current practice, including the use of selective coatings on windows.* These coatings are chosen on the basis of the local climate, to reduce thermal transmission (by increasing the thermal-infrared emissivity and reflectivity of the window). They are most effective on west- and east-facing windows in climates requiring cooling or in transitional climates where an efficient shell can obviate the need to buy an air conditioner.
- *Moving ducts into the conditioned space for new construction, and reducing leakage through on-site pressure testing in both new and existing homes.*
- *Improving heating and cooling systems themselves,* for example, by using programmable thermostats, by using higher-efficiency furnaces that condense water vapor produced by the combustion of methane (or other fuels) to extract additional energy and achieve efficiencies over

90 percent, by using variable-speed and higher-efficiency motors and fans for air circulation, and by using ground-source heat pumps (for electric heating) or gas-fired heat pumps.

- *Upgrading equipment for cooling,* focusing on better heat transfer from evaporators and condenser coils in air conditioners and employing variable-speed drives that allow units to operate efficiently at partial loads (rather than turning on and off frequently). This measure can control humidity more effectively, as well as save energy.
- *Changing ventilation systems to provide sufficient fresh air to a system that uses the proper amount of mechanical ventilation while sealing the home to nearly airtight standards.* Controlling ventilation can greatly mitigate indoor air quality and mold problems while also offering the opportunity to recover both latent and sensible heat from the exhaust airstream.
- *Using evaporative cooling.* While once-through evaporative coolers work well only in desert climates, indirect systems that transfer sensible heat from the humidified airstream can provide comfort in a much broader zone of climate while using about one-quarter or less of the energy of compression-based cooling.
- *Making greater use of passive solar heating and cooling,* although this design technique has not yet found widespread acceptance in the marketplace owing to the difficulties of custom designing the orientation and thermal characteristics of each home.

After space heating and cooling, the next-largest user of energy in residences is water heating. Water-heating energy use can be reduced both by improving the efficiency of the water-heating device itself and by reducing the demands for hot water, including for clothes washing and bathing, throughout a home. Substantial gains have been made in the best-performing clothes washer and showerhead products compared with standard products. For example, the highest specification for utility incentives for washing machines has a modified energy factor (MEF) of 2.2; current stock has an MEF of about 0.85. Heat-pump water heaters, which have become very popular in Japan, can reduce electricity use by two-thirds relative to an electric-resistance water heater. Older showerheads use 3.5 or more gallons per minute; newer ones meeting current standards use 2.5 or fewer gallons per minute, and a few newer models use about half this level of waterflow to provide a comfortable shower (Harrod and Hain, 2007). Similar lists of technologies for residential lighting and appliances are found in most studies of efficiency potential. Beyond technologies themselves, efficiency can be improved through residential lighting design that raises the ratio of productive light output (lux on the visual task) to power use in homes to a level comparable to that in office buildings.

The major sources of energy use in commercial buildings are heating, ventilation, cooling, and lighting. Studies of energy efficiency potential usually look at specific measures within these categories, such as improving the rated efficiency of rooftop air conditioners by 20–30 percent or substituting 100 lumen per watt lamp-ballast combinations for existing product combinations that provide fewer than 70 lumens per watt.

2.4.3 Approach by Individual "Widgets" or Detailed Energy Efficiency Measures

The energy end-use approach to estimating potential savings suffers from the limited ability of readers to review critically the assumptions that are made and the models that are used to derive the costs and savings for specific energy efficiency measures. This problem is accompanied by often limited guidance to program administrators about how best to achieve the savings—that is, what types of equipment or designs should be promoted.

In contrast, "widget"-based supply curves look at technologies and measures at a very detailed level. They can involve a spreadsheet of many hundreds or thousands of lines that tabulates detailed technical measures. They can cover both retrofits and new buildings by establishing separate sets of rows in the spreadsheets for retrofits compared with new buildings. This is the approach that relates most closely to the policies and programs used to obtain energy savings through improved energy efficiency.

Widget-based analyses look at the same types of technologies and measures as those considered in whole-building- or end-use-based analyses, but they also include the following for residential buildings:

- More efficient appliances, by efficiency rating;
- More efficient heating and cooling equipment, by efficiency rating;
- Additions of insulation (increasing "R" values) to ceilings, walls, and floors; and
- The substitution of CFLs and light-emitting diodes (LEDs) for incandescent lightbulbs.

For commercial buildings, the more detailed approaches include the following:

- More efficient lamps, ballasts, and luminaires;
- The substitution of more efficient lighting sources for less efficient ones (such as compact fluorescent lamps for general-service incandescents or downlights, or ceramic metal halide lamps for incandescent reflectors, or the use of infrared-reflective incandescent reflector lamps instead of conventional ones, or the use of LEDs for colored light sources);
- Controls to reset air-conditioning system temperatures;
- Variable-speed fans/drives and pumps;
- Lower-pressure fan systems; and
- Occupancy sensors for lighting and air quantities.

While widget-based analyses are easier to review and interpret, they tend to exclude many cost-effective options for systems integration, such as the following:

- Using lighting designs that optimize the distribution of light so that it is brightest where the most light is desired and less intense elsewhere;
- Using envelope designs that permit daylighting, especially in commercial buildings;
- Using envelope measures that are intended to reduce the size or complexity of the HVAC system;
- Using separate ventilation systems in which the benefits include occupant satisfaction and the ability to control the system under nontypical operating conditions; and
- Changing the building's orientation to take advantage of passive heating or cooling.

The amount of efficiency available at any particular cost from a widget-based, detailed end-use-and-technology analysis is generally lower than what would be estimated by a whole-building-based analysis. However, the results are easier to review and validate and may thus be more credible. The discussion of whole-building-based analysis noted that a number of commercial buildings achieve 50 percent savings with no increase in first cost. But buildings achieving such 50 percent savings are not normally included on supply curves, in part because buildings that achieve this can be seen as unrepresentative of the savings across the sector.

2.5 CONSERVATION SUPPLY CURVES

This section presents conservation supply curves for residential and commercial buildings developed in 2008 by researchers from the Lawrence Berkeley National Laboratory (LBNL; Brown et al., 2008). The analysis starts with the reference case from the EIA's Annual Energy Outlook (AEO) 2007 as a business-as-usual (BAU) scenario, with disaggregation by fuel and end-use (EIA, 2007a). The researchers adjusted the published AEO end-use consumption values for 2030 to allocate some of the consumption in the "other uses" category (mainly cooking and electronics) to the traditional end-uses in which that consumption appropriately belongs. This reallocation was based on data published by the Department of Energy (DOE/EERE, 2007). Tables 2.9 and 2.10 show the revised AEO reference case that is used as the BAU scenario, with energy consumption and cost of conserved energy (CCE) presented in terms of electricity and natural gas.

The analysis considers only electricity and natural gas, which together account for approximately 92 percent of the primary energy used in U.S. buildings. Petroleum products—distillate fuel oil and liquefied petroleum gas, or LPG—account for most remaining energy use in buildings, with approximately 12 percent of homes using one of these two petroleum products as the primary heating fuel (EIA, 2007b). The analysis of the natural gas space-heating-savings potential presented below most likely applies to homes heated by fuel oil and LPG as well, but this was not explicitly analyzed by Brown et al. (2008).

The BAU scenario, which includes some level of energy efficiency improvement driven by market forces as well as by codes and standards, assumes that residential electricity use increases 1.4 percent per year and that commercial electricity use increases 1.9 percent per year on average during 2006–2030. For comparison, residential electricity use increased 2.4 percent per year and commercial use 2.8 percent per year on average over the period 1990–2006 (EIA, 2007b). With respect to the use of natural gas, the BAU scenario assumes growth rates of 0.8 percent per year in the residential sector and 1.6 percent per year in the commercial sector over the period 2006–2030. It should be noted that the effects of the Energy Independence and Security Act of 2007 (EISA; Public Law 110-140) are not included in the BAU scenario.

2.5.1 Methodology and Efficiency Measures

To calculate cost-effective energy-savings potential in 2030, Brown et al. (2008) compiled percentage savings estimates by end-use, drawn from several prior stud-

TABLE 2.9 Summary of Residential Building Energy Consumption, Savings Potential, and Efficiency Measure Costs in 2030, by End-Use

Fuel and End-Use	Business as Usual (BAU) 2030 U.S. Consumption[a]	Technoeconomic Potential: % Savings Relative to BAU Case	Technoeconomic Potential: Consumption Savings	Cost of Conserved Energy[b]
Electricity	*(TWh)*		*(TWh)*	*(2007¢/kWh)*
Space heating[c]	164	17	28	3.5
Space cooling[c]	328	27	89	5.3
Water heating[c,d,e]	149	27	39	2.0
Refrigeration[c]	121	31	38	4.6
Cooking[c,e]	103	0	0	N/A
Clothes dryers[c,e]	103	0	0	N/A
Freezers[c]	42	21	9	7.4
Lighting[c]	338	50	169	1.2
Clothes washers[c]	9	50	4	2.3
Dishwashers[f]	11	11	1	5.8
Color televisions[c]	267	25	67	0.9
Personal computers[f]	68	57	39	4.3
Furnace fans[f]	40	25	10	3.7
Other uses[c]	154	48	74	1.9
Total electricity	1896	30	567	2.7
Natural Gas	*(Quads)*		*(Quads)*	*(2007$/million Btu)*
Space heating[g]	3.89	30	1.15	5.5
Space cooling[g]	0.00	0	0.00	N/A
Water heating[g]	1.20	29	0.35	11.8
Cooking	0.26	0	0.00	N/A
Clothes dryers[g]	0.09	3	0.00	2.9
Other uses[g]	0.04	10	0.00	1.1
Total natural gas	5.47	28	1.51	6.9

Note: A corresponding table for 2020 can be found in Brown et al. (2008).

[a]2007 AEO reference case (EIA, 2007a) end-use consumption for the "Other uses" end-use was reallocated to match the 2007 DOE Buildings Energy Databook 2007 (DOE/EERE, 2007) end-use shares.

[b]End-uses with cost of conserved energy (CCE) listed as N/A were not analyzed by the LBNL researchers.

[c]Source for potential savings and CCE is CEF study Table D-1.1 (IWG, 2000). Values for CCE are from the CEF Advanced Case, calculated using a real discount rate of 7 percent and lifetimes as shown in CEF study Appendix C-1.

[d]CCE for electric water heating was incorrect in the original CEF study (IWG, 2000) and has been corrected here.

[e]CEF study results (IWG, 2000) were adjusted to remove fuel switching (electric to gas) as a measure for water heaters, cooking, and clothes dryers.

[f]Source for potential savings and CCE is the updated LBNL analysis documented in Brown et al. (2008).

[g]Source for potential savings and CCE is the New York State natural gas potential study (Mosenthal et al., 2006).

Source: Brown et al., 2008.

TABLE 2.10 Summary of Commercial Building Energy Consumption, Savings Potential, and Efficiency Measure Costs in 2030, by End-Use

		Technoeconomic Potential		
Fuel and End-Use	Business as Usual (BAU) 2030 U.S. Consumption[a]	% Savings Relative to BAU Case	Consumption Savings	Cost of Conserved Energy
Electricity	*(TWh)*		*(TWh)*	*(2007¢/kWh)*
Space heating[b]	77	39	30	0.5
Space cooling[b]	238	48	115	2.8
Water heating[b]	59	11	6	1.2
Ventilation[b]	131	45	59	0.5
Cooking[c]	11	30	3	8.3
Lighting[b]	543	25	137	5.2
Refrigeration[b]	89	38	34	1.3
Office equipment—PCs[c]	120	60	71	3.9
Office equipment—non-PCs[c]	271	25	68	3.2
Other uses[b]	523	35	182	1.4
Total electricity	2062	34	705	2.7
Natural Gas	*(Quads)*		*(Quads)*	*(2007$/million Btu)*
Space heating[b]	2.30	47	1.09	1.9
Space cooling[b]	0.06	38	0.02	4.1
Water heating[b]	1.06	15	0.16	2.3
Cooking[c]	0.47	31	0.14	7.3
Other uses[b]	0.47	20	0.09	1.9
Total natural gas	4.36	35	1.51	2.5

Note: A corresponding table for 2020 can be found in Brown et al. (2008).

[a]AEO reference case (EIA, 2007a) end-use consumption for the "Other uses" end-use was reallocated to match the 2007 DOE Buildings Energy Databook 2007 (DOE/EERE, 2007) end-use shares.

[b]Source for potential savings and CCE is CEF Table D-1.1 (IWG, 2000). Values for CCE are from the CEF Advanced Case, calculated using a real discount rate of 7 percent and lifetimes as shown in CEF report Appendix C-1.

[c]Source for potential savings and CCE is the updated LBNL analysis documented in Brown et al. (2008).

Source: Brown et al., 2008.

ies, and applied them to the BAU scenario. The approach was to consider specific energy efficiency measures for each end-use, as explained below. For most end-uses, *Scenarios for a Clean Energy Future* (IWG, 2000; hereafter referred to as the CEF study), sponsored by the DOE, was used to estimate savings potential (Koomey et al., 2001). The CEF study[8] adjusted the energy-savings potential for

[8]See also Box 4.2, "The *Scenarios for a Clean Energy Future* Study" in Chapter 4.

different end-uses in order to account for energy efficiency improvements in the baseline scenario and thereby avoid double counting the energy savings. For the residential natural gas end-uses, a recent study of natural-gas-savings potential in New York was used as the principal reference (Mosenthal et al., 2006). For selected end-uses that were not analyzed in the CEF study (IWG, 2000), Brown et al. (2008) compiled technical data to estimate savings percentages and the CCE. The specific data source for each end-use is identified in Tables 2.9 and 2.10.

To provide a better sense of the technologies that were used to estimate these potentials, Tables 2.11 and 2.12, respectively, list the principal residential-building and commercial-building measures or efficiency improvement assumptions used for each end-use. For the most part, the technologies are widely available in the marketplace today and are well proven. Technologies such as CFLs, T8 lamps and electronic ballasts, ENERGY STAR® appliances, horizontal-axis clothes washers, and high levels of building thermal integrity have been implemented by some consumers, but they have not been implemented in all applications in which they are technically and economically feasible owing to a wide range of market barriers and failures (see Section 2.7) A few of the technologies, such as heat-pump water heaters, are still produced on a limited scale and are considered near-term emerging technologies. However, other emerging technologies, such as LED lights, solar water heaters, and very high efficiency new buildings, are not included in the supply curves. Other excluded technologies include passive solar heating and cooling, gas-fired and geothermal heat pumps, the separation of ventilation from heating and cooling systems, products that must be special ordered (e.g., R-7 windows), non-air-based heating and cooling distribution systems in commercial buildings, and the redesign of building envelopes for daylighting and cooling load minimization.

To estimate aggregate savings potential in 2030, Brown et al. (2008) multiplied the percentage energy-savings potential shown by end-use in Tables 2.9 and 2.10 by the estimates of energy use by end-use in the BAU scenario. The CCE is reported as the levelized annual cost of the efficiency measures over their lifetime divided by the estimated annual energy savings. The CCE accounts for the costs of incremental measures only; no cost is included for public policies or programs aimed at stimulating the adoption of a measure. Consistent with the CEF study, a real discount rate of 7 percent was used to calculate these values. Cost-of-conserved-energy values from the CEF and New York studies were inflated to 2007 dollars using the GDP implicit price deflator.

TABLE 2.11 Residential Building Measures Included in the Conservation Supply Curve Analysis

Fuel and End-Use	Efficiency Measure Description
Electricity	
Thermal shell	Existing electric-heated homes: no efficiency measures; new homes: up to 40% savings compared to 2006 International Energy Conservation Code
Space heating equipment	Electric furnace switched to heat pump, improved heat-pump efficiency
Space cooling equipment	Improved-efficiency central and room air conditioners, variable speed room air conditioners
Water heating	Reduced standby-loss electric-resistance water heater, heat pump water heater, horizontal axis clothes washer
Refrigeration	Best-in-class ENERGY STAR® refrigerator, 2008
Freezers	Best-in-class ENERGY STAR® freezer, 2008
Lighting	Compact fluorescent fixtures, halogen-infrared lamps, reduced-wattage incandescents, motion sensors
Clothes washers	Horizontal-axis washer with improved motor
Dishwashers	Dishwasher with improved pump design and improved motor
Color televisions	Reduced standby power use
Personal computers	ENERGY STAR®-rated PC and monitor, power-management-enabled
Furnace fans	Electronically commutated permanent magnet furnace-fan motor, single-speed operation
Other uses	More efficient motors in ceiling fans, pool pumps, and other small motors; improved fan and pump design; reduced standby power use in set-top boxes and other electronics; improved insulation for water beds, spas, and other small heating loads
Natural Gas	
Thermal shell	Air sealing, R-19 floor insulation, R-21 wall insulation, R-49 attic insulation, integrated design for new construction (SF 30% > code, MF 50% > code), triple-pane low-e windows, insulated attic hatch
Space heating equipment	Insulated/sealed/balanced ducts, ducts placed within thermal shell condensing furnace, sensible heat recovery ventilation, direct-vent fireplace, direct-vent boiler, programmable thermostat, boiler pipe insulation
Space cooling equipment	Not applicable
Water heating	On-demand water heater, 0.63 EF gas water heater, low-flow plumbing fittings, ENERGY STAR® clothes washer, reduced water heater tank temperature, gray water heat exchanger/GFX, pipe insulation
Cooking	Not applicable
	Humidity sensor control
Other uses	Pool and spa covers

Source: Brown et al., 2008.

TABLE 2.12 Commercial Building Measures Included in This Analysis

Fuel and End-Use	Efficiency Measure Description
Electricity	
Thermal shell	No efficiency measures
Space heating equipment	Up to 55% savings in existing buildings from improved HVAC equipment and controls
Space cooling equipment	Up to 55% savings in existing buildings from improved HVAC equipment and controls
Water heating	20% savings compared to frozen efficiency baseline
Ventilation	Up to 55% savings in existing buildings from improved HVAC equipment and controls
Cooking	ENERGY STAR®-rated dishwasher, fryer, hot-food-holding cabinet, and steamer; more efficient broilers, griddles, and ovens
Lighting	T-8 lamps and electric ballasts; 32% combined savings from occupancy controls, daylight dimming, and improved lighting design
Refrigeration	20–45% savings compared to frozen efficiency baseline
Office equipment—PCs	ENERGY STAR®-rated personal computer and monitor, power-management-enabling software
Office equipment—non-PCs	ENERGY STAR®-rated copies and printers
Other uses	More efficient motors in ceiling fans, pool pumps, and other small motors; improved fan and pump design; reduced standby power use in electronics; improved insulation; small heating loads; up to 55% reduction in district services due to improved shell, equipment, and controls
Natural Gas	
Thermal shell	No efficiency measures
Space heating equipment	Up to 55% savings in existing buildings from improved HVAC equipment and controls
Space cooling equipment	Up to 55% savings in existing buildings from improved HVAC equipment and controls
Water heating	10% savings compared to frozen efficiency baseline
Cooking	ENERGY STAR®-rated fryer and steamer; more efficient broilers, griddles, and ovens
Other uses	10% reduction in miscellaneous gas use; up to 55% reduction in district services due to improved shell, equipment, and controls

Source: Brown et al., 2008.

2.5.2 Information Sources

Technology costs were drawn by Brown et al. (2008) from the CEF advanced case, which assumed a greater penetration of more advanced efficiency technologies than in the moderate case. In using these savings potentials to estimate the national savings potential in 2030, the researchers assumed that the CEF savings potential estimated for 2000–2020 would still be applicable for the 2020–2030 time period. While some efficiency measures such as CFLs, more efficient lighting devices for commercial buildings, and ENERGY STAR® personal computers and other electronic devices have been adopted to a significant degree, new efficiency measures have entered the marketplace since 2000, and others are under development and expected to be commercialized in the near future. Thus, while today's energy efficiency baseline has improved since 2000, Brown et al. (2008) assumed that the number of efficiency technologies and practices being developed and not yet adopted have kept pace with this improvement, keeping the overall efficiency potential roughly constant.

Because the CEF study did not model the savings potential of building-shell retrofits to existing homes, the LBNL researchers instead used estimates of residential natural-gas savings derived from a recent New York study (Mosenthal et al., 2006). The applicability of that study to the national context rests on the assumption that the *percentage* savings (relative to baseline energy use) in New York is representative of the country as a whole. The CCE, however, depends on the absolute energy savings for a given measure, so Brown et al. (2008) scaled the CCEs to account for differences in heating degree-days between New York and the national average.[9] The CCEs were calculated using a 7 percent discount rate, to be consistent with the other end-uses in this analysis. The Brown et al. (2008) study provides the details of this analysis.

Several end-uses were not analyzed in the CEF study, and savings and cost data for them were not available from other studies at that time. These end-uses were commercial office equipment, commercial cooking, residential office equipment, residential furnace fans, and residential dishwashers. For these end-uses, LBNL researchers compiled data on technology performance and cost and developed savings-potential estimates specifically for this analysis. The summary results of this analysis are shown in Tables 2.9 and 2.10; details can be found in Brown et al. (2008).

---

[9]The researchers used the potential savings estimates for "downstate" New York (New York City and immediate vicinities) for this study. The climate scaling increased the CCEs by about 15 percent and was only applied to the space-heating end-use.

2.5.3 Results

Figures 2.8 and 2.9 show the potential for electricity (Figure 2.8) and natural gas (Figure 2.9) efficiency improvements over the 2010–2030 period in the residential and commercial sectors. The x-axis shows the total reduction in 2030 energy use, and the Y-axis shows the CCE in fuel-specific units. Each step on the graphs in these two figures represents the total savings for a given end-use for all the cost-effective efficiency measures analyzed for that end-use. These are referred to as "supply curves" because they indicate how much energy savings is available for a given cost. The CCE is calculated as the savings-weighted average for all the measures in that end-use cluster. End-uses that do not have technology costs reported in Tables 2.9 and 2.10 are not included in these plots (e.g., residential cooking). It should be noted that the space-heating and the space-cooling steps in Figures 2.8 and 2.9 include efficiency improvements in both the thermal shell and the HVAC equipment, analyzed in an integrated manner.

Each of the supply curves indicates that the projected BAU energy consumption in 2030 can be reduced by about 30–35 percent at a cost less than current retail energy prices. Table 2.13 compares the weighted-average cost of conserved energy from each supply curve with national average retail energy prices as of 2007. The data in the table show that the average CCE is well below the retail energy price in all areas, meaning that adopting these efficiency measures is cost-effective for households and businesses. In fact, the average CCE for these electricity-savings measures is only about one-quarter of the average retail electricity price. Of course, factors such as local energy prices and weather will influence cost-effectiveness in any particular location.

Table 2.14 presents data on the aggregate costs and benefits of efficiency technologies for the entire buildings sector. The cumulative capital investment required to achieve these savings between 2010 and 2030 is about \$440 billion.[10] The value of annual energy-bill savings in 2030 is nearly \$170 billion. Thus, these efficiency measures in aggregate have a 2.6 year simple payback period on average, or savings over the life of the measures that are nearly 3.5 times larger than

[10]The investment includes both the full "add-on" cost for new equipment or measures (e.g., attic insulation) and the incremental cost of purchasing an efficient technology (e.g., a high-efficiency boiler) compared with purchasing its conventional-technology equivalent (e.g., a standard boiler). These investments would be made by the individuals and private entities making the purchases. The costs of programs to support, motivate, or require these improvements are not included.

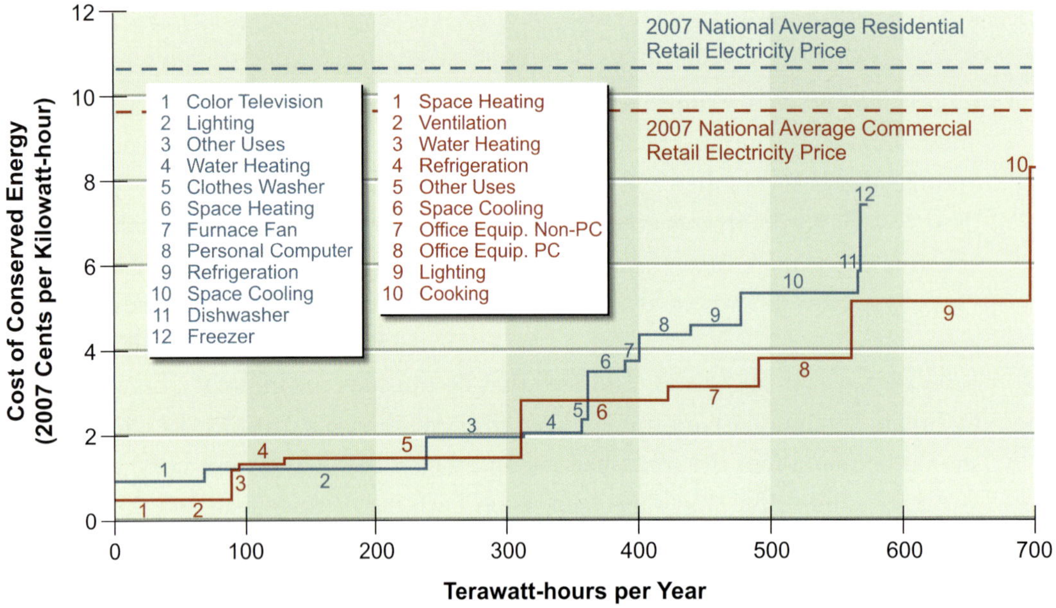

FIGURE 2.8 *Estimates of the cost of conserved energy (CCE) and energy-savings potential for electricity efficiency technologies in buildings in 2030. The CCEs for potential energy efficiency measures (numbered) are shown versus the ranges of potential energy savings for these measures. The total savings potential is 567 TWh per year in the residential sector (blue solid line) and 705 TWh per year in the commercial sector (red solid line). For comparison, the national average 2007 retail price of electricity in the United States is shown for the residential sector (blue dashed line) and the commercial sector (red dashed line). For many of the technologies considered, on average the investments have positive payback without additional incentives. CCEs include the costs for add-ons such as insulation. For replacement measures, the CCE accounts for the incremental cost—for example, between purchasing a new but standard boiler and purchasing a new high-efficiency one. CCEs do not reflect the cost of programs to drive efficiency. All costs are shown in 2007 dollars.*
Source: Data from Brown et al., 2008.

the investment required on a discounted net present value basis. These averages are based on combining efficiency measures with CCE values ranging from less than 1¢/kWh to 8¢/kWh in the case of electricity saving measures, and $1/million Btu to $12/million Btu in the case of natural-gas-saving measures. There is an up-front cost to achieving substantial energy savings, but this cost is paid back a number of times over the lifetime of the energy efficiency measures.

A few other studies have developed conservation supply curves or, equivalently, the cost of reducing carbon dioxide (CO_2) emissions for the United States. A recent study prepared by McKinsey and Company has received considerable attention (Creyts et al., 2007). The panel was unable to verify (owing to lack of

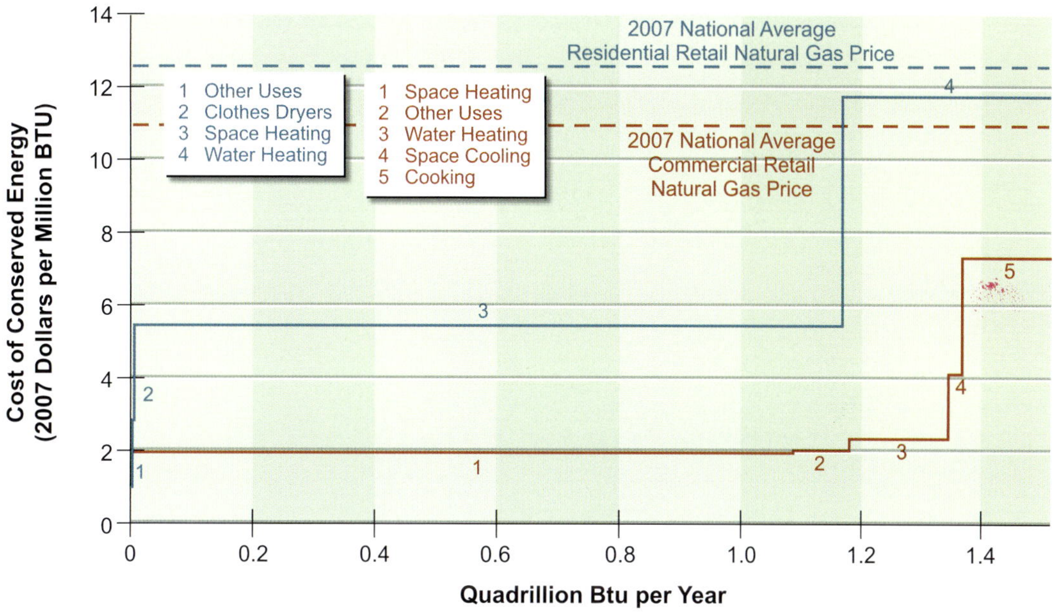

FIGURE 2.9 *Estimates of the cost of conserved energy (CCE) and energy savings potential for natural gas efficiency technologies in buildings in 2030. The CCEs for potential energy efficiency measures (numbered) are shown versus the ranges of potential energy savings for these measures. The total savings potential is 1.5 quads per year in the residential sector (blue solid line) and 1.5 quads per year in the commercial sector (red solid line). For comparison, the national average 2007 retail price of natural gas in the United States is shown for the residential sector (blue dashed line) and the commercial sector (red dashed line). For many of the technologies considered, on average the investments have positive payback without additional incentives. CCEs include the costs for add-ons such as insulation. For replacement measures, the CCE accounts for the incremental cost—for example, between purchasing a new but standard boiler and purchasing a new high-efficiency one. CCEs do not reflect the cost of programs to drive efficiency. All costs are shown in 2007 dollars.*
Source: Data from Brown et al., 2008.

data) the assumptions and results regarding energy efficiency potential in buildings. Nevertheless, the results of the McKinsey and Company study generally parallel those of the studies that the panel reviewed in terms of the magnitude and cost of saved energy in buildings.

2.5.4 Limitations

Owing to time and resource constraints, the Brown et al. (2008) analysis relied on data from previous efficiency-potential studies. As a result, the analysis can be improved on in several respects, some of which are highlighted below.

TABLE 2.13 Estimated Average Cost of Conserved (Saved) Energy in Residential and Commercial Buildings Compared with National Average Retail Energy Prices, 2007

	Average Cost of Conserved (Saved) Energy	National Average Retail Energy Price
Residential		
Electricity	2.7¢/kWh	10.6¢/kWh
Natural gas	$6.9/million Btu	$12.7/million Btu
Commercial		
Electricity	2.7¢/kWh	9.7¢/kWh
Natural gas	$2.5/million Btu	$11.0/million Btu

Note: For the specific savings, see Figures 2.8 and 2.9.
Source: Brown et al., 2008. Energy prices are from EIA, 2008c.

TABLE 2.14 U.S. Efficiency Investment and Savings by 2030 (for the Buildings Sector)

Sector and Energy Type	Cumulative Capital Investment (billion 2007 $)	Annual Utility Bill Savings in 2030 (billion 2007 $)[a]	Simple Payback Time (years)
Residential			
Electricity	137	64	2.1
Natural gas	104	19	5.5
Commercial			
Electricity	163	68	2.4
Natural gas	38	17	2.3
Total	442	168	2.6

[a]Assumes 2007 retail electricity and natural gas prices.
Source: Brown et al., 2008.

- The end-use technology data used in this study are mostly drawn from the CEF study (IWG, 2000), which reflects technology and market conditions in the late 1990s. Clearly, many factors have changed since then, including new technologies becoming available as well as costs falling for some energy efficiency measures owing to improved manufacturing processes, increased volumes, and the relocation of manufacturing facilities to countries where costs are lower. For example, one study found that the average retail price of CFLs dropped about 75 percent between 1999 and 2007 (Itron, 2008).

- As explained in Section 2.5.2, it is assumed that new efficiency measures compensate for the loss of savings potential due to measures adopted since 2000. This is a simplifying assumption that introduces uncertainty in the point estimates of the savings potential presented above.
- Energy prices have risen significantly since the CEF study (IWG, 2000), which increases the number of energy efficiency technologies that are cost-effective, thus increasing the energy efficiency potential.
- For the residential natural-gas end-uses, the New York study (Mosenthal et al., 2006) is only a rough approximation of savings potential across the country. A national study that includes all relevant technologies (including shell retrofits) would add considerable value to a study extrapolated from New York.
- The effect of the Energy Independence and Security Act of 2007 is considered part of the remaining efficiency potential; that is, the effect of EISA is not included in the baseline. This assumption probably has the largest effect on the lighting end-use, because EISA contains aggressive provisions for lighting efficiency.
- The results of Brown et al. (2008) are point estimates of savings potential that ignore uncertainty about how energy use in the building sector will evolve during the next 20+ years. Some of the major areas of uncertainty include energy prices, the availability and price of efficiency technologies, and potential changes in consumer behavior. They also include the policy context—for example, whether or not limits on greenhouse gas emissions are enacted, and if so, with what degree of stringency.
- Studies of efficiency potential, such as the CEF and New York studies, are highly aggregated analyses that tend to ignore the great variability in the building stock with respect to climate, building configuration, equipment ownership, building occupancy and use, and other factors. Future studies should be conducted at a greater level of disaggregation to address variability in the building stock.

As noted above, the conservation supply-curve approach to estimating potential savings itself has limitations. The initial models of the cost of saving energy did not account for the life of energy-using equipment or for whether the new technology would fit into the space available and perform the same functions, and these models generally lacked the detail to enable a determination of whether a particular technology would actually be attractive in a particular setting. The models assumed that the existing equipment needed to be replaced and could be replaced with a more efficient technology. The models were static in assuming that customers would want to buy the best technology today, instead of waiting until a better technology was available. The models did not account for the time and costs of disseminating information about the new technologies, the availability of capital to acquire the often more expensive equipment, the risks to existing production, and other barriers. Some economists expressed skepticism about the results—especially when businesses and consumers did not take advantage of the new, better technologies that promised large economic benefits. (See, for example, Stavins et al., 2007; Jaccard et al., 2003; Sutherland, 2000; Jacoby, 1999; and Jaccard and Montgomery, 1996.)

While current models are more sophisticated in considering these issues, they are still aggregate models that do not consider the specific circumstances of each energy-efficient replacement. They still do not fully account for educational and dissemination costs and other barriers. Thus, on the one hand, they tend to overstate the economic attractiveness of some new technologies; on the other hand, they are only dealing with a small number of energy-efficient technologies, so they neglect many other attractive alternatives, thus understating some potential benefits.

2.6 ADVANCED TECHNOLOGIES AND INTEGRATED APPROACHES

The conservation supply curves presented in Section 2.5 do not take into account a number of newer technologies and whole-building-design approaches. These technologies and approaches add to the energy-savings potential identified in the conservation supply curves; thus, the panel judges that these supply curves represent lower estimates of energy-savings potential.

This section reviews some of the advanced technologies that are the most promising for further improving the energy efficiency of buildings. These include discrete technologies such as solid-state lighting, advanced windows, and high-efficiency air-conditioning equipment, as well as the full integration of the technologies into new, highly efficient whole buildings, both residential and commercial. These technologies demonstrate that energy efficiency is a dynamic resource—new and improved technologies now under development will reach the marketplace in the future, thereby increasing the potential for energy efficiency and energy savings.

The review below is not comprehensive; it does not address many advanced technologies related to building materials, design, and appliances.

2.6.1 Solid-State Lighting

Solid-state lighting is an important emerging technology for energy savings, given that lighting accounts for about 18 percent of primary energy use in buildings. CFLs are a major improvement over incandescent lamps with respect to efficacy (about 60 lumens per watt versus 15 lumens per watt), but they contain mercury, are difficult to dim, are not a point light source,[11] and are not "instant-on." Light-emitting diodes do not suffer from these disadvantages. As shown in Figure 2.10, the performance of white LED lamps has improved greatly in recent years. The best white LEDs are now more efficient than fluorescent lamps, with further gains expected within the next 5 years. Advances will come from improvements in the ratio of injected electrons to emitted photons in the active region, the efficiency of extracting generated photons out of the packaged part, phosphors, thermal efficiency, and scattering efficiency (Azevedo et al., 2009). The expectation is that white LEDs will reach 150 lumens per watt (Craford, 2008). LEDs last longer than fluorescent lamps do, are dimmable, and are becoming available in warm white with excellent color rendering. The dimmability and "instant-on" capability of LEDs make them especially appealing for applications such as streetlighting, gas station lighting, and display case lighting in retail stores where occupancy sensors can be used to dim them or turn them off when people are not present.

The primary issue with LEDs is cost, but their cost is decreasing rapidly. A 1000-lumen LED source that costs around $25 (wholesale cost) in 2008 is projected by the DOE to cost $2 in 2015 (DOE, 2008b). At a $50 retail cost, the CCE for an LED with a 20,000-hour lifetime and an efficacy of 60 lumens per watt replacing an incandescent lamp is about $0.13/kWh. The CCE would fall to $0.008/kWh if the cost goal of $2 wholesale ($4 retail) is achieved and performance improves to 150 lumens per watt along with a 50,000-hour lifetime. Given

[11]Point light sources are easier to focus, which is important in some applications—for example, in retail stores.

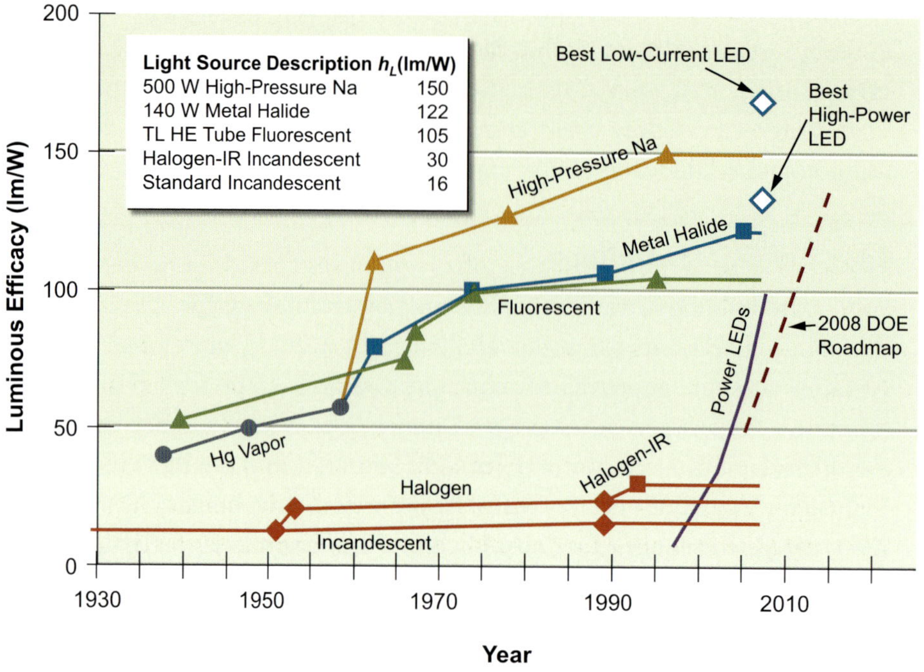

FIGURE 2.10 *White light source performance: light-emitting diodes (LEDs) compared with conventional light sources. Luminous efficacy is the ratio of luminous flux (in lumens) to power (in watts).*
Note: TL HE = tube light, high-efficiency; IR = infrared.
Source: Craford, 2008.

expected improvements, LEDs are projected to become competitive with CFLs by 2012 even if the lights are used only 2 hours a day and a 20 percent annual discount rate is assumed (Azevedo et al., 2009).

High-quality LED replacement lamps are now becoming available, and over the next 5 years a wide variety of higher-power lamps is expected. The penetration rate into the illumination market is difficult to predict and will be different for different market segments. DOE has modeled several segments as well as the overall market penetration rate (Navigant Consulting, 2006). The model projects that LEDs will yield a 12 percent savings in lighting energy use in 2017 and a 33 percent savings by 2027, relative to projected lighting energy use without LEDs. The projected 2027 electricity savings are greater than the energy consumed to illuminate all the homes in the United States today.

The penetration rate for organic LEDs (OLEDs), an alternative form of solid-state illumination, has also been independently modeled and also yields 33 percent lighting-energy savings by 2027. However, the OLED savings will occur later than the LED savings will (less than 6 percent in 2017 for OLED savings) and are much more speculative in view of impending technology hurdles and cost challenges.

2.6.2 Advanced Cooling

Cooling is one of the largest uses of energy in residential and commercial buildings, responsible for about 10 percent of total U.S. electricity use and 25–30 percent of total peak electricity demand (EIA, 2007a; Koomey and Brown, 2002). Significant potential exists for reducing cooling demand in buildings—or eliminating it entirely in some climates—through strategies that combine measures to reduce building cooling requirements and peak loads (e.g., highly efficient building envelopes, shading, reflective surfaces and roofs, reductions in heat gains from lights and other equipment, natural ventilation, and thermal storage) with emerging cooling technologies. These technologies are designed to supplement or replace vapor compression-based cooling with low-energy, thermally driven cooling approaches. They include indirect evaporative cooling and indirect-direct evaporative cooling (IDEC), solar thermal cooling (STC) systems, advanced controls and low-lift cooling strategies, and thermally activated desiccants. Each of the technologies has already been used commercially as individual components, but further research and development and commercial demonstration projects are needed to develop the technologies as integrated systems and to optimize their performance.

Indirect evaporative cooling systems use an evaporative process to cool a building's interior without adding moisture to the indoor air. Various indirect evaporative systems are currently entering the marketplace, including systems that couple cooling towers with floor slabs and radiant ceilings. IDEC lowers air temperature by first passing air across a heat exchanger surface whose other side is cooled by evaporation. This precooled air then passes through a direct evaporative process, where it is cooled and humidified. Where applicable, IDEC units are capable of reducing cooling energy demand by 70–80 percent (PG&E, 2006). These advanced evaporative systems are now applicable in the dry climates of the western United States, with minimal net increase in total household water use (Kinney, 2004).

STC combines solar thermal technologies with traditional chiller-based systems to produce hot water that drives heat-driven absorption, and absorption or adsorption chillers that generate chilled water for space conditioning. Currently, the systems are most cost-effective in commercial and industrial buildings with large roof areas, such as one- to three-story commercial buildings, hospitals, and food and beverage processing plants (Burns et al., 2006).

Low-lift cooling increases efficiency by reducing the temperature difference and thus the work performed by the cooling system. Options for low-lift cooling include the use of a dedicated outdoor air supply with enthalpy heat recovery from exhaust air, radiant cooling panels or floor systems, low-lift vapor compression equipment, and advanced controls. The technical energy savings potential is estimated to be 60–74 percent for temperate to hot and humid climates and 30–70 percent in milder climates (Jiang et al., 2007).

Desiccants allow the independent control of temperature and humidity within the HVAC system, thereby providing greater control over humidity loads within the building, which can also improve indoor air quality. For more than a decade, desiccant systems have been used in combination with conventional HVAC systems in specialized markets where humidity control is important, such as in supermarkets and hotels (DOE, 2004b). In humid regions, desiccant-based dehumidification could reduce residential electricity demand by 25 percent, because less energy is used to achieve dehumidification (PNNL, 1997). Combining desiccants with STC systems could achieve additional energy savings (Stabat et al., 2003). Researchers at DOE's National Renewable Energy Laboratory are developing heat-driven liquid desiccant systems that are capable of being powered by solar thermal energy or through heat recovery from reciprocating engines, microturbines, and fuel cells (Lowenstein et al., 2006).

The energy-savings potential of these technologies is substantial. It is estimated that they could reduce total cooling-energy demand in residential and commercial buildings by 15 percent (0.6 quad) in 2020 and 33 percent (1.5 quads) in 2030, as shown in Figure 2.11. Incorporating advanced building-design practices that minimize cooling loads could save an additional 0.75 quad annually in 2030.

Within the next decade, the most economically competitive cooling strategies include thermal envelope improvements to reduce cooling loads, desiccant dehumidification in humid climates, and indirect-direct evaporative cooling in hot, dry climates. STC systems will become increasingly competitive with electrically driven vapor compression systems as the cost of solar collectors declines. By 2030, thermally driven cooling systems that use renewable energy or waste heat for cooling could replace conventional cooling technologies in new buildings and could be used to retrofit existing buildings.

There is also potential to use advanced building sensors and controls to

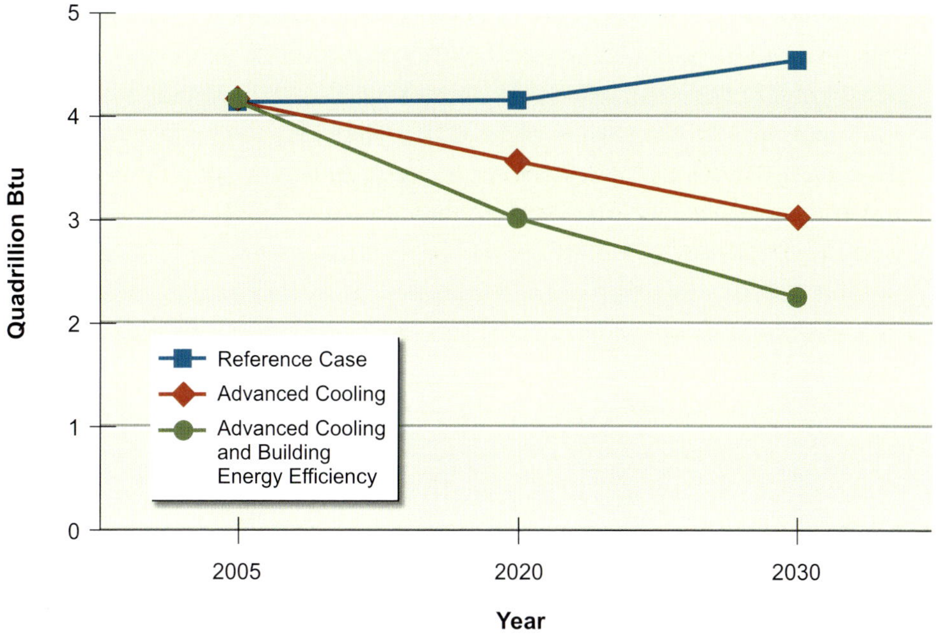

FIGURE 2.11 *Potential reduction in cooling demand in U.S. buildings using advanced technologies.*
Source: Courtesy of S. Dunn, Southwest Energy Efficiency Project, Boulder, Colo., November 2007; based on data from the Annual Energy Outlook 2007 (EIA, 2007a) and the Buildings Energy Data Book 2007 (DOE/EERE, 2007).

reduce HVAC and lighting energy consumption. Currently, most advanced control approaches have a very small market share. These approaches include occupancy sensors, demand-controlled ventilation, photosensor-based lighting control, and continuous commissioning—that is, the ongoing testing of building equipment and systems to detect and diagnose faults. Brambley et al. (2005) estimate that the widespread adoption of advanced sensors and controls could reduce primary energy use by commercial buildings by about 6 percent.

2.6.3 Technologies to Reduce Energy Consumption in Home Electronics

Consumer electronics—that is, the products dealing with the processing of information—are responsible for about 13 percent of residential electricity use (Roth and McKenney, 2007; EIA, 2008a). This consumption is likely to increase simply because the growth in the number of products in homes shows little sign of abating (EIA, 2008a). Numerous efficiency improvements have been incorporated into—and ENERGY STAR® specifications have been adopted for—a wide range of products, but energy use continues to increase in a few important products, notably flat-panel televisions and set-top boxes. In March 2009, the Energy Information Administration projected that electricity use by televisions and set-top boxes would increase by 65 percent between 2006 and 2030 (DOE, 2009). New products are also appearing, such as digital picture frames, which will contribute to further increases in energy consumption. Many other kinds of appliances, such as dishwashers, furnaces, and water heaters, have electronic controls that are responsible for a small but noticeable fraction of those products' overall energy consumption.

Electronic products typically consume energy both while active and while switched off. In many cases, the energy consumption while switched off exceeds that while in use, owing to the limited number of hours that many electronic devices are switched on (IEA, 2001). The average California home now contains more than 40 products that are continuously drawing power greater than 110 watts (Meier et al., 2008). An increasing number of electronic products are also connected to networks (WiFi, Ethernet, USB, Bluetooth, and others). With present designs, many products must remain in a power-intensive, fully-on mode so as not to be disconnected from the network. As with standby power use, the network connectivity of new products may become the standard situation rather than the exception.

At least five strategies exist to reduce the energy use of consumer electronics, but they are not yet widely used. First, improvements in power supplies could reduce electricity use in all power modes. Second, many products can be redesigned to exploit smaller and more efficient circuitry, which usually results in lower energy use. Third, some products can incorporate an auto-power-down feature. This feature is already required in new ENERGY STAR® specifications for digital television adapters. Fourth, protocols can be employed to allow products on a network to operate with a low-power sleep level without losing network connectivity. Finally, "power strips" can be designed more cleverly to manage energy consumption in clusters of products. In such systems, the "power" switch has migrated from individual products to a much smarter device able to take into account many other variables. For example, such a device can sense when a computer is turned off and can shut off the electricity flow to the computer and to other devices plugged into the strip at the same time. These strategies are already employed in a few products but have had minimal impact on energy use to date.

2.6.4 Technologies to Reduce Energy Consumption in Servers and Data Centers

Servers and data centers were responsible for about 61 billion kWh of electricity use in 2006, 1.5 percent of total national electricity use and more than double the level of 2000 (EPA, 2007b). The bulk of this electricity consumption is for site infrastructure, including cooling systems (50 percent) and what are termed "volume servers" (34 percent). The installed base of servers and external hard drives is increasing very rapidly. If current trends in server and data-center expansion and energy efficiency continue, the EPA (2007b) projects that servers and data centers will consume approximately 107 TWh by 2011, 75 percent more than in 2006.

There is large potential for cutting the electricity consumption of servers and data centers, and to do so cost-effectively. The techniques for improving efficiency include the following:

- Virtualization, which allows data processing to be accomplished with fewer servers;
- Improved microprocessors with higher performance per watt;
- Servers with more efficient power supplies, fans, and microprocessors;
- More efficient data-storage devices; and
- More efficient cooling techniques, uninterruptible power supplies, and other "site infrastructure" systems.

The EPA (2007b) estimates that with the more widespread adoption of cost-effective energy efficiency technologies and practices already in use in some servers and data centers today, the overall electricity use of servers and data centers could be limited to 48 TWh in 2011, 55 percent less than in the current trends scenario. Furthermore, the EPA estimates that if state-of-the-art technologies and practices were fully adopted in all servers and data centers, overall electricity use could be limited to about 34 TWh in 2011, nearly a 70 percent reduction from what is projected under current trends.

In addition to saving the owners of servers and data centers money on their utility bills, utilities would benefit from reducing server and data-center electricity use because many servers and data centers are concentrated in areas such as New York City and San Francisco, which have congested transmission and distribution grids. Server and data-center energy efficiency is starting to be addressed by utility and government energy efficiency programs such as the ENERGY STAR® program, but these efforts are still in their infancy; for example, an objective and credible energy performance rating for data centers, as well as ENERGY STAR® specifications for data-center equipment, are still under development. On May 15, 2009, the EPA published its initial specification for ENERGY STAR® computer servers.[12]

2.6.5 Advanced Window Technologies

Windows are responsible for about 2.7 quads of energy use annually in homes and about 1.5 quads in the commercial sector, and they impact another 1.0 quad of potential lighting-energy savings (Apte and Arasteh, 2006). Advances have been made over the last two decades primarily in reducing the heat-transfer coefficient (U-value) of windows through the use of low-emissivity (low-E) coatings and by reducing the solar heat gain coefficient (SHGC) by means of the use of spectrally selective low-E coatings. The window U-value is the primary determinant of winter heat loss; the window SHGC is the primary determinant of summer cooling loads. Window U-values have changed little in mainstream markets in recent years, having become "stuck" at ENERGY STAR® values that typify performance values achieved starting in the 1980s. Two new window technology advances are now available in niche markets. They currently have higher-than-acceptable cost, but they could have far-reaching implications if they could become mainstream products and systems.

The first advance is highly insulating "superwindows" that achieve U-values in the range of 0.1–0.2, compared to a typical U-value of 0.5 for double glazing and 0.35–0.4 for ENERGY STAR® windows currently being sold in cold climates (Apte and Arasteh, 2006). Such windows are available in limited quantity in Europe and in the United States but are not yet mass-produced. While research efforts continue with highly insulating aerogel and vacuum glazings, each of these approaches requires fundamental changes in glazing and window assembly and massive investment in facilities for start-up. An alternative approach is to develop a family of highly insulating window systems based on the use of two low-E-coated glazing layers in a triple-glazed assembly, with gas fills and improved edges. This combination of measures is capable of achieving overall glazing performance of U = 0.1–0.2 using existing production facilities and could have a price only

---

[12]See http://www.energystar.gov/index.cfm?c=archives.enterprise_servers.

marginally above current gas-filled, double-glazed low-E units. These highly insulating glazings will also require a new series of companion insulating sashes and frames so that the whole window product meets target thermal performance requirements. These products are likely to have primary value in the residential sector but will find applications in commercial buildings as well.

The second opportunity is a new generation of dynamic products that reduces cooling loads in climates with substantial cooling loads and makes daytime lighting controls the preferred solution for reducing electric lighting in commercial buildings (Apte and Arasteh, 2006). The requirement here is for more dynamic control—the ability to modulate solar gain transmitted through the windows (thus minimizing cooling load), but also to admit sufficient daylight to reduce electric lighting in commercial buildings while controlling glare. This can be achieved in two ways: (1) for example, using "smart" electrochromic coatings that can reversibly switch properties in response to sensor input and that are capable of optimizing between cooling and lighting and can provide glare control as well; and (2) employing dynamic optical control using automated, motorized shading systems in buildings, including automated shades, blinds, shutters, and so on. These systems are commonplace in many European buildings but are rarely found in the United States. Fully automated, motorized roller shades with dimmable, addressable fluorescent lighting to capture daylight benefits were recently installed in a new 52-story office building in New York City.

Apte and Arasteh (2006) show that taken together, these two classes of emerging window technologies could produce large energy savings in the U.S. building stock if widely deployed in new and existing buildings. Tables 2.15 and 2.16 provide energy-savings estimates for the residential and commercial sectors, respectively, if the specified set of window technologies was fully deployed in the building stock. The tables show that the full penetration of ENERGY STAR® windows in residential and similar windows in commercial buildings would provide nearly 1.8 quads of energy savings per year. However, if the advanced technologies described above are commercialized and fully penetrate the buildings stock, the full technical potential for energy savings would increase to nearly 3.9 quads per year. Since the stock today accounts for approximately 4.2 quads of energy use per year, these improvements shift the role of windows in buildings to being approximately "energy neutral." This is achieved by greatly reducing the unwanted seasonal losses and gains in all buildings in all climates and then by providing useful solar heat in winter in homes and useful daylight year round in commercial buildings. It will take many years, of course, to replace the existing

TABLE 2.15 Annual Energy Savings Potential of New Residential Window Technologies

Scenario	Energy Savings over Installed Window Stock in 2005 (quads)		
	Heat	Cool	Total
Sales (business as usual)[a]	0.49	0.37	0.86
ENERGY STAR® (Low-e)	0.69	0.43	1.12
Dynamic Low-e	0.74	0.75	1.49
Triple Pane Low-e	1.20	0.44	1.64
Mixed Triple, Dynamic	1.22	0.55	1.77
High-R Superwindow	1.41	0.44	1.85
High-R Dynamic	1.50	0.75	2.25

[a] The average properties of residential windows sold in 2004; used as a "business as usual" scenario.
Source: Apte and Arasteh, 2006.

TABLE 2.16 Annual Energy Savings Potentials of New Commercial Window Technologies

Scenario	Energy Savings over Installed Stock in 2005 (quads)		
	Heat	Cool	Total
Sales (business as usual)[a]	0.03	0.17	0.20
ENERGY STAR® (Low-e)	0.33	0.32	0.65
Dynamic Low-e	0.45	0.53	0.98
Triple Pane Low-e	0.71	0.31	1.02
High-R Dynamic	1.10	0.52	1.62

[a] The average properties of residential windows sold in 2004; used as a "business as usual" scenario.
Source: Apte and Arasteh, 2006.

stock with these new technologies, but a transition to "zero-net-energy windows" could provide enormous benefits eventually (Arasteh et al., 2006).

2.6.6 Low-Energy and Zero-Net-Energy New Homes

It is possible to construct homes that combine high levels of energy efficiency in the building envelope, heating and cooling systems, and appliances, along with passive and active solar features, in order to approach zero-net-energy consumption.[13] This is the objective of the Department of Energy's Zero Energy Homes program. Although the market share for zero-energy homes (ZEHs) is still very low, the level of awareness and commitment among builders is growing rapidly.

A growing number of new homes being built use 50 percent as much energy as that used by typical new homes, or even less, as evidenced by the fact that more than 23,000 new homes met this performance criterion and qualified for a federal tax credit as of 2007. The following examples demonstrate the techniques used to achieve very low conventional energy use in new homes.

Two highly instrumented homes were built with the same floor plan in Lakeland, Florida, in 1998 (Parker et al., 2000). One of these was of conventional construction and served as the project control. The experimental building included an interior duct system, a high-efficiency heat pump, better wall insulation, a white reflective roof, solar water heating, efficient appliances, and efficient lighting. Over 1 year, the experimental home used 6,960 kWh of electricity and had photovoltaic (PV) system production of 5,180 kWh. For the same year, the control home used 22,600 kWh; that is, the experimental home consumed 70 percent less. Including the PV production, the experimental home's net energy use was only 1,780 kWh, a 92 percent reduction relative to the control home.

A 3079-square-foot (286-square-meter) ZEH was built in Livermore, California, in 2002 (Parker and Chandra, 2008). The home featured fairly high levels of insulation; an innovative, computerized night cooling system (*NightBreeze*®) using outside air introduced by the duct system; high-performance windows with window shading; an attic radiant barrier; and highly efficient appliances and lighting. Heating was provided by a hydronic loop using a tankless gas water heater. Cooling was provided by the *NightBreeze*® with compressor cooling backup. In 2004, the 3.6 kW PV system produced more electricity (4890 kWh) than the house used (4380 kWh) so that net electricity consumption was negative: –510 kWh. Very little compressor cooling was ever needed. However, natural gas consumption totaled 700 million Btu per year—likely due to excess heat loss in a hot-water circulation loop.

In Lenoir City, Tennessee, the Oak Ridge National Laboratory constructed five ZEHs within a Habitat for Humanity development (Christian et al., 2004).

---

[13]A home with zero-net-energy consumption may at times produce more energy than it consumes (for example, through photovoltaic panels on the roof), and at other times it may consume more energy than it produces.

The project focused on small, affordable homes while evaluating a variety of efficient building methods and technologies such as the following:

- A heat-pump water heater linked to the refrigerator for heat recovery,
- A crawl space with thermostat-controlled ventilation to assist with space cooling and dehumidification in the summer,
- Ground source heat pumps using foundation heat recovery,
- Structural insulated panels,
- An interior duct system,
- High-performance windows,
- Efficient appliances, and
- A gray water waste-heat recovery system.

Conventional energy use in these homes was reduced by 35–60 percent. Nonetheless, because a number of innovative technologies were tested in these experimental homes, the cost per unit of energy savings was relatively high (Parker and Chandra, 2008).

Another ZEH Habitat for Humanity home was built in Wheat Ridge, Colorado (Norton and Christiansen, 2006). The small home was superinsulated with R-60 ceiling, R-40 double-stud walls, and R-30 floor insulation. Ventilation was provided by a small heat-recovery ventilator. Very-high-performance, low-E solar glass with argon fill and a U-value of 0.2 was used for the east, west, and north faces, with a higher transmission U-value 0.3 glass used for the south exposure. For hot water, the home used a 9-square-meter solar collector with 757 liters of storage, backed up by a tankless gas water heater. The home was mated with a 4 kW rooftop PV system.

During a recent year, the PV system in the Wheat Ridge home produced 1,542 kWh more than the electricity used in the home. It is interesting to note that some 60 percent of the electricity use in the home was for nonappliance, nonlighting miscellaneous electric loads. Only 5.7 million Btu of natural gas were used during this period. Thus, the home was a net energy producer on both a site and a source basis. The total incremental cost of the project was $42,500, including $32,000 for the PV system and $7,100 for the solar water-heating system. The incremental cost of the efficiency measures was only about $3,400, due in part to the elimination of a full-size furnace.

Premier Gardens is a community-level project in Sacramento, California.

This project saw 95 entry-level homes constructed with high levels of energy efficiency: R-38 ceiling insulation and R-13 to R-19 wall insulation, tankless gas water heaters, high-efficiency gas furnaces, tightly sealed ducts buried in the attic insulation, and fluorescent lighting in all permanent fixtures (Parker and Chandra, 2008). The Premier Gardens homes also included 2.2 kW of PV on each house. Across the street from the development, a similar housing project was constructed without the energy efficiency measures or solar power. Performance monitoring in both developments showed that the Premier Gardens homes averaged 34 percent lower gas consumption and 16 percent lower electricity use. With the PV generation included, the homes averaged 54 percent lower net electrical demand and even greater savings during summer peak periods. The incremental cost of the Premier Gardens homes (not including the California PV buy-down) averaged about \$19,000.

Figure 2.12 compares the performance of various ZEHs. The graph shows the energy performance as well as the incremental cost of different homes or housing projects. The costs for efficiency features are generally modest, but the solar power systems add \$20,000 to \$50,000 or more to project costs. Efficiency measures also reduce energy use more cost-effectively than solar systems do, as indicated by the steeper first set of data points relative to the second set of data points in Figure 2.12.

Figure 2.13, which shows the cost of saving or supplying energy for a set of homes similar to those discussed in Figure 2.12, indicates that efficiency measures are often cost-effective with a cost of saved energy at less than \$0.10/kWh. However, the combination of solar PV and efficiency measures may not be cost-effective owing to the still-high cost of the solar systems, suggesting that efficiency measures should be emphasized first in low-energy homes.

Recent analysis shows that the source-energy use of a typical 2,000-square-foot, two-story house can be reduced by 50 percent at an incremental first cost of approximately \$13,000. If this incremental cost is financed through a 30-year mortgage at 7 percent interest, the annualized cost for greater energy efficiency is just two-thirds of the annual utility bill savings, leading to net annual savings for the homeowner of about \$450 on average (Anderson and Roberts, 2008). The efficiency measures that can be used to achieve this level of energy efficiency include additional insulation, tight sealing of the building envelope, highly efficient heating and cooling equipment, a tankless gas hot-water heater, ENERGY STAR® appliances, and fluorescent lamps in most light fixtures.

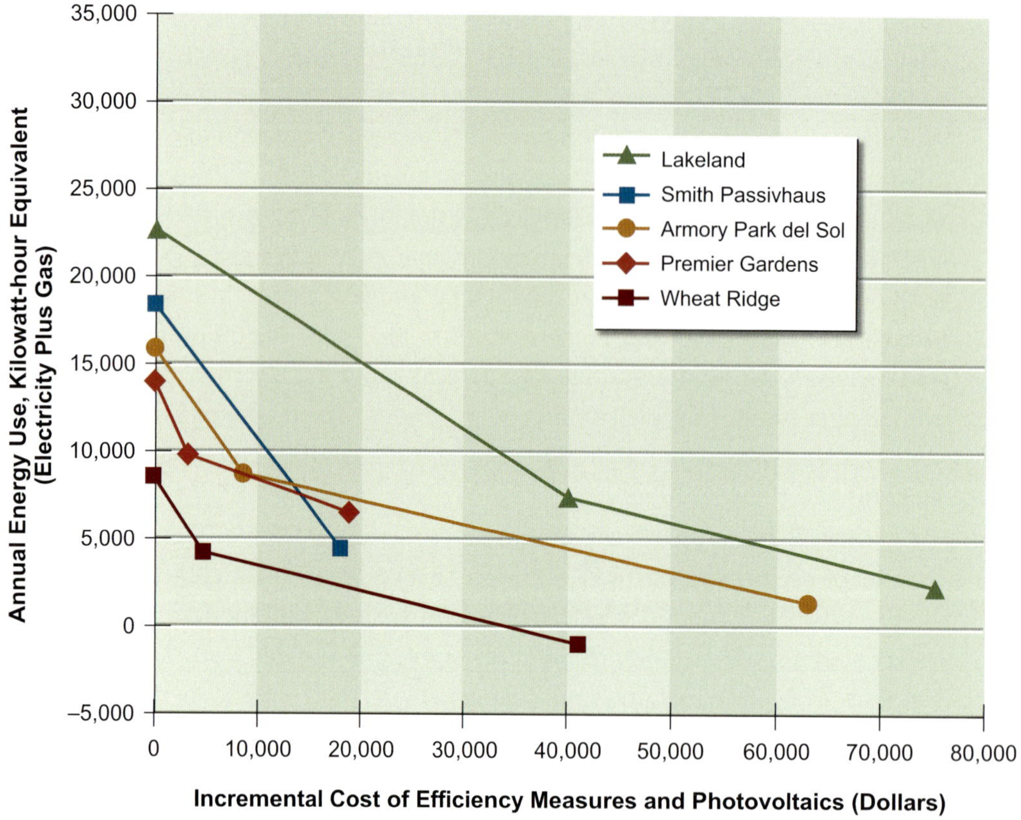

FIGURE 2.12 *Energy use versus incremental first cost of five zero-energy and near-zero-energy homes. Baseline homes (actual or estimated) are the first point of each line (i.e., intersecting the Y-axis). The first drop in (conventional) energy use and first increase in incremental cost are a result of energy efficiency measures. The second drop in (conventional) energy use and second increase in incremental cost are due to the solar photovoltaic (PV) energy systems. The houses were constructed in Sacramento, California (Premier Gardens); Wheat Ridge, Colorado; Tucson, Arizona (Armory Park del Sol); Urbana, Illinois (Smith Passivhaus); and Lakeland, Florida.*
Source: Courtesy of D. Parker, Florida Solar Energy Center, Cocoa, Fla.

2.6.7 Low-Energy New Commercial Buildings

Technical innovation in the building sector will continue to drive improvements in building performance, but there is a significant gap between the potential of existing building technologies and the effective adoption of these strategies by the building sector. This gap represents a huge opportunity for improvements in building performance that will generate substantial performance improvements in the near term. New buildings as complete structures last many decades, but major

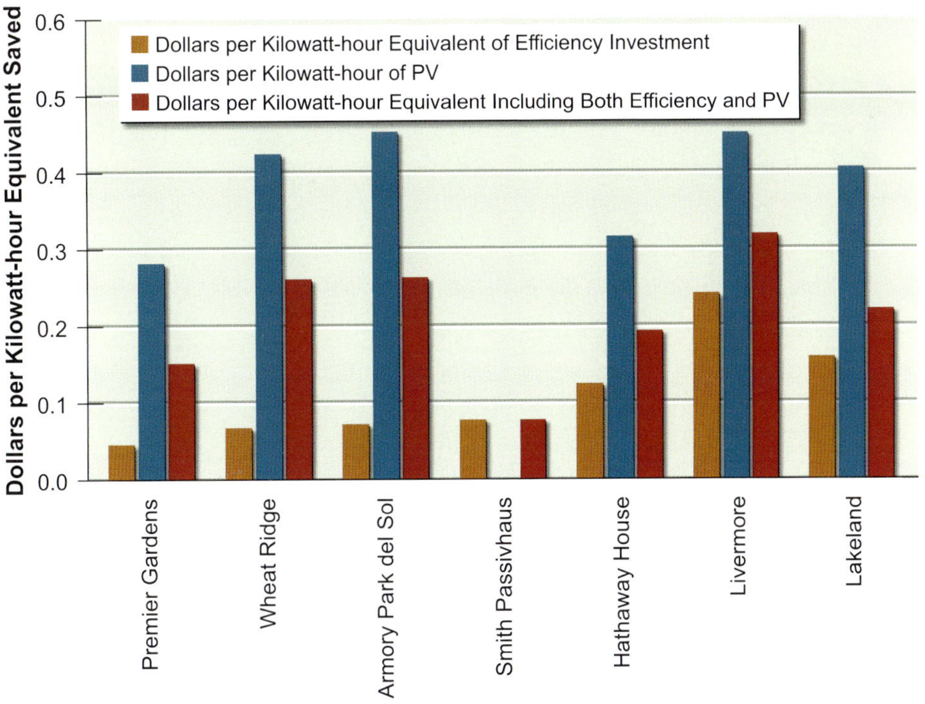

FIGURE 2.13 *Cost of saved energy and solar photovoltaic (PV) energy supply in seven low-energy homes (those shown in Figure 2.12 plus two additional projects: one in Washington, D.C. [Hathaway House] and one in Livermore, California). Costs are shown on an annualized basis using a fixed-charge rate of 0.06 and with gas savings converted to kilowatt-hour equivalent.*
Source: Courtesy of D. Parker, Florida Solar Energy Center, Cocoa, Fla.

energy-using subsystems within such buildings are often redesigned on 5-year or 20-year cycles, so improved subsystems could be applied at least partially to existing buildings, particularly in the commercial sector.

A review of the best-performing new buildings in the country suggests that buildings that achieve energy-use reductions of 50 percent or more below standard practice do not typically incorporate cutting-edge technologies, but instead successfully integrate multiple "state-of-the-shelf"[14] technologies to achieve these performance levels (Turner and Frankel, 2008).[15] This approach represents a huge

[14] "State-of-the-shelf" technologies represent the state-of-the-art selection of technologies that are widely available (on the shelf) today.

[15] Also see http://www.gettingtofifty.org—a searchable database of information about projects whose energy performance targets 50 percent beyond ASHRAE 90.1-2001; NBI (2008).

opportunity for improved energy performance using existing available technologies (Griffith et al., 2007).

The main difference between high-performing buildings and conventional buildings is essentially an attention to integration, interaction, and quality control throughout the design, construction, and operation of the building. This process, typically referred to as *integrated design,* represents a transformation not in technology but in conceptual thinking about how building systems can most effectively work together and the successful implementation of design intent (Torcellini et al., 2006).

One aspect of building performance that will grow tremendously in the next decade is the incorporation of more robust monitoring tools for building performance. A critical limitation of the ability of building designers and operators to improve building performance is the lack of good information about the impacts of design and operating decisions on actual building performance. A review of energy modeling results for 80 recently constructed Leadership in Energy and Environmental Design (LEED) buildings suggests a wide variation in the accuracy of energy modeling in predicting actual energy use (Turner and Frankel, 2008). In current practice, there is almost no mechanism through which the design community can receive real feedback on the effectiveness of its design strategies; no way to separate operational issues from design-based performance characteristics; and very little actionable feedback to building operators on real-time building performance. To address this problem, a host of efforts are currently under way for developing more effective tools to monitor and manage building operational performance using real-time data and intuitive data visualization.[16]

2.7 BARRIERS TO IMPROVING ENERGY EFFICIENCY IN BUILDINGS

Proponents of energy efficiency point to a wide range of market failures or barriers that inhibit greater investment in energy efficiency measures, including the following:

---

[16]Building performance measurement protocols are currently under development by the Center for Neighborhood Technology, the New Buildings Institute, the Green Building Alliance, the U.S. Green Building Council, the American Society of Heating, Refrigerating, and Air-Conditioning Engineers, and a host of private companies.

- *Limited supply and availability* of some energy efficiency measures;
- *Consumers lacking information or having incomplete information* about energy efficiency opportunities, and the high transaction costs for obtaining such information;
- *Users of energy lacking the capital* to invest in energy efficiency measures;
- *Fiscal or regulatory policies that discourage energy efficiency investments,* often inadvertently;
- *Decision making that does not consider or value energy efficiency*;
- *Perceived risk* associated with the performance of relatively new energy efficiency measures;
- *Split incentives* whereby the party designing, constructing, or purchasing a building or piece of equipment does not pay the operating costs; and
- *Energy prices that do not reflect the full costs imposed on society* by energy production and use (externalities).[17]

It is important to recognize there is no single market for energy efficiency. The energy efficiency "market" consists of many end-uses, a myriad of intermediaries, hundreds of millions of energy users, and millions of decision points. In addition, it is useful to distinguish between what are generally viewed as market failures and market barriers (Box 2.1). Market failures occur if there is a flaw in the way that markets operate. Market barriers are not flaws in the way that markets operate, but they limit the adoption of energy efficiency measures nonetheless.

2.7.1 Market Failures

Environmental externalities are one of the most important and frequently cited examples of unpriced costs and benefits. Energy prices do include costs associated with meeting environmental standards, but other adverse environmental impacts, such as emissions of mercury or CO_2, land disruption, or legal water contamination, are not factored in to energy prices. Likewise, the cost paid by society to protect and defend sources of oil and other energy imports is not included in energy

---

[17]No attempt was made to rank the various market failures or barriers by importance in this list or in the subsequent discussion.

BOX 2.1 ***Market Failures and Market Barriers Inhibiting Greater Energy Efficiency***

Market Failures

- Unpriced costs and benefits
- Distortional fiscal and regulatory policies
- Misplaced or split incentives
- Insufficient and inaccurate information

Market Barriers

- Low priority of energy issues
- Incomplete markets for energy efficiency
- Lack of access to capital
- Fear of not getting value from next buyer

Sources: Brown, 2001; IEA, 2007.

prices. As a result of failing to include these costs in market prices, more fossil energy is consumed than is socially optimal (Brown et al., 2007).

Many energy economists acknowledge that not including environmental and social costs in energy prices is a problem. For example, Jaffe and Stavins (1994) stated, "While much controversy surrounds the magnitude of the value of the environmental damages associated with energy use, the direction of the effect is unambiguous . . . consumers face incentives to use more energy than is socially desirable if they do not bear the full costs of the pollution their energy use fosters."

There are also barriers to recognizing and taking into account the full benefits of energy efficiency measures in consumer decision making. For example, at peak times, small reductions in demand can have a disproportionate effect in reducing price. The benefits of reducing demand accrue not only to the customers who reduce their load but to all customers, since all pay a lower price for their electricity. Because the majority of benefits accrue to customers who take no action, no customer realizes the full benefit of reducing his or her own load, and so there is less motivation to reduce demand than would be the case if a customer could realize the full benefit (Elliott and Shipley, 2005; Wiser et al., 2005; Spees and Lave, 2007).

Various types of fiscal policies discourage investments in energy efficiency. For example, capital investments in commercial buildings must be depreciated over more than 30 years, while energy purchases can be fully deducted from taxable income the year in which they occur (Brown, 2001). This puts energy efficiency investments and upgrades at a disadvantage in terms of cost-effectiveness. Likewise, most states charge sales tax on energy efficiency measures and projects but do not tax electricity and fuel sales (Brown et al., 2007). In addition, conventional energy sources such as oil and natural gas production receive tax subsidies such as depletion allowances.

Some regulatory policies also inhibit investments in energy efficiency in buildings and industry. In particular, regulatory policies that allow public utilities to increase their profits by selling more electricity or natural gas are a disincentive to effective utility energy efficiency programs (Carter, 2001). Many utilities also have adopted tariffs and interconnection standards that discourage end users from adopting energy-efficient combined heat and power (CHP) systems (Brooks et al., 2006). The variability in the stringency and enforcement of building energy codes across states and localities is cited as another barrier to energy efficiency in buildings (Brown et al., 2007).

Misplaced incentives, also known as split incentives or principal-agent problems, exist in numerous situations. The most visible example of misplaced incentives is in rental markets, where building owners are responsible for investment decisions but tenants pay the energy bills. Nearly one-third of U.S. households rent their homes, and 40 percent of privately owned commercial buildings rent or lease their space (Brown et al., 2005). A number of studies have revealed lower levels of energy efficiency in dwellings occupied by renters compared to those occupied by owners in the United States. For example, a survey in California found that insulation, energy-efficient windows, programmable thermostats, and other energy efficiency measures are less common in rental housing compared to owner-occupied dwellings (see Figure 2.14).

Misplaced incentives also are found in new construction markets in which decisions about building designs and features are made by people who are not responsible for paying the energy bills. Architects, builders, and contractors have an incentive to minimize first cost in order to win bids and maximize their profits (Koomey, 1990; Brown et al., 2007). Also, commercial leases are often structured so that the landlord allocates energy costs to tenants on the basis of square footage leased. In that case, neither the landlord nor any of the tenants has the incentive to invest in cost-effective efficiency (Lovins, 1992).

Split incentives also exist for certain appliances and end-uses (IEA, 2007). When a homeowner installs a cable or satellite television box in his or her home, the box is purchased by the service provider, but the electricity bill is paid by the homeowner. When a retail store owner installs a beverage vending machine on

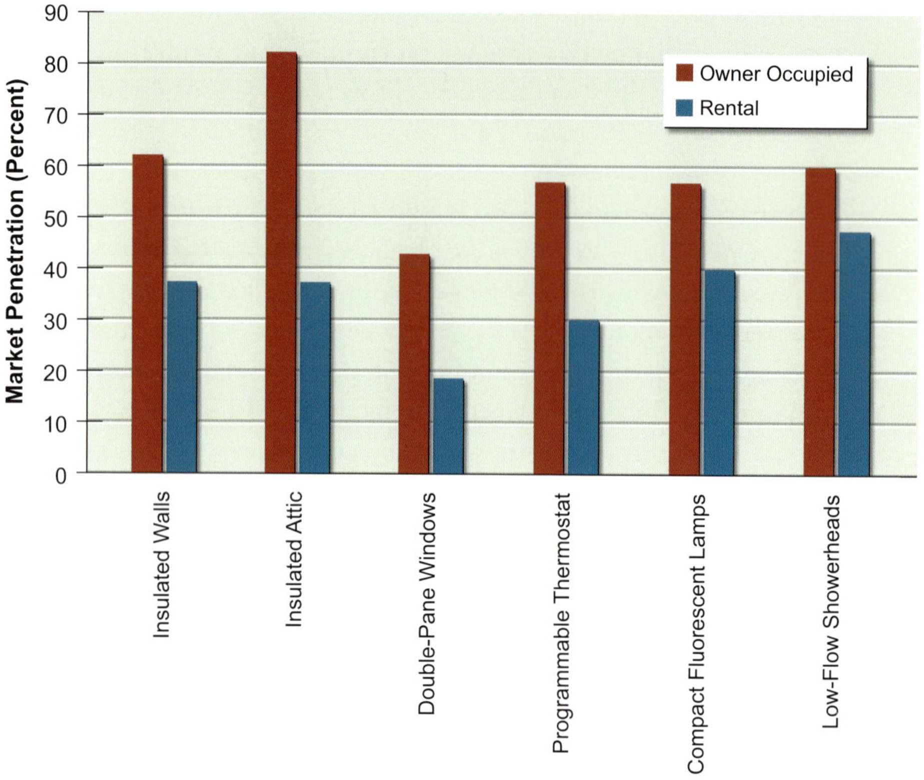

FIGURE 2.14 *Comparison of the market penetration of energy efficiency measures in owner-occupied and rental housing in California. "Penetration" is the percentage of U.S. households with an energy-saving feature such as insulation.*
Source: CEC, 2004. Reprinted with permission.

his or her property, the machine is specified by the bottler but the utility bill is paid by the retailer. In these cases, the owner of the equipment has no incentive to spend extra money to improve energy efficiency.

In some cases the split in incentives is not between different economic actors but between different centers of responsibility within a single organization. In larger companies, energy efficiency investment decisions are often made by financial officers in charge of capital budgets, but the energy savings accrue to the division responsible for operating a particular piece of energy-efficient equipment. The operating division does not have access to capital or authority to make investment decisions, and the financial officers may end up ignoring cost-savings opportunities in the utilities area.

A large body of research documents that consumers are often poorly informed about technology characteristics and energy efficiency opportunities. Some consumers do not know where to find credible information on energy efficiency options. Consumers may know how much more an energy-efficient air conditioner or water heater costs, but they would not know how much they will save per year by purchasing the more efficient technology, despite the fact that these technologies may carry mandatory energy efficiency labels. In addition, it can take many years to inform and educate a large majority of households and businesses about energy efficiency options. For example, after nearly 8 years of active promotion of—and incentives for—CFLs, nearly one-third of households surveyed in the Pacific Northwest in late 2004 were still unaware of this energy efficiency measure (Rasmussen et al., 2005). Indeed, many owners or managers of large commercial buildings have no knowledge of the size of the energy bills of their properties, despite the fact that energy is the largest cost component of net operating income, typically coming in at 15 percent. Some owners or managers even conceive of energy as a fixed cost beyond their control.

This lack of information is an even greater problem, and harder to fix, for individual end-uses. For example, when a tenant of a commercial building buys office equipment, the electricity usage of this equipment will not be metered, either at the user level or even on a level that would allow a rational decision to be made about the efficiency of the equipment. And not a single end-use in homes is ever metered separately. Thus homeowners have no direct information as to whether their computer, or videogame box, or hair dryer is a big energy user or a trivial one.

Likewise, consumers or businesses often lack the ability or time to process and evaluate the information that they do have, a situation sometimes referred to as "bounded rationality" (Koomey, 1990; Golove and Eto, 1996). And consumers often have difficulty using information on energy labels or calculating the payback period for a more efficient appliance (Sanstad and Howarth, 1994). Even when performance ratings are available (such as ENERGY STAR® labeling), consumers may not know how the energy-efficient device will function and how much energy and money will be saved in their own homes or businesses.

Consumers or businesses may perceive (rightly or wrongly) that energy efficiency technologies do not perform as well as the standard, less-efficient products they are used to.[18] For example, consumers may believe that energy-efficient fluo-

---

[18]Usually the assumption is not only wrong but reversed: the energy-efficient option performs better than what the consumer is used to. So while the popular press often reports negative con-

rescent lamps provide poorer-quality light compared to that from incandescent lamps; that energy-efficient homes have poorer air quality and are less healthful than leaky, inefficient homes; or that energy-efficient furnaces or air conditioners are less reliable than "low-tech" standard efficiency models (Jaffe and Stavins, 1994). Likewise, businesses may be concerned that energy-efficient devices are less reliable and could lead to costly downtime. In some cases, such concerns were legitimate when a technology was first introduced (e.g., for CFLs), but are no longer valid today.

Recent research has shown that human decision making departs from rationality in certain consistent ways, and several of those ways impact energy efficiency investments. Decisions reflect risk aversion and loss aversion, in that a $1 gain represents less positive utility than a $1 loss does negative utility. It takes a gain of $2.50 to balance a loss of $1 as a result of risk aversion, and a similarly biased ratio for loss aversion (Thaler et al., 1997). Decision makers confronted with a complicated choice will tend to leave things the way they are rather than risking failure or optimizing the situation as a decision maker with a fresh perspective would. This situation is also known as "status quo bias."

These problems might represent a relatively minor failure of the market when the decision maker is acting on his or her own behalf, but they are serious problems when that person is an agent for others, for example, a business manager who is not the owner. Corporate shareholders do not want irrational behavior, and such behavior certainly is not consistent with the way to maximize profit (Goldstein, 2007).

The lack of information is often a market barrier rather than a market failure. In some cases, however, it rises in significance because the problem is not merely the dissemination of existing information but rather the generation of information in the first place. For example, televisions are a growing source of energy use, but information on the energy use of a particular set is unavailable because (until 2008) there was no standard on how to test a television's energy consumption. This type of problem often occurs because the product's trade association wants to set test standards or, in some cases, prefers that there not be any test standards.

---

sumer reactions to CFL color, a national survey of consumers as part of the McGraw-Hill annual construction survey found that more consumers than not consider the light quality of CFLs better than that of incandescent lamps (LeBlanc, 2007).

2.7.2 Market Barriers

Turning to the market barriers, businesses tend to pay limited attention to energy-use and energy-savings opportunities if energy costs are a small fraction of the total cost of owning or operating the business or factory, or if energy efficiency is not viewed as a priority by the company management. Energy costs represent less than 2 percent of the total cost of operating a factory or commercial business in many (but not all) cases. Furthermore, businesses are most concerned with developing new products, maintaining production, and increasing sales; energy consumption is usually a secondary or tertiary concern. As a result of these and other factors, many businesses limit energy efficiency investments to projects with payback periods of no more than 2 or 3 years (DeCanio, 1993; Geller, 2003).

Many individual consumers also do not value the lifetime energy savings provided by more efficient appliances, vehicles, or other energy efficiency measures. For example, consumers on average expect vehicle fuel-efficiency improvements to pay back their first cost in 3 years or less even though vehicles remain in use for about 14 years on average (Greene and Schafer, 2003). Chapter 5 discusses the strategies for overcoming this barrier as well as other market barriers discussed in this section.

Regarding incomplete markets for energy efficiency, some measures are relatively new and are still not widely available in the marketplace or not well supported by product providers (Hall et al., 2005). These include measures such as highly efficient light fixtures, reflective roofing materials, heat-pump water heaters, and modern evaporative coolers. The limited availability of these products results in a lack of consumer awareness and demand, and the lack of demand makes it difficult to expand availability. Also, some very effective energy efficiency services such as duct testing and sealing and the recommissioning of existing buildings are not widely available or marketed in many parts of the country.[19]

Regarding a lack of capital to invest in energy efficiency measures, this is particularly a problem for low-income households that have limited resources and limited access to credit. In addition, some businesses (particularly small businesses) have insufficient capital or borrowing ability.

---

[19]It should be noted, however, that some energy efficiency measures such as insulation, compact fluorescent lamps, or ENERGY STAR® appliances are readily available.

Detailed studies of particular markets have found multiple and substantial barriers inhibiting the adoption of cost-effective energy efficiency improvements (IEA, 2007). In the motors market, for example, motor suppliers may fail to stock high-efficiency motors; buyers may lack accurate information on motor efficiency or other opportunities for cost-effectively saving energy in motor systems; facility managers often shy away from newer technologies, fearing reliability problems; and motors may be replaced on an emergency basis, resulting in little or no time to consider energy efficiency (Nadel et al., 2002). Also, many motors are purchased by original equipment manufacturers (OEMs), companies that assemble pumps, blowers, air-conditioning systems, and other items. OEMs generally purchase motors based on lowest first cost since they are not responsible for paying operating costs, another example of split incentives.

2.7.3 Implications of Market Barriers

Particular policy and program remedies are available for many of the market failures and barriers described above (Hall et al., 2005). These include educating consumers and businesses, increasing the supply and visibility of energy-efficient products and services in retail establishments, offering consumers and businesses financial incentives to get their attention and stimulate greater willingness to adopt efficiency measures, removing inefficient products or buildings from the marketplace through codes and standards, and reforming pricing and regulatory policies.

But other market failures or barriers are deeper and harder, if not impossible, to correct. Diffuse decision making, risk aversion, loss aversion, and status quo bias, in particular, would seem difficult to solve, even conceptually. Moreover, there are transactions costs related to educating consumers, addressing the split incentives problem, or convincing households or businesses to invest in energy efficiency to a greater degree. The real question is whether policy and program interventions are cost-effective mechanisms for stimulating greater investment in energy efficiency measures—that is, whether the value of the energy savings, peak-demand reduction, and nonenergy benefits (see below) exceed the costs (both for the efficiency measures and policy or program implementation). As discussed in detail in Chapter 5, many types of energy efficiency policies and programs offer effective and economically attractive ways of removing or reducing the market failures and barriers described above.

2.8 MARKET DRIVERS

A number of factors—economic, environmental, and business related—are serving to overcome the barriers and market failures discussed above. Likewise, many energy efficiency measures provide multiple benefits, including nonenergy benefits that help to drive the adoption of these measures. In addition, numerous public policies, including state and utility efficiency programs, building energy codes, and appliance efficiency standards are stimulating a greater adoption of efficiency measures. These policies and programs are discussed further in Chapter 5.

2.8.1 Economic, Environmental, and Business Considerations

The costs of electricity, natural gas, and petroleum products are volatile but are generally rising (see Figure 2.15). Corrected for inflation, the average retail price for electricity paid by households in the United States increased 34 percent between 1975 and 1985, but then fell 24 percent during 1985–2002 (EIA, 2008b). However, the national average price increased 10 percent in real dollars during 2002–2007. Retail natural gas prices paid by residential consumers increased 45 percent on average during 2002–2007. The downturn in the world economy apparent at the time of this writing will mitigate growth in energy prices for a while, but the underlying determinants of demand growth remain in place. Higher energy prices improve the cost-effectiveness of energy efficiency measures and stimulate greater interest in and willingness to adopt such measures on the part of consumers (AIA, 2007). This is clear from the high demand for fuel-efficient vehicles and the heavy drop-off in demand for large sport-utility vehicles and other gas guzzlers in response to rapidly rising gasoline prices in late 2007 and early 2008.

Environmental awareness and concern about energy security and global climate change are also on the rise. A national survey administered by researchers from the Massachusetts Institute of Technology in 2003 and 2006 showed that citizens ranked global warming as the nation's most pressing environment concern in 2006, whereas in 2003 global warming was ranked at number 6 out of 10 (Curry et al., 2007). Furthermore, the same survey found that support for action to address global warming is rising, as is the willingness to pay more for electricity in order to address global warming. In particular, the amount that respondents indicated they were willing to pay to address global warming increased 50 percent between 2003 and 2006. Many states and cities are adopting greenhouse gas emissions reduction goals and action plans, thereby increasing the awareness of and support for the adoption of energy efficiency measures.

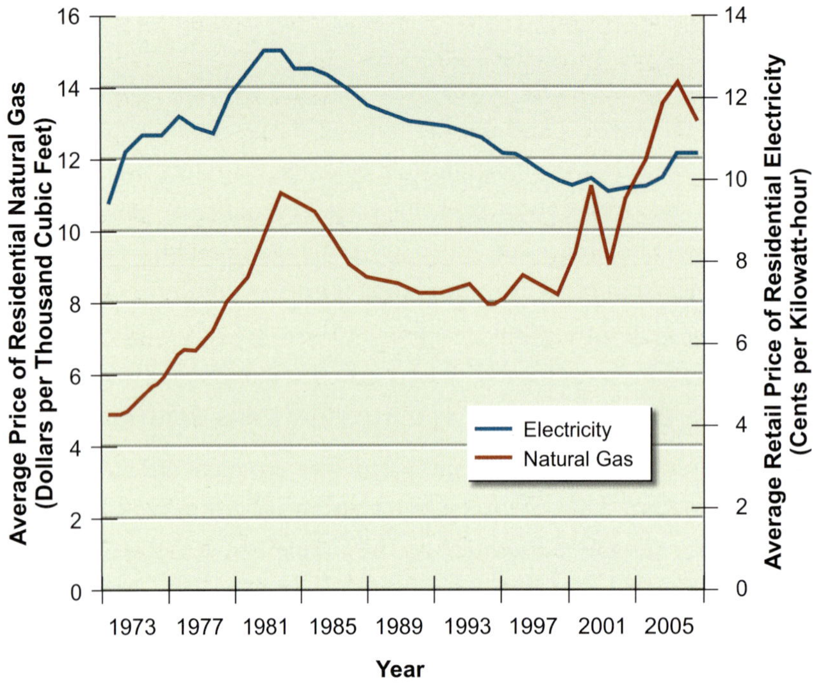

FIGURE 2.15 *Average price of residential natural gas and average retail price of residential electricity.*
Source: EIA, 2008b.

Growing awareness of the ENERGY STAR® label and its increasing use in energy-related purchase decisions are other indicators that interest in energy efficiency is rising (EPA, 2008). ENERGY STAR® is promoted in part as a way for consumers to reduce polluting emissions and protect the environment. Likewise, the awareness and adoption of "green building" practices are rapidly rising. For example, 2.3 billion square feet of commercial buildings were registered or certified under the LEED rating program at the end of 2007, up more than 500 percent in 2 years (Makower, 2008).

Many corporations with forward-looking agendas are making commitments to increase energy efficiency and/or to reduce their greenhouse gas emissions and are achieving impressive results (Hoffman, 2006). These commitments often pertain to energy use and greenhouse gas emissions within the company itself. However, some large corporations such as Wal-Mart, Hewlett-Packard, and General Electric are expanding their production and/or marketing of energy- and resource-efficient products. Wal-Mart, for example, exceeded its goal of selling

100 million compact fluorescent lamps in 2007 and announced that it would launch its own brand of CFLs (Makower, 2008). The retailer also announced plans to cut the energy use of its stores and help increase the energy efficiency of its entire supply chain. Hewlett-Packard has introduced a wide range of more efficient personal computers that meet the ENERGY STAR® specifications issued in 2007. And General Electric recently announced that it would start manufacturing and marketing energy-efficient tankless and heat-pump water heaters, in conjunction with the DOE's announcing ENERGY STAR® criteria for residential water heaters. Such efforts increase consumer awareness and the adoption of energy-efficient products.

2.8.2 Nonenergy Benefits

Energy efficiency is often defined in terms of using technology to reduce the amount of energy consumed in providing a given level of service. However, energy-efficient technologies in many cases provide a higher level of service as a result of "nonenergy" benefits. In some cases, the value of these nonenergy benefits exceeds the value of the energy saved over the lifetime of the product. It is also possible that energy-efficient technologies can reduce the level or quality of service—that is, they can result in nonenergy costs. But empirically, the number of cases in which this is true is small.

The CFL offers one example of nonenergy benefits. The nonenergy benefits of CFLs are primarily the increased lifetime. Compact fluorescent lamps meeting the ENERGY STAR® specifications have a minimum lifetime of 6,000 hours, but increasing numbers of products have lifetimes of 10,000 hours or more. For comparison, the lifetime for incandescent bulbs ranges from 750 hours to 2,000 hours, with 1,000 hours being most typical. For applications where lights are hard to change or where staff must be paid to change the bulbs, the value of reduced maintenance greatly exceeds the value of the energy savings.[20]

The color rendition of CFLs, which is different from that of incandescents, in some cases is a benefit. Incandescents are only available in a limited range of color temperatures from about 2500 K to 3000 K, with the low end of the range

[20]For example, to change incandescent lamps providing illumination to a three-story atrium, maintenance crews must set up scaffolding to climb up three stories in order to change the bulbs. The cost of avoiding doing this every 1,000 hours—two or three times a year for typical usage patterns of office buildings—exceeds by an order of magnitude the value of lifetime energy savings of the lamp, which is on the order of \$50–\$100.

obtainable only through dimming and the high end of the range obtainable only through halogen lamps. CFLs are available at a choice of color temperatures ranging from about 2700 K to about 6500 K; commonly available products are at 2700 K, 3500 K, and 4100 K.

LEDs used for traffic signals provide substantial cost savings to the municipalities that use them because of their longer lifetime; these savings go far beyond the value of energy savings. Longer lifetime means less expense for maintenance crews to replace burned-out lamps. LED traffic signals also have a safety advantage—if an incandescent signal fails, the entire red light or green light goes out, whereas if LEDs fail, they fail one lamp at a time, and the driver can still see a red light, green light, or amber light even if the pattern is not perfectly circular. And, because of their much lower energy use, it is technically feasible to operate LED signals with battery backups, so that traffic signals can function even in a blackout.

Energy-efficient (and water-efficient) domestic clothes washers are marketed on the basis of their superior cleaning ability and the fact that they cause less wear to fabrics, in addition to the value of the savings of energy, water, or even detergent. Many of these products can also, through their energy-efficient design, handle larger garments or those that might previously have required dry cleaning. These factors are much more important than energy savings in the marketing of these clothes washers.

Natural daylighting is another energy-saving strategy with significant non-energy benefits. High-quality daylighting has been shown to have a positive association with better student performance in schools, higher retail sales in stores, and productivity improvements in offices and other workplaces (Romm, 1999; Heschong and Wright, 2002; Kats, 2006).

Improved insulation and fenestration systems in buildings not only reduce energy use for air-conditioning and cooling but also provide greater comfort. In these cases, this benefit is a consequence of the physics of heat transfer: better-insulated walls or windows have interior surface temperatures that are closer to room temperature than do poorly insulated ones; this creates a radiant heat-transfer environment for the human body, which is more comfortable. In hotter climates, windows that reflect near-infrared solar heat result in less solar heat gain on clothes or skin than would be the case with conventional windows. Homes that are sealed to prevent air leakage from or to the outside and are better insulated also provide better acoustic isolation (i.e., less interior noise). And sealing air ducts in buildings not only saves energy but also can provide more even heating and cooling within the building—for example, avoiding rooms at the end of the duct system that are not adequately heated in the winter or cooled in the summer.

The panel is unaware of any comprehensive study looking at nonenergy impacts (costs and benefits) from energy efficiency measures. This informal review finds numerous examples of nonenergy benefits (some of which are larger than the energy benefits themselves) and a few examples of nonenergy costs. Such a study would be useful, particularly if it attempted to quantify these costs and benefits.

2.9 FINDINGS

The following findings derive from the panel's analysis of energy efficiency in buildings summarized in this chapter.

B.1 Studies assessing the potential for energy savings in buildings take several different approaches, looking at whole-building results as well as results by end-use and technology. Nevertheless, their results tend to be consistent.

B.2 The potential for large, cost-effective energy savings in buildings is well documented. Median predictions of achievable, cost-effective savings are 1.2 percent per year for electricity and 0.5 percent per year for natural gas, amounting to a 25–30 percent energy savings for the buildings sector as a whole over the next 20–25 years. If this level of savings were to be achieved, it would offset the EIA (2008a) projected increase in energy use in this sector over the same period.

B.3 Studies of energy efficiency potential are subject to a number of limitations and biases. On the one hand, factors such as not accounting for new and emerging energy efficiency technologies can lead such studies to underestimate energy-savings potential, particularly in the midterm and long term. On the other hand, some previous studies were overly optimistic about the cost and performance of certain efficiency measures, thereby overestimating energy-savings potential, particularly in the short term. Although these limitations must be acknowledged, they do not affect the panel's overall finding that the potential for energy savings in buildings is large.

B.4 Many advanced technologies under development and likely to become commercially available within the next decade—including LED lamps, innovative window systems, new types of cooling systems, and power-saving electronic devices—will further increase the energy-savings potential in buildings. In addition, new homes and commercial buildings with relatively low overall energy use have been demonstrated throughout the country. With appropriate policies and programs, they could become the norm in new construction.

B.5 Despite substantial barriers to widespread energy efficiency improvements in buildings, a number of countervailing factors could drive increased energy efficiency, including rising energy prices, growing concern about global climate change and the resulting willingness of consumers and businesses to take action to reduce emissions, a movement toward "green buildings," and growing recognition of the significant nonenergy benefits offered by energy efficiency measures.

2.10 REFERENCES

ACEEE (American Council for an Energy-Efficient Economy). 2007. The Potential for Electricity Conservation in New York State, September 1989. Prepared for the New York State Energy Research and Development Authority, Niagara Mohawk Power Corporation, and the New York State Energy Office.

AHAM (Association of Home Appliance Manufacturers). 2008. Data compiled by the Association of Home Appliance Manufacturers. Washington, D.C.: AHAM. Available at http://www.aham.org.

AIA (American Institute of Architects). 2007. As home energy costs remain high, residential architects report that sustainable design motivates homeowners. AIArchitect This Week. Volume 14. September 7. Available at http://info.aia.org/aiarchitect/thisweek07/0907/0907n_econres.cfm.

Amann, J.T., A. Wilson, and K. Ackerly. 2007. Consumer Guide to Home Energy Savings. 9th Edition. Washington, D.C.: American Council for an Energy-Efficient Economy.

Anderson, R., and D. Roberts. 2008. Maximizing Residential Energy Savings: Net Zero Energy Home (ZEH) Technology Pathways. NREL/TP-550-44547. Golden, Colo.: National Renewable Energy Laboratory. November.

Apte, J., and D. Arasteh. 2006. Window-Related Energy Consumption in the U.S. Residential and Commercial Building Stock. LBNL-60146. Berkeley, Calif.: Lawrence Berkeley National Laboratory. Available at http://gaia.lbl.gov/btech/papers/60146.pdf.

Arasteh, D., S. Selkowitz, J. Apte, and M. LaFrance. 2006. Zero energy windows. Proceedings of the 2006 ACEEE Summer Study on Energy Efficiency in Buildings. Washington, D.C.: American Council for an Energy-Efficient Economy.

Ashdown, B.G., D.J. Bjornstad, G. Boudreau, M.V. Lapsa, S. Schexnayder, B. Shumpert, and F. Southworth. 2004. Heat Pump Water Heater Technology: Experiences of Residential Consumers and Utilities. ORNL/TM-2004/81. Oak Ridge, Tenn.: Oak Ridge National Laboratory. June. Available at http://www.ornl.gov/sci/engineering_science_technology/ eere_research_reports/appliances/water_heaters/heat_pump_water_heaters/ornl_tm_2004_81/ornl_tm_2004_81.pdf.

ASHRAE (American Society of Heating, Refrigerating and Air-Conditioning Engineers). 2008. Advanced Energy Design Guides. Atlanta, Ga.: ASHRAE. Available at http:// www. ashrae.org/technology/page/938.

Azevedo, I.L., M.G. Morgan, and F. Morgan. 2009. The transition to solid-state lighting. Proceedings of the IEEE 97(3):481-510.

Battles, S., and B. Hojjati. 2005. Two Decades of U.S. Household Trends in Energy-Intensity Indicators: A Look at the Underlying Factors. Washington, D.C.: Energy Information Administration. Available at http://www.eia.doe.gov/emeu/efficiency/2005_IAEE.pdf.

BEA (Bureau of Economic Analysis). 2008. Implicit Price Deflators for Gross Domestic Product. Washington, D.C.: BEA, Department of Commerce. February. Available at http://www.bea.gov/national/index.htm. Accessed March 5, 2008.

Belzer, D.B. 2007. Estimates of U.S. Commercial Building Electricity Intensity Trends: Issues Related to End-Use and Supply Curves. PNNL-16820. Richland, Wash.: Pacific Northwest National Laboratory.

Brambley, M.R., P. Haves, S.C. McDonald, P. Torcellini, D.G. Hansen, D. Holmberg, and K. Roth. 2005. Advanced Sensors and Controls for Building Applications: Market Assessment and Potential R&D Pathways. PNNL-15149. Richland, Wash.: Pacific Northwest National Laboratory.

Brooks, S., B. Elswick, and R.N. Elliott. 2006. Combined Heat and Power: Connecting the Gap Between Markets and Utility Interconnection and Tariff Practices (Part 1). Washington, D.C.: American Council for an Energy-Efficient Economy. March.

Brown, M.A. 2001. Market failures and barriers as a basis for clean energy policies. Energy Policy 29(14):1197-1207.

Brown, M.A., F. Southworth, and T.K. Stovall. 2005. Towards a Climate-Friendly Built Environment. Arlington, Va.: Pew Center on Global Climate Change.

Brown, M.A., J. Chandler, M.V. Lapsa, and B.K. Sovacool. 2007. Carbon Lock-In: Barriers to Deploying Climate Change Mitigation Technologies. ORNL/TM-2007/124. Oak Ridge, Tenn.: Oak Ridge National Laboratory. November.

Brown, R., S. Borgeson, J. Koomey, and P. Biermayer. 2008. U.S. Building-Sector Energy Efficiency Potential. LBNL-1096E. Berkeley, Calif.: Lawrence Berkeley National Laboratory. September. Available at http://repositories.cdlib.org/lbnl/LBNL-1096E.

Burns, S., M. Goggin, D. Hinrichs, and K. Lee. 2006. Technical and economic assessment of solar thermal absorption cooling systems in small commercial buildings. Presented at the 2006 World Energy Engineering Congress. Available at http://www.naturalenergytechnologies.com/resources/WEEC_Paper_Hinrichs.pdf.

Carter, S. 2001. Breaking the consumption habit: Ratemaking for efficient resource decisions. Electricity Journal 14(10):66-74.

CEC (California Energy Commission). 2004. California Statewide Residential Appliance Saturation Study. Final Report. CEC-400-04-009. Sacramento, Calif.: CEC. June.

Christian, J.E., D. Beal, and P. Kerrigan. 2004. Toward simple affordable zero energy houses. Proceedings of the Thermal Performance of the Exterior Envelopes of Buildings IX. Clearwater, Fla.: American Society of Heating, Refrigerating and Air-Conditioning Engineers. December.

Craford, M.G. 2008. High power LEDs for solid state lighting: Status, trends, and challenges. Journal of Light and Visual Environment 32(2):58-62.

Creyts, J., A. Derkach, S. Nyquist, K. Ostrowski, and J. Stephenson. 2007. Reducing U.S. Greenhouse Gas Emissions: How Much at What Cost? McKinsey and Company. December.

Curry, T.E., S. Ansolabehere, and H. Herzog. 2007. A Survey of Public Attitudes Towards Climate Change and Climate Change Mitigation Technologies in the United States: Analyses of 2006 Results. Cambridge, Mass.: Laboratory for Energy and the Environment, Massachusetts Institute of Technology. April.

DeCanio, S.J. 1993. Barriers within firms to energy-efficient investments. Energy Policy 21(9):906-914.

DeCotis, P.A., Lawrence J. Pakenas, J.M. Tarantino. 2004. Improving energy productivity: Successes and remaining potential. Proceedings of the 15th National Energy Services Conference and Exposition. Jupiter, Fla.: Association of Energy Service Professionals International.

DOE (Department of Energy). 2004a. Whole-House Approach Benefits Builders, Buyers, and the Environment. Washington, D.C.: DOE. Available at http://apps1.eere.energy.gov/buildings/publications/pdfs/building_america/34867.pdf.

DOE. 2004b. Cooling Technologies for Buildings. Washington, D.C.: DOE, Building Technologies Program. Available at http://www.eere.energy.gov/buildings/info/components/hvac/cooling.html.

DOE. 2008a. Energy Intensity Indicators in the U.S. Washington, D.C.: DOE, Office of Energy Efficiency and Renewable Energy. Available at http://www1.eere.energy.gov/ba/pba/intensityindicators/trend_data.html.

DOE. 2008b. Solid State Lighting Research and Development Multi-Year Program Plan FY'09-FY'13. Washington, D.C.: DOE, Building Technologies Program.

DOE. 2009. 2006 Manufacturing Energy Consumption Survey. Washington, D.C. Available at http://www.eia.doe.gov/emeu/mecs/contents.html.

DOE/EERE (Energy Efficiency and Renewable Energy). 2007. Buildings Energy Data Book 2007. Washington, DC: DOE/EERE. Available at http://buildingsdatabook.eere.energy.gov/.

DOE/EERE. 2008. Buildings Energy Data Book 2008, Section 2.1.5. Washington, D.C.: DOE/EERE. Available at http://buildingsdatabook.eere.energy.gov/.

Dunn. S. 2007. High Performance Homes in the Southwest: Savings Potential, Cost Effectiveness and Policy Options. Boulder, Colo.: Southwest Energy Efficiency Project. November. Available at http://www.swenergy.org/hph/High_Performance_Homes.pdf.

EETF (Energy Efficiency Task Force). 2006. Energy Efficiency Task Force Report. Prepared for the Clean and Diversified Energy Initiative, Western Governors' Association, Denver, Colo. January. Available at http://www.westgov.org/wga/initiatives/cdeac/Energy%20Efficiency-full.pdf.

EIA (Energy Information Administration). 2007a. Annual Energy Outlook 2007, with Projections to 2030. DOE/EIA-0383(2007). Washington, D.C.: DOE, Energy Information Administration. February. Available at http://www.eia.doe.gov/oiaf/archive/aeo07/index.html.

EIA. 2007b. Annual Energy Review 2006. DOE/EIA-0384(2006). Washington, D.C.: Department of Energy, Energy Information Administration.

EIA. 2008a. Annual Energy Outlook 2008, with Projections to 2030. DOE/EIA-0383(2008). Washington, D.C.: DOE, Energy Information Administration. February. Available at http://www.eia.doe.gov/oiaf/aeo/pdf/0383(2008).pdf.

EIA. 2008b. Annual Energy Review 2007. DOE/EIA-0384(2007). Washington, D.C.: DOE, Energy Information Administration.

EIA. 2008c. Monthly Energy Review (April). DOE/EIA-0035(2008/04). Washington, D.C.: Department of Energy, Energy Information Administration.

Elliott, R.N., and A.M. Shipley. 2005. Impacts of Energy Efficiency and Renewable Energy on Natural Gas Markets: Updated and Expanded Analysis. Washington, D.C.: American Council for an Energy-Efficient Economy.

Energy Innovations. 1997. Energy Innovations: A Prosperous Path to a Clean Environment. Washington, D.C.: Alliance to Save Energy, American Council for an Energy-Efficient Economy, Natural Resources Defense Council, Tellus Institute, and Union of Concerned Scientists.

ENERGY STAR®. 2008. Information on Compact Fluorescent Light Bulbs (CFLs) and Mercury. Washington, D.C.: ENERGY STAR® Program. June. Available at http://www.energystar.gov/ia/partners/promotions/change_light/downloads/Fact_Sheet_Mercury.pdf.

EPA (Environmental Protection Agency). 2007a. ENERGY STAR® Qualified Homes Market Indices for States. Washington D.C.: EPA, ENERGY STAR® Program. Available at http://www.energystar.gov/index.cfm?fuseaction=qhmi.showHomesMarketIndex.

EPA. 2007b. Report to Congress on Server and Data Center Energy Efficiency—Public Law 109-431. Washington, D.C.: EPA, ENERGY STAR® Program. August 2. Available at http://www.energystar.gov/ia/partners/prod_development/downloads/EPA_Datacenter_Report_Congress_Final1.pdf.

EPA. 2008. National Awareness of ENERGY STAR® for 2007: Analysis of CEE Household Survey. Washington, D.C.: EPA, ENERGY STAR® Program. Available at http://www.energystar.gov/ia/news/downloads/nat_awareness_0408.pdf.

Geller, H. 2003. Energy Revolution: Policies for a Sustainable Future. Washington, D.C.: Island Press.

Geller, H., and S. McGaraghan. 1998. Successful government-industry partnership: The U.S. Department of Energy's role in advancing energy-efficient technologies. Energy Policy 26:167-177.

Geller, H., A. de Almeida, B. Barkovich, C. Blumstein, D. Goldstein, A. Meier, P. Miller, O. de la Moriniere, A. Rosenfeld, and L. Schuck. 1986. Residential Conservation Power Plant Study. Phase 1—Technical Potential. Washington, D.C.: American Council for an Energy-Efficient Economy.

Goldstein, D.B. 2007. Saving Energy, Growing Jobs: How Environmental Protection Promotes Economic Growth, Profitability, Innovation, and Competition. Point Richmond, Calif.: Bay Tree Publishing.

Goldstein, D.B. 2008. Extreme efficiency: How far can we go if we really need to? Proceedings of the 2008 ACEEE Summer Study on Energy Efficiency in Buildings. Washington, D.C.: American Council for an Energy Efficient Economy. August.

Goldstein, D.B., and M.G. Hoffman. 2004. Forecasting an increasing role for energy efficiency in meeting global climate goals. Proceedings of the 2004 ACEEE Summer Study on Energy Efficiency in Buildings. Washington, D.C.: American Council for an Energy Efficient Economy. August.

Golove, W.H., and J.H. Eto. 1996. Market Barriers to Energy Efficiency: A Critical Reappraisal of the Rationale for Public Policies to Promote Energy Efficiency. LBNL-38059. Berkeley, Calif.: Lawrence Berkeley National Laboratory.

Greene, D.L., and A. Schafer. 2003. Reducing Greenhouse Gas Emissions for U.S. Transportation. Arlington, Va.: Pew Center on Global Climate Change.

Griffith, B., N. Long, P. Torcellini, D. Crawley, and J. Ryan. 2007. Energy efficiency improvements that use the best available technologies and practices and integrated whole-building design approaches can, on average, reduce energy consumption by 43%. In Assessment of the Technical Potential for Achieving Net Zero Energy Buildings in the Commercial Sector. NREL/TP-550-41957. Golden, Colo.: National Renewable Energy Laboratory. Available at http://www.nrel.gov/docs/fy08osti/41957.pdf.

Hall, N., C. Best, J. Roth, P. Jacobs, and L. Lutzenhiser. 2005. Assessing markets to design programs that more fully attack key market barriers and take advantage of market opportunities—Why do we continue to miss so many opportunities? Pp. 697-713 in Reducing Uncertainty Through Evaluation. Brooklyn, N.Y.: International Energy Program Evaluation Conference. August.

Harrod, J., and L. Hain. 2007. Showerheads! Home Energy. November/December, pp. 40-41.

Heschong, L., and R.L. Wright. 2002. Daylighting and human performance: Latest findings. Pp. 8.91-8.104 in Proceedings of the 2002 ACEEE Summer Study on Energy Efficiency in Buildings. Washington, D.C.: American Council for an Energy-Efficient Economy.

Hoffman, A.J. 2006. Getting Ahead of the Curve: Corporate Strategies That Address Climate Change. Arlington, Va.: Pew Center on Global Climate Change. October.

IEA (International Energy Agency). 2001. Things That Go Blip in the Night: Standby Power and How to Reduce It. Paris, France: International Energy Agency.

IEA. 2006. Energy Technology Perspectives 2006 in Support of the G8 Plan of Action: Scenarios and Strategies to 2050. Paris, France: IEA.

IEA. 2007. Mind the Gap: Quantifying Principal-Agent Problems in Energy Efficiency. Paris, France: IEA.

Itron. 2008. California Residential Efficiency Market Share Tracking: Lamps 2007. Prepared for Southern California Edison. Rosemead, Calif.: Itron Inc. September 15.

IWG (Interlaboratory Working Group). 1997. Scenarios of U.S. Carbon Reductions: Potential Impacts of Technologies by 2010 and Beyond. LBNL-40533 and ORNL/CON-444. Prepared by the Interlaboratory Working Group on Energy-Efficient and Low-Carbon Technologies. Oak Ridge, Tenn.: Oak Ridge National Laboratory; Berkeley, Calif.: Lawrence Berkeley National Laboratory. September. Available at http://enduse.lbl.gov/projects/5lab.html.

IWG. 2000. Scenarios for a Clean Energy Future. LBNL-44029. ORNL/CON-476. Prepared by the Interlaboratory Working Group on Energy-Efficient and Clean Energy Technologies. Oak Ridge, Tenn.: Oak Ridge National Laboratory; Berkeley, Calif.: Lawrence Berkeley National Laboratory. November. Available at http://www.ornl.gov/sci/eere/cef/.

Jaccard, M., and W. Montgomery. 1996. Costs of reducing greenhouse gas emissions in the USA and Canada. Energy Policy 24(10):889-898.

Jaccard, M., J. Nyboer, C. Bataille, and B. Sadownik. 2003. Modeling the cost of climate policy: Distinguishing between alternative cost definitions and long-run cost dynamics. Energy Journal 24(1):49-73.

Jacoby, H. 1999. The uses and misuses of technology development as a component of climate policy. In Climate Change Policy: Practical Strategies to Promote Economic Growth and Environmental Quality—Proceedings of a Symposium sponsored by the American Council for Capital Formation Center for Policy Research, National Press Club, Washington, D.C., September 23, 1998 (C. Walker, M. Bloomfield, and M. Thorning, eds.). Washington, D.C.: American Council for Capital Formation Center for Policy Research.

Jaffe, A.B., and R.N. Stavins. 1994. The energy efficiency gap: What does it mean? Energy Policy 22(10):804-810.

Jiang, W., D.W. Winiarski, S. Katipamula, and P.R. Armstrong. 2007. Cost-Effective Integration of Efficient Low-Lift Base-Load Cooling Equipment. PNNL-17157. Richland, Wash.: Pacific Northwest National Laboratory. December.

Kats, G. 2006. Greening America's Schools: Costs and Benefits. Washington, D.C.: Capital E, LLC.

Kinney, L. 2004. New Evaporative Cooling Systems: An Emerging Solution for Homes in Hot Dry Climates with Modest Cooling Loads. Boulder, Colo.: Southwest Energy Efficiency Project. Available at http://www.swenergy.org/pubs/Evaporative_Cooling_Systems.pdf.

Koomey, J. 1990. Energy Efficiency Choices in New Office Buildings: An Investigation of Market Failures and Corrective Policies. Doctoral Dissertation. University of California, Berkeley. Available at http://enduse.lbl.gov/info/JGKdissert.pdf.

Koomey, J., and R.E. Brown. 2002. The Role of Building Technologies in Reducing and Controlling Peak Electricity Demand. LBNL-49947. Berkeley, Calif.: Lawrence Berkeley National Laboratory. September.

Koomey, J.G., C.A. Webber, C.S. Atkinson, and A. Nicholls. 2001. Addressing energy-related challenges for the U.S. buildings sector: Results from the Clean Energy Futures Study. Energy Policy 29(14):1209-1222.

LeBlanc, B. 2007. A Unique Survey Opportunity. Presentation by Bill LeBlanc, ESource, Boulder, Colo., September 27.

Lovins, A. 1992. Energy-Efficient Buildings: Institutional Barriers and Opportunities. Boulder, Colo.: E-Source. December.

Lowenstein, A., S. Slayzak, and E. Kozubal. 2006. A zero carryover liquid-desiccant air conditioner for solar applications. Draft. ASME/SOLAR06. July 8-13, 2006, Denver, Colo. ISEC2006-99079. Available at http://www.ailr.com/PDF/ASME-Solar06%20LDAC%20ISEC2006-99079.pdf.

Makower, J. 2008. State of Green Business 2008. Oakland, Calif.: Greener World Media, Inc. Available at http://stateofgreenbusiness.com/files/StateOfGreenBusiness2008.pdf.

Meier, A., B. Nordman, J. Busch, C. Payne, R. Brown, G. Homan, M. Sanchez, and C. Webber. 2008. Low Power Mode Energy Consumption in California Homes. CEC-500-2008-035. Sacramento, Calif.: California Energy Commission, Public Interest Energy Research Program.

Messenger, M. 2008. The use of conservation supply curves in California planning and implementation: Comparing estimates of achievable savings versus reported program savings. Draft final paper prepared by Itron, Inc., for the National Academies. June.

Miller, P.M, J.H. Eto, and H. Geller. 1989. The Potential for Energy Conservation in New York State. Washington, D.C.: American Council for an Energy-Efficient Economy.

Mills, E. 2009. Building Commissioning: A Golden Opportunity for Reducing Energy Costs and Greenhouse-gas Emissions. Berkeley, Calif.: Lawrence Berkeley National Laboratory. July. Available at http://cx.lbl.gov/2009-assessment.html.

Mosenthal, P.R. R.N. Elliott, D. York, C. Neme, P. Chernick, and K. Petak. 2006. Natural Gas Energy Efficiency Resource Development Potential in New York. Albany, N.Y.: New York State Energy Research and Development Authority. October 31. Available at http://www.nyserda.org/energy_Information/otherdocs.asp#NaturalGas.

Nadel, S., and H. Geller. 2001. Smart Energy Policies: Saving Money and Reducing Pollutant Emissions Through Greater Energy Efficiency. Washington, D.C.: American Council for an Energy-Efficient Economy. September. Available at http://aceee.org/pubs/e012full.pdf.

Nadel, S., R.N. Elliott, M. Shepard, S. Greenberg, G. Katz, and A.T. de Almeida. 2002. Energy-Efficient Motor Systems: A Handbook on Technology, Program, and Policy Opportunities. Second Edition. Washington, D.C.: American Council for an Energy-Efficient Economy.

Nadel, S., A.M. Shipley, and R.N. Elliott. 2004. The technical, economic, and achievable potential for energy efficiency in the United States: A meta-analysis of recent studies. Proceedings of the 2004 ACEEE Summer Study on Energy Efficiency in Buildings. Washington, D.C.: American Council for an Energy-Efficient Economy.

Navigant Consulting Inc. 2006. Energy Savings Potential of Solid State Lighting in General Illumination Applications. Final Report prepared for Lighting Research and Development Building Technologies Program Office of Energy Efficiency and Renewable Energy U.S. Department of Energy. Washington, D.C.: DOE. December.

NBI (New Buildings Institute). 2008. Getting to Fifty Web site. White Salmon, Wash.: NBI. Available at http://www.newbuildings.org/gtf.

Norton, P., and C. Christiansen. 2006. A Cold Climate Case Study for Affordable Zero Energy Homes. NREL- CP-550-39678. Golden, Colo.: National Renewable Energy Laboratory. Available at http://www.nrel.gov/buildings/zero_energy.html.

NPCC (Northwest Power and Conservation Council). 2005. The Fifth Northwest Electric Power and Conservation Plan. Portland, Oreg.: NPCC.

NRC (National Research Council). 2001. Energy Research at DOE: Was it Worth It? Washington, D.C.: National Academy Press.

Optimal Energy, Inc. 2003. Energy Efficiency and Renewable Energy Resource Development Potential in New York State. American Council for an Energy-Efficient Economy, Vermont Energy Investment Corporation, and Christine T. Donovan Associates. Report prepared for New York State Energy Research and Development Authority, Albany, N.Y. August.

Parker, D., and S. Chandra. 2008. Very Low Energy Homes in the United States: Perspectives on Performance from Measured Data. Cocoa, Fla.: Florida Solar Energy Center.

Parker, D.S., J.P. Dunlop, S.F. Barkaszi, J.R. Sherwin, M.T. Anello, and J.K. Sonne. 2000. Towards zero energy demand: Evaluation of super efficient building technology with photovoltaic power for residential housing. Pp. 1.207-1.223 in Proceedings of the 2000 ACEEE Summer Study on Energy Efficiency in Buildings. Washington, D.C.: American Council for an Energy-Efficient Economy.

PG&E (Pacific Gas and Electric). 2006. Laboratory Evaluation of the OASys™ Indirect/Direct Evaporative Cooling Unit. Application Assessment Report #0510. Available at http://www.etcc-ca.com/database/docs/ETCC_ProjectDoc_7.pdf.

PNNL (Pacific Northwest National Laboratory). 1997. Desiccant Dehumidification and Cooling Systems: Assessment and Analysis. PNNL-11694. Richland, Wash.: PNNL.

Rasmussen, T., V. Goepfrich, and K. Horkitz. 2005. Drivers of CFL purchase behavior and satisfaction: What makes a consumer buy and keep buying? Pp. 897-910 in 2005 International Energy Program Evaluation Conference. Available at http://www.iepec.org/2005PapersTOC/papers/096.pdf.

Romm, J.J. 1999. Cool Companies: How the Best Businesses Boost Profits and Productivity by Cutting Greenhouse Gas Emissions. Washington, D.C.: Island Press.

Rosenfeld, A.H. 1985. Residential energy efficiency: Progress since 1973 and future potential. Pp. 92-121 in Energy Sources: Conservation and Renewables (D. Hafemeister, H. Kelly, and B. Levi, eds.). New York, N.Y.: American Institute of Physics.

Roth, K., and K. McKenney. 2007. Energy Consumption by Consumer Electronics in U.S. Residences. Report No. D5525. Cambridge, Mass.: TIAX, LLC.

Rufo, M., and F. Coito. 2002. California's Secret Energy Surplus: The Potential for Energy Efficiency. Report prepared by Xenergy, Inc., for the Energy Foundation and the Hewlett Foundation. San Francisco, Calif.: Energy Foundation.

Sandahl, L.J, T.L. Gilbride, M.R. Ledbetter, H.E. Steward, and C. Calwell. 2006. Compact Fluorescent Lighting in America: Lessons Learned on the Way to the Market. PNNL-15730. Richland, Wash.: Pacific Northwest National Laboratory.

Sanstad, A.H., and R.B. Howarth. 1994. Normal markets, market imperfections, and energy efficiency. Energy Policy 22(10):811-818.

Scheckel, P. 2007. The Home Energy Diet: How to Save Money by Making Your House Energy-Smart. Gabriola Island, British Columbia: New Society Publishers.

Spees, K., and L. Lave. 2007. Demand response and electricity market efficiency. Electricity Journal 20(3):69-85.

Stabat, P., S. Ginestet, and D. Marchio. 2003. Limits of feasibility and energy consumption of desiccant and evaporative cooling in temperate climates. Proceedings of 2003 CIBSE/ASHRAE Conference. Available at http://www.cibse.org/pdfs/4dstabat.pdf.

Stavins, R., J. Jaffe, and T. Schatzki. 2007. Too good to be true? An examination of three economic assessments of California climate change policy. National Bureau of Economic Research (NBER) Working Paper No. 13587. Cambridge, Mass.: NBER, Inc. November.

Suozzo, M. J. Benya, M. Hyderman, P. DuPont, S. Nadel, and R.N. Elliott. 2000. Guide to Energy-Efficient Commercial Equipment. 2nd edition. Washington, D.C.: American Council for an Energy-Efficient Economy.

Sutherland, R. 2000. "No cost" efforts to reduce carbon emissions in the U.S.: An economic perspective. Energy Journal 21(3):89-112.

Thaler, R., A. Tversky, D. Kahneman, and A. Schwartz. 1997. The effect of myopia and loss aversion on risk taking: An experimental test. Quarterly Journal of Economics 112(2):647-661.

Torcellini, P., S. Pless, M. Deru, B. Griffith, N. Long, and R. Judkoff. 2006. Lessons Learned from Case Studies of Six High-Performance Buildings. NREL/TP-550-37542. Golden, Colo.: National Renewable Energy Laboratory. June.

Turner, C., and M. Frankel. 2008. Energy Performance of LEED for New Construction Buildings. White Salmon, Wash.: New Buildings Institute. Available at http://www.newbuildings.org/downloads/Energy_Performance_of_LEED-NC_Buildings-Final_3-4-08b.pdf.

Wiser, R., M. Bolinger, and M. St. Clair. 2005. Easing the Natural Gas Crisis: Reducing Natural Gas Prices Through Increased Deployment of Renewable Energy and Energy Efficiency. LBNL-56756. Berkeley, Calif.: Lawrence Berkeley National Laboratory.

3 Energy Efficiency in Transportation

Energy efficiency in the U.S. transportation sector merits special attention from the standpoint of energy security and the environment because this sector is almost solely dependent on a single fuel—petroleum—about 60 percent of which is imported. Moreover, the transportation sector is responsible for about 30 percent of U.S. emissions of greenhouse gases.

3.1 SCOPE AND CONTENT OF THIS CHAPTER

This chapter describes the U.S. transportation system and its energy consumption. It identifies near-term (through 2020) opportunities for energy efficiency and the technologies that could capitalize on them. (See Box 3.1 for definitions of fuel efficiency, fuel economy, and fuel consumption.) It considers technologies that could improve energy efficiency in the medium term (through 2030–2035), as well as longer-term opportunities that could stem from technologies that are now at an early stage of research and development (R&D). Finally, it touches on the possibilities for broader changes in transportation systems.

Reflecting the charge to the Panel on Energy Efficiency Technologies, the transportation technologies covered here are described in terms of their performance (improvements in energy efficiency and fuel consumption), their costs, and their effects on the environment (mainly reductions in greenhouse gas emissions). This review is not an in-depth study of all the factors that could improve technology performance, cost, or deployment, or associated environmental effects. Hence, the potential improvements discussed here should be considered as first-step technology assessments rather than as forecasts.

BOX 3.1 ***Fuel Efficiency, Fuel Economy, and Fuel Consumption***

"Energy efficiency" in transportation is generally discussed using terminology specific to this sector of the economy, as defined below. The primary terms used to quantify the fuel consumed by a vehicle as it is driven are "fuel economy" (or "fuel efficiency") and "fuel consumption."

Fuel efficiency is a relative term used to describe how effectively fuel is used to move a vehicle. Thus, a heavy and a light vehicle, using the same technology in the same ways, would have the same fuel efficiency but very different fuel economy. Note that fuel-efficiency improvements do not necessarily result in increased fuel economy, as they are often offset by the negative effects of increases in vehicle power and weight. Thus, fuel efficiency is related to the amount of useful work that is derived from the combustion of fuel. Whether that useful work is applied to increase the number of miles that can be traveled per gallon of fuel or to provide other amenities (such as size and power) is a separate question.

Fuel economy is expressed as miles per gallon of fuel consumed; it is the term most commonly used in the United States in discussing vehicle fuel consumption.

Fuel consumption is the inverse of fuel economy. It refers to the fuel consumed by the vehicle as it travels a given distance. Widely used in the Europe (expressed in liters per 100 km), this metric is a clearer measure of fuel use than is fuel economy. The amount of fuel consumed in driving from one place to another (say, New York City to Washington, D.C.) is what matters to consumers. In U.S. units, fuel consumption is usually expressed as gallons per 100 miles.

To illustrate: A vehicle with fuel economy of 50 miles per gallon (mpg), which corresponds to fuel consumption of 2 gallons per 100 miles, is twice as fuel-efficient as a vehicle of the same size, weight, and power that gets 25 mpg, corresponding to 4 gallons per 100 miles.

Passengers and freight are transported by land vehicles, aircraft, and waterborne vessels through vast networks of land, air, and marine infrastructure. For this report, the panel partitioned this sector into passenger transport and freight transport and separated each of these into highway transportation and nonhighway transportation.

Highway transportation is responsible for 75 percent of the energy used in transportation and has the greatest potential for energy efficiency; it is therefore the focus of this chapter. Nonetheless, nonhighway modes of transportation (aviation, railroad, and marine) together account for about 17 percent of the energy used in the sector and are an important potential source of energy savings collectively. Efficiency improvements in these transport modes are also discussed. Transport modes such as mass transit and intercity rail have important roles in bringing about more energy-efficient passenger and freight transportation, particularly if they shift traffic to modes that can be more energy-efficient. However, they are not treated in detail here.

Section 3.2 outlines energy use for U.S. transportation overall. Passenger and freight transport are covered in Sections 3.3 and 3.4, respectively. Section 3.5 briefly discusses the effects of alternative fuels on the efficiency of highway vehicles.

Much of the discussion in this chapter is "vehicle-centric" in the sense that it focuses on opportunities for boosting energy efficiency through the engineering of highway vehicles (and aircraft) and their subsystems and equipment (e.g., engines, transmissions, body designs, and tires). Indeed, a great deal of R&D attention has been given to vehicle engineering for energy efficiency.

The energy required for transportation, however, is greatly influenced by the performance of the systems in which these vehicles operate. "Systems" refer to the physical networks of infrastructure through which vehicles move, as well as the underlying logistic, institutional, commercial, and economic considerations that influence the mix of vehicles used, how they are used, and how the infrastructure itself performs. For example, congestion management that allows vehicles to operate at more constant speeds, with fewer starts and stops and less idling, could increase overall transportation efficiency. System energy efficiency can also be improved through the more direct routing of trucks and aircraft, the optimization of operating speeds, more intense use of infrastructure, and changes in land-use density and patterns. Similarly, energy use in air transportation is influenced by air-traffic-management requirements. The degree to which the underlying systems operate effectively, therefore, can foster—or in some cases, hinder—energy efficiency. Some of these system-level issues are discussed briefly in Section 3.6.

Energy use in transportation is also influenced by factors that give rise to the demand for travel and that affect the amount or type of travel. Change in these areas, however, is a complex topic that can only be touched on in this chapter. The demand for transportation comes from individuals and businesses pursuing social and economic activities. Reducing these activities may save energy, but may or may not be otherwise desirable. The panel did not examine possibilities for saving energy by reducing the activities that spur demand for transport. It focused instead on the use of more energy-efficient modes of transportation as a means of achieving energy savings (for example, using mass transit or freight rail in place of individual vehicles or trucks), although this chapter does at times point to some of the system-level requirements, such as changes in land-use patterns and density, that may be needed to support such improvements.

In examining opportunities for energy efficiency in transportation, the panel considered the three time periods set out in the America's Energy Future (AEF) project:

1. Early deployment: through 2020;
2. Medium-range deployment: 2020 through 2030–2035;
3. Longer-range deployment: beyond 2030–2035.

Current technologies offer many improvements in fuel economy that become increasingly competitive and attractive as fuel prices rise. For the early period of its assessment (through 2020), the panel focused primarily on opportunities to improve the energy efficiency of mainstream power trains and vehicles. Reductions in fleet fuel consumption through 2020 are likely to come primarily from improving today's spark-ignition engine, compression-ignition (diesel) engine, and hybrid-electric vehicles fueled with petroleum, biofuels, and other nonpetroleum hydrocarbon fuels.[1] Annual, incremental improvements in engines and transmissions are expected to continue. When coupled with changes in the deployment fractions of these propulsion systems, as well as substantial vehicle weight reductions, these improvements could reduce average vehicle fuel consumption steadily over this time period.

In the medium-term (2020 through 2030–2035), changes in power-train and vehicle technologies that go beyond incremental changes become feasible. Plug-in hybrid-electric vehicles using electricity plus any of the above fuels may well become a significant fraction of new-vehicle sales. Their deployment may be followed by substantial numbers of (fully) battery-electric vehicles.

Over the longer term (beyond 2030–2035), major sales of hydrogen fuel-cell vehicles and the necessary hydrogen supply and distribution infrastructure may develop.

---

[1]Note that biofuels and other nonpetroleum hydrocarbon fuels are covered not in this report, but in Liquid Transportation Fuels: Technological Status, Costs, and Environmental Impacts—the report of the AEF Panel on Alternative Liquid Transportation Fuels (NAS-NAE-NRC, 2009b).

Only when a sizable fraction of in-use vehicles incorporates efficiency improvements and new power trains can the effect of these changes on the nation's energy use and greenhouse gas emissions reach significant levels. The panel constructed some illustrative scenarios of vehicle and technology deployment as a means of estimating the potential overall effects of improved passenger vehicles and technologies on fuel consumption and the environment. The results are discussed in Section 3.3.

Finally, Section 3.7 outlines the challenges that will have to be met and the impediments that will have to be overcome to improve energy efficiency in transportation, and Section 3.8 presents the panel's findings for the sector.

3.2 ENERGY USE IN TRANSPORTATION

Energy use for transportation in the United States has experienced tremendous growth over the past several decades, although the trend registered brief pauses during the economic recessions of 1974, 1979–1982, 1990–1991, and 2001.

In 2007 the United States consumed 29 quads (quadrillion British thermal units, or Btu) of energy for transportation, or about 28 percent of total U.S. energy use. Moreover, the sector used more than 70 percent of the petroleum consumed in the United States.

Energy use in each mode of transportation reflects its degree of use as well as its energy efficiency characteristics. Figure 3.1 breaks down total U.S. transportation energy use into components, by mode, for the year 2003. As shown, passenger travel is dominated by automobiles and by air transport for longer distances.[2] Mass transit and scheduled intercity rail and bus services have important roles in some locations but account for only a small proportion of total passenger-miles.[3] On the freight side, the major transport modes are by truck, rail, water, pipeline, and air. Trucking dominates in terms of tons and value of shipments.

In 2006, petroleum accounted for 96 percent of the energy used for transportation; gasoline accounted for 62 percent of the energy used (EIA, 2006).

---

[2]Bureau of Transportation Statistics, National Transportation Statistics, *Transportation Energy Data Book,* available at http://www.bts.gov/publications/national_transportation_statistics/.

[3]One passenger carried for 1 mile is referred to as a "passenger-mile." For example, an automobile carrying four people 8 miles is responsible for 32 passenger-miles of travel.

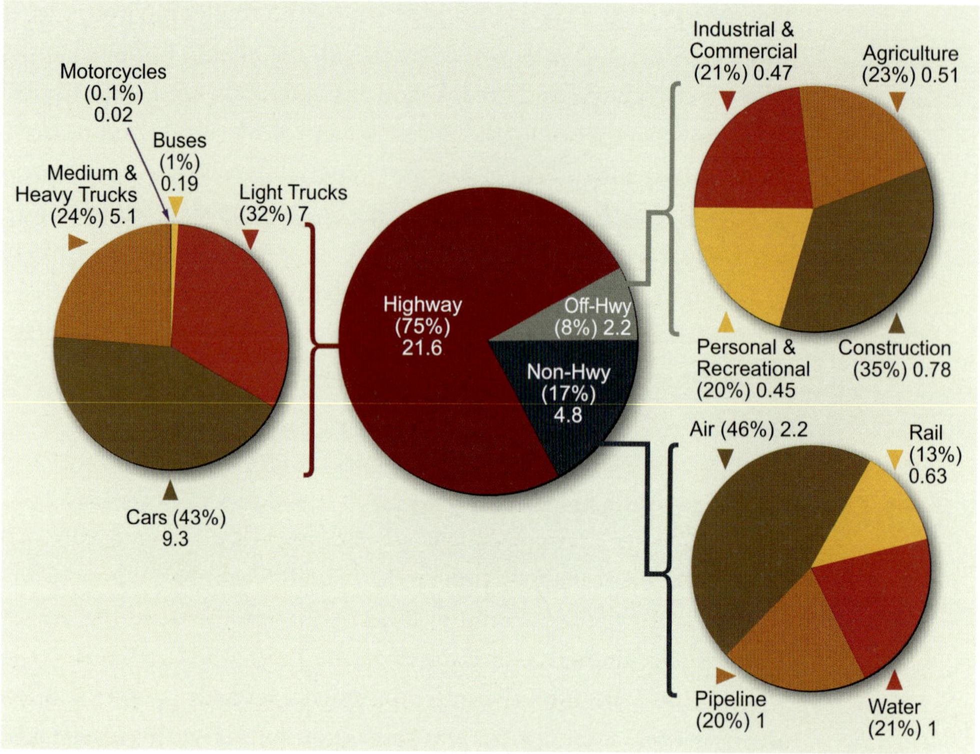

FIGURE 3.1 *U.S. transportation energy consumption (quads) by mode and vehicle in 2003.*
Note: Total U.S. energy consumption = 98.2 quads.

Light-duty vehicles (LDVs), defined as passenger cars and light trucks, are the primary users of gasoline. Heavy-duty vehicles (HDVs), defined as heavy trucks and buses and medium-duty trucks, accounted for most of the diesel fuel consumed, about 17 percent of the energy used. LDVs and HDVs accounted for about 60 percent and 20 percent of transportation sector carbon dioxide (CO_2) emissions, respectively.

3.2.1 Public Transit

Although public transit consumes a relatively small fraction of overall transportation energy, it serves several important roles in urban transportation systems.

Today, the energy required for mass transit in the United States is less than 2 percent of the total energy used for transport (Davis et al., 2008). However, transit buses on average consume about the same amount of energy per passenger-mile as that consumed by LDVs, largely because of low average ridership, especially during non–rush hours (Davis et al., 2008). Rail transit is somewhat better in terms of energy use per passenger-mile, but apart from New York City and a few other densely populated cities that have heavy ridership during both peak and nonpeak hours, transit rail is also characterized by light usage for much of the day and thus high average energy use per rider. These averages mask the specific times (rush hours) and corridors during which public transit uses less energy per passenger-mile than passenger cars do and where targeted promotion of transit use could contribute to a reduction in total energy use. It merits noting that the run-up in gasoline prices in 2007–2008 coincided with increases in public transit ridership and that public transit ridership has grown by one-third in the United States over the past 12 years (APTA, 2008).

Although the panel did not analyze the potential for energy efficiency gains in public transit per se, it does consider in Section 3.6 how energy efficiency can be improved through system-level improvements in the provision, use, and operation of transportation systems. In so doing, the panel mentions how changes in the provision of transit services and in the operation of the highway and aviation infrastructure can boost energy efficiency.

3.2.2 Commercial Versus Private Transportation

The drivers for energy efficiency in commercial transportation differ from those for private transportation. Lifetime operating costs, and thus energy efficiency, are important to companies supplying passenger and freight transportation. The commercial transportation sector is so highly competitive that even small cost differentials among firms can have a major influence on their relative profitability and growth.

In contrast, fuel is a small fraction of the lifetime cost of owning a motor vehicle for private transportation. For many consumers, vehicle comfort, style, and operating performance are more important than fuel consumption. The time period over which the costs of driving are considered (relative to initial vehicle costs) also tends to be shorter for passenger transport than for freight transport.

3.3 THE POTENTIAL FOR ENERGY EFFICIENCY IMPROVEMENT IN PASSENGER TRANSPORTATION

Automobiles, light trucks, and aviation are the main modes of passenger transportation in the United States. Private automobiles account for the vast majority of local and medium-distance passenger-trips[4] (those under 750 miles), and airlines account for longer trips (BTS, 2006).

3.3.1 Light-Duty Vehicles—Efficiency Trends

The major factors driving vehicles to become more (or less) fuel-efficient are the price of fuel (including taxes), regulations, and consumer preferences for particular vehicle attributes. This section reviews experience with these factors.

3.3.1.1 International Experience

Europe has historically had high fuel and vehicle taxes that raise owner and user costs. Moreover, diesel fuel has often been taxed at a rate lower than that for gasoline. High prices have pushed consumers to demand fuel-efficient vehicles, giving a larger market share to diesel engines. The average fuel economy of new light-duty vehicles in Europe today approaches 40 miles per gallon (mpg), 60 percent higher than in the United States.[5]

Vehicle fuel economy in Japan is similar to that in Europe. In 2006, Japan revised its fuel economy standard to 47 mpg, to be achieved by 2015 (ICCT, 2007b).

3.3.1.2 U.S. Experience

In contrast to the trend in Europe, from 1980 until recently, real gasoline prices had been falling in the United States, encouraging consumers to buy larger, heavier, more powerful vehicles and to drive more, rather than to seek greater fuel economy. During periods of high fuel prices (such as those prevailing in mid-2008), U.S. consumers have demonstrated more interest in fuel economy. The average fuel economy of recently sold new vehicles in the United States is about 25 mpg. The U.S. Energy Independence and Security Act of 2007 (EISA; Public

---

[4]One passenger taking one trip, regardless of trip length, is referred to as a passenger-trip.

[5]Note that "real-world" fuel economy values are lower—in Europe real-world fuel economy is about 28 mpg; in the United States it is about 21 mpg. See Schipper (2006).

Law 110-140) requires that corporate average fuel economy (CAFE) standards be set for LDVs for model years 2011 through 2020 which will ensure that, by 2020, the industry-wide CAFE for all new passenger cars and light trucks, combined, is at least 35 mpg.[6] This is a 40 percent increase over today's average fuel economy.

Although fuel economy has not improved, fuel efficiency *has* improved. Owing to relatively low fuel prices and static fuel-economy standards between the mid-1980s and the early 2000s, increases in vehicle size and performance have offset energy efficiency gains in vehicles (Lutsey and Sperling, 2005; An and DeCicco, 2007). As a result, average new-vehicle fuel-economy levels have stagnated for nearly two decades, and total vehicle fleet use and greenhouse gas emissions have increased steadily owing to the increasing fleet size and vehicle-miles traveled (VMT).[7]

Before discussing specific energy-saving technologies for LDVs, the panel notes that energy prices will have a significant impact on the pace of the development and introduction of these technologies. The effects of fuel costs on driving and demand for motor fuel have been studied extensively. Recent work by Small and Van Dender (2007) suggests that, as U.S. incomes have risen, the dominant effect of increases in fuel prices has been more demand for vehicle fuel economy rather than reduced driving. In other words, people are more likely to drive vehicles with higher fuel economy than to sacrifice making trips in the face of rising gasoline prices. Small and Van Dender (2007) estimate that, in the short term, each 10 percent increase in fuel costs results in a 0.1 percent reduction in VMT but a 0.3 percent increase in realized miles per gallon, often achieved through more intensive use of the vehicles with the highest fuel economy in the household.[8] In the longer term, when consumers have time to choose among alternative vehicle technologies and types, realized fuel economy increases by a full 2 percent for each 10 percent increase in fuel prices, while travel falls by 0.5 percent. (Figure 3.2 illustrates how recent increases in fuel prices have increased the fraction of new vehicles sold that are automobiles rather than light trucks, since cars average 27.5 mpg compared with 22.3 mpg for light trucks, and how the subsequent fall in prices has reversed this trend.) The response in the direction of

---

[6]The Obama administration has recently proposed that these requirements, specified by Subtitle A of EISA (Public Law 110-140), be accelerated.

[7]One vehicle-mile is one vehicle traveling 1 mile (regardless of the number of passengers).

[8]Hughes et al. (2008) drew the same conclusion.

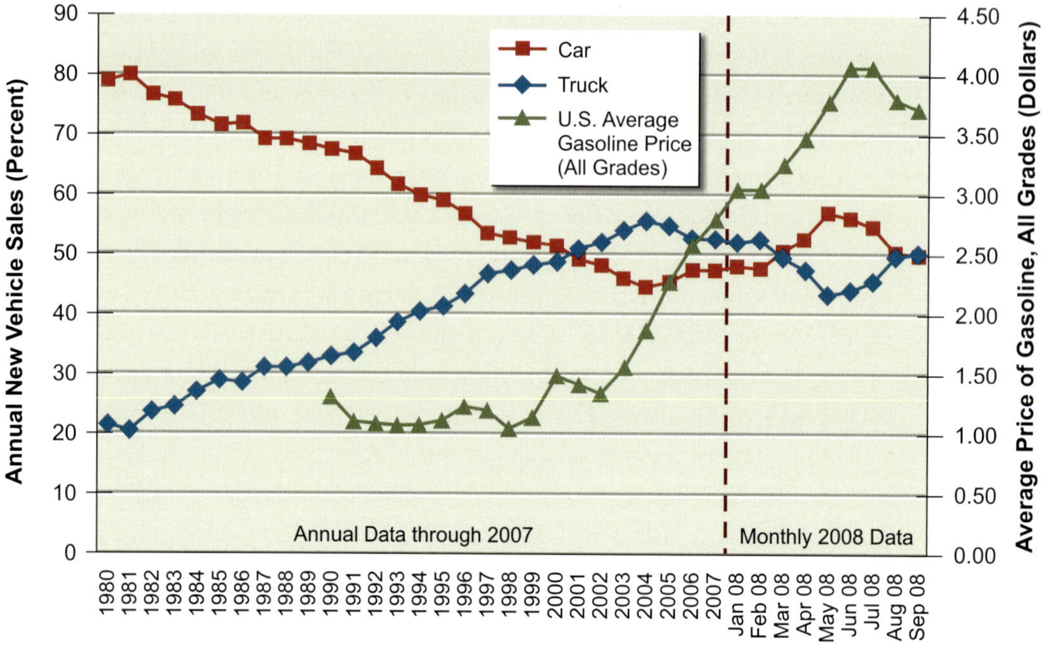

FIGURE 3.2 *U.S. car and light truck percentage of new vehicle sales versus average price of gasoline (all grades).*
Source: Gasoline prices (2007$) for 1980–2007 are from Table 5.24 in EIA (2008). For 2008, the monthly nominal gas prices are from EIA (2009). Light truck and car percentages from 1980 to 2006 are from Table 4.6 of the Transportation Energy Data Book, Edition 27 (Davis et al., 2008). Light truck and car percentages for 2007 and 2008 are from Ward's Automotive Group, a division of Prism Business Media, Inc., Ward's U.S. Light Vehicle Sales Summary, Ward's AutoInfoBank, available at http://wardsauto.com.

improved fuel economy suggests that as incomes and energy prices rise, they will spur demand for the kind of energy-saving technologies discussed next.

3.3.2 Light-Duty Vehicles—Technologies

Long-standing concern with oil imports and greenhouse gas emissions has led to several studies by the National Research Council (NRC), examining ways to reduce both of these. In particular, three studies have had as their main focus technologies that could improve fuel efficiency in light-duty vehicles (NRC, 1992, 2002, 2008b). Two other studies on hydrogen technologies and the hydrogen economy (NRC, 2008c and 2004a, respectively) also considered conventional vehicle technologies that could have an effect by 2020. Moreover, energy centers at leading U.S. universities and federal laboratories, as well as private consultants, have recently carried out in-depth studies of transportation energy efficiency technologies and their potential to reduce petroleum use and greenhouse gas emissions through about 2030 (e.g., Bandivadekar et al., 2008; Lutsey, 2008). The following review of vehicle efficiency technologies draws on these and other studies; their quantitative estimates are in general agreement.

The review is organized as follows: engine improvements, transmission improvements, and other (nonpropulsion system) improvements. A brief overview of the possibilities for each of these to increase fuel efficiency in LDVs is given first. This is followed by a summary of the overall decrease in petroleum consumption that could result from these changes.[9] Many of these technologies are already being used in a limited number of vehicles, but their use is expected to expand to satisfy the higher fuel-economy standards specified in EISA.

3.3.2.1 Engine Improvements

Gasoline Spark-Ignition Engine

The gasoline spark-ignition (SI) engine efficiency improvements that could be deployed in large volume in the next decade include but are not limited to the following: variable valve timing, two- and three-step variable valve lift, cylinder deactivation, direct injection turbocharging with engine downsizing, engine-friction reduction, and smart cooling systems. Many of these are already in low-volume production. In the medium-term (15–20 years), technologies such as camless valve actuation, continuously variable valve lift, and homogeneous-charge compression ignition (HCCI)[10] could be deployed in increasing numbers. A survey of recent technology assessments shows that the above technologies have the potential to reduce vehicle fuel consumption, on average, by approximately 10–15 percent in the new-vehicle sales mix[11] in the nearer term (by 2020) and by

---

[9]Note that this review does not cover the effects of biofuels on petroleum consumption.

[10]HCCI combines features of spark-ignition and compression-ignition (diesel) engines by making it possible to ignite gasoline and other hydrocarbon fuels using compression.

[11]These reduction percentages indicate the fuel consumption of a new, state-of-the-art vehicle on the date stated, relative to its current-technology-equivalent vehicle. These percentages are panel estimates of what can realistically be expected from engine, drivetrain, and vehicle improvements in the near- and midterm future, based on estimates in the references given. Vehicle performance levels are assumed to be comparable to today's values. The economic and regulatory context is assumed to place greater but not extreme market emphasis on lowering vehicle fuel consumption.

an additional 15–20 percent in the medium term (by 2030) (EEA, 2007; Kasseris and Heywood, 2007; Ricardo, Inc., 2008; NRC, 2008b). Turbocharged, downsized gasoline engines, which are some 10–15 percent more efficient than equal-performance, naturally aspirated gasoline engines, are expected to steadily replace a significant fraction of naturally aspirated (nonturbocharged) gasoline engines, improving energy efficiency and contributing to meeting future fuel-economy standards.

Diesel Compression-Ignition Engine

Owing to their high compression ratios and reduced pumping losses, turbocharged diesel engines currently offer approximately a 20–25 percent efficiency benefit over gasoline SI engines when adjusted for the higher energy density of diesel fuel. Efficiency improvements in compression-ignition (CI) engines are likely to come primarily from increased power density, improved engine-system management, more sophisticated fuel-injection systems, and improved combustion processes. New technologies are emerging for after-treatment to reduce emissions of particulate matter and oxides of nitrogen to levels comparable to those of SI engines. The primary challenges for diesel engines in the United States are the added costs and fuel penalties (of about 3–6 percent) associated with the after-treatment systems required to reduce these emissions (Bandivadekar et al., 2008; Johnson, 2008, 2009; Ricardo, Inc., 2008). By 2020, improvements in energy and after-treatment technologies have the potential to reduce the fuel consumption of new diesel-engine vehicles relative to current diesel vehicles by about 10 percent, and by an additional 10–15 percent by 2030.

Gasoline Hybrid-Electric Vehicle

Hybrid vehicles combine an internal combustion engine (ICE) with electric drive from a battery-electric motor/generator system. Usually both systems can drive the vehicle, and the ICE recharges the batteries. (Hence, these vehicles are also called "charge-sustaining" hybrids.) The primary fuel-consumption benefits of a gasoline hybrid-electric vehicle (HEV) derive from regenerative braking, engine downsizing, the active management of energy use to maintain the most efficient engine operating conditions, and the elimination of idling.

Hybrid vehicles are increasingly being classified on the basis of the extent of the functions offered by the electric motor/generator. Relative to equivalent gasoline SI engines, belt-driven starter-generator systems can eliminate engine idling, reducing fuel consumption by 4–6 percent; integrated starter-generator systems that can recover energy from regenerative braking, along with eliminating engine idling (a mild hybrid), can reduce fuel consumption by 10–12 percent. A parallel full hybrid with power assist, such as Honda's Integrated Motor Assist system, can increase this benefit to more than 20–25 percent, whereas more complex systems using two motors, such as Toyota's Hybrid Synergy Drive, can reduce fuel consumption by more than 30 percent.

Some prototype diesel HEVs are under development and could be in limited production volumes within a few years. These could have about 10 percent higher efficiency (which corresponds to 20 percent lower diesel fuel consumption due to higher fuel density) than that of an equivalent gasoline hybrid. The cost for a diesel HEV would be significantly higher than for a gasoline-fueled version.

Plug-In Hybrid Electric Vehicle

Plug-in hybrid-electric vehicles (PHEVs) are hybrid vehicles that can be recharged from an external source of electricity. The liquid-fuel savings that can be realized is directly related to the amount of electricity stored in the battery.

PHEVs require substantially larger battery packs than those used in conventional HEVs. Depending on the nature of the HEV being redesigned as a PHEV, the redesigned vehicle will likely require a larger electric motor and higher-capacity power electronics. The larger battery and, if needed, larger components increase propulsion system size, weight, and cost. As with the charge-sustaining hybrids discussed above, they also use an onboard ICE to recharge the batteries.

Batteries for these vehicles are usually sized to obtain an all-electric driving range of 20 to 60 miles. Compared with a gasoline SI-engine-powered vehicle, PHEVs could reduce petroleum consumption by up to 75 percent, depending on the onboard battery size and (thus) range, and on how these vehicles are driven (Kromer and Heywood, 2008). The corresponding reduction in greenhouse gas emissions depends on the greenhouse gas intensity of the electricity used to charge the battery. Although PHEVs are likely to be introduced in modest numbers into the U.S. market over the next 5 years, the development of a mass market for PHEVs will require low-cost, lightweight batteries that can store the needed electricity and last for 10 years or more (Anderman, 2007). The current status of battery performance and development is summarized in Box 3.2.

BOX 3.2 *Status of Advanced Battery Technology*

Lead acid batteries were invented in the 19th century and are still the standard battery technology in vehicles today. The GM EV1, a production battery-electric vehicle (BEV), used this battery technology as recently as 1999, and then transitioned to the nickel-metal hydride (NiMH) battery.

The next generation of batteries, based on lithium-ion chemistry, is widely deployed in consumer electronic devices. Of course, the power and energy storage requirements of these devices are much smaller than those of electric vehicles.

Hybrid-electric vehicles (HEVs) require batteries with high power (commonly stated in units of watts per kilogram). Plug-in HEVs (PHEVs) and BEVs require significant energy storage (along with sufficient power). Today's batteries have an energy storage capacity of 150–200 Wh/kg. A typical vehicle consumes approximately 0.25 kWh per mile in all-electric mode. Typical electric motors that can propel a vehicle require power ranging between 50 and 150 kW.

Chemistries

Table 3.2.1 summarizes the promising advanced battery chemistries and their performance characteristics. Significant amounts of research and development are being devoted to promising new versions of the chemistries of cathode materials, anode materials, and electrolytes, as well as to manufacturing processes.

TABLE 3.2.1 Lithium-ion Battery Cathode Chemistries

	Lithium Cobalt Oxide	Lithium Manganese Spinel	Lithium Nickel Manganese Cobalt	Lithium Iron Phosphate
Automotive status	Limited auto applications (due to safety concerns)	Pilot	Pilot	Pilot
Energy density	High	Low	High	Moderate
Power	Moderate	High	Moderate	High
Safety	Poor	Good	Poor	Very good
Cost	High	Low	High	High
Low-temperature performance	Moderate	High	Moderate	Low
Life	Long	Moderate	Long	Long

Source: Adapted from Alamgir and Sastry, 2008.

Performance and Cost Targets

The U.S. Advanced Battery Consortium (USABC) has established a set of long-term performance goals for electrochemical energy storage devices:

- The target for PHEV batteries is an energy storage capacity of 11.6 kWh with an energy density of 100 Wh/kg and a unit cost of stored energy of $35/kWh.
- The target for BEV batteries is an energy storage capacity of 40 kWh with an energy density of 200 Wh/kg and a unit cost of stored energy of $100/kWh.

In addition, goals were established for battery life in terms of the number of 80 percent discharge cycles. Meeting these goals is likely to be required for widespread commercialization of electrically powered vehicles.

Lithium-ion batteries currently lead in energy density (Wh/kg) metric and have an average annual improvement rate of 3.7 percent. Lead-acid batteries lead in the cost of stored energy ($/kWh) at $50/kWh and have an average annual reduction rate of around 3 percent. However, lead-acid batteries are unable to satisfy the battery life requirements for PHEVs and BEVs. Today's lithium-ion batteries that have the cycle life desired for automotive applications cost between $500/kWh and $1000/kWh.

The cost target (in $/kWh) is currently viewed as the greatest challenge for lithium-ion battery technology.

Industry Developments

The lithium-ion consumer electronics market is currently at around 2 billion units annually. The volume of lithium-ion batteries in automotive applications, however, is very small. Frost & Sullivan (2008) predict a 19.6 percent compound annual growth rate for shipments of HEV batteries, as well as a smaller but rapidly growing market for PHEV and BEV batteries.

An auto battery alliance has been promoted by the U.S. Department of Energy's Argonne National Laboratory and includes 3M, ActaCell, All Cell Technologies, Altair Nanotechnologies, EaglePicher, EnerSys, Envia Systems, FMC, Johnson Controls-Saft, MicroSun, Mobius Power, SiLyte, Superior Graphite, and Townsend Advanced Energy.

All major vehicle manufacturers have partnered with major battery manufacturers: Ford with Johnson Controls-Saft, General Motors with LG Chem, Chrysler with General Electric, Toyota with Panasonic/Sanyo, Nissan with NEC via the Automotive Energy Supply joint venture, and Honda with GS Yuasa.

Specialists anticipate that it may be 10 to 20 years before advanced battery technology can reach the USABC performance and cost targets.

Battery-Electric Vehicle

The successful development and deployment of PHEVs enabled by developments in advanced battery technology might also lead to batteries suitable for battery-electric vehicles (BEVs) (see Box 3.2).

Although several models of BEVs are being introduced into the market today in limited production volumes, in the near-term those BEVs that are commercially viable are likely to be small cars with modest performance capabilities, such as "city BEVs."

Hydrogen Fuel-Cell Vehicle

Fuel-cell technology, in a hybrid system with hydrogen as the fuel, offers the promise of significantly higher propulsion system efficiency than that of ICE technology, as well as zero vehicle tailpipe greenhouse gas emissions. Several scientific, engineering, and business challenges must be overcome before hydrogen fuel-cell vehicles (HFCVs) can be commercialized successfully (NRC, 2004a, 2008c, 2008d; Crabtree et al., 2004). As discussed in the studies just cited and in other references, the principal challenges are increasing the durability and lowering the costs of fuel cells, achieving cost-effective storage of hydrogen in fueling stations and on board the vehicle, and reducing the environmental impacts from deploying a hydrogen supply and fueling infrastructure with low greenhouse gas emissions.

The NRC's recent study on the transition to a hydrogen-based transportation system (NRC, 2008c) discusses scenarios for the introduction of HFCVs. As these scenarios show, there is significant potential for reducing oil imports and CO_2 emissions with HFCVs in the long term (2035–2050), but little opportunity for impact in the near term (by 2020) because of the time required to overcome existing technical challenges, to provide the fueling infrastructure, and to ramp up to high-volume vehicle production.

3.3.2.2 Transmission Improvements

Automatic transmissions are popular in the United States primarily because of their ease of use and smooth gearshift. Transmission efficiency is likely to improve in the near- to midterm through increasing the number of gears and reducing losses in bearings, gears, sealing elements, and the hydraulic system. While four-speed transmissions have dominated the U.S. market, five-speed transmissions are becoming standard as well (EPA, 2007). Six-speed automatic transmissions, as well as automated manual transmissions, are already used in some cars and

TABLE 3.1 Expected Transmission System Efficiencies

Transmission	Efficiency (%)
Current automatic transmission (4- and 5-speed)	84–89
Automatic transmission (6- or 7-speed)	93–95
Dual clutch transmission (wet-clutch)	86–94
Dual clutch transmission (dry-clutch)	90–95
Continuously variable transmission (CVT)	87–90

Source: Ricardo, Inc., 2008, and EEA, 2007.

are likely to become more widely used over the next decade. Manufacturers have begun incorporating seven- and eight-speed transmissions into some luxury vehicles, and the penetration of these transmissions can be expected to increase in the midterm (2020–2035). Energy and Environmental Analysis, Inc., estimates that each additional gear results in a retail price increase of approximately $50 (EEA, 2007).

Table 3.1 lists the efficiencies that can be expected from various transmission systems in the near- to midterm. As shown, improvements of 2–9 percent are realizable and provide equivalent percentage reductions in vehicle fuel consumption. Note that, although a continuously variable transmission (CVT) allows the engine to operate near its maximum efficiency, the estimated efficiency of CVTs is lower than the corresponding estimate for six- and seven-speed automatic transmissions.[12] CVTs have been in low-volume production for well over a decade.

3.3.2.3 Nonpropulsion System Improvements

Vehicle Weight and Size Reduction

Reducing vehicle weight is one obvious way to reduce fuel consumption. A commonly used rule of thumb is that a 10 percent reduction in vehicle weight can reduce fuel consumption by 5–7 percent, when accompanied by appropriate engine downsizing at constant performance (Bandivadekar, 2008). Vehicle simulation results suggest that the relative benefits of weight reduction may be smaller than this in some types of hybrid vehicles because the hybrid propulsion system

[12]The CVT has a slightly lower torque-transmitting efficiency owing to its higher frictional losses. However, when coupled with the engine in the vehicle, its extra flexibility improves the overall engine-plus-CVT efficiency. Not much difference is apparent among the several "best" transmission technologies.

actively manages engine use to stay in areas of higher and more uniform efficiency and also to recoup vehicle energy during braking (An and Santini, 2004; Wohlecker et al., 2007).

Weight reduction can be achieved by substituting lighter-weight materials (such as aluminum) for heavier ones, by redesigning vehicles, and by downsizing vehicles and components. For example, downsizing a passenger car by one Environmental Protection Agency (EPA) size class (e.g., from large to midsize) can reduce vehicle weight by between 9 and 12 percent (Cheah et al., 2007). Unlike vehicle weight, however, vehicle size is an attribute that consumers value.

The cost of reducing vehicle weight through the use of lighter-weight materials is estimated to be about $3/kg ($1.40/lb) (Bandivadekar et al., 2008). Secondary weight reduction (e.g., using a smaller engine because the vehicle is lighter) and weight reduction due to vehicle redesign are usually assumed to occur when a vehicle is redesigned, so the cost is assumed to be small.

Rolling Resistance Reduction

A recent NRC report on tires and passenger-vehicle fuel economy (NRC, 2006) agrees with earlier estimates in the literature (Schuring and Futamura, 1990) that each reduction of 0.001 in the coefficient of the rolling resistance of passenger tires—equivalent to a 10 percent reduction in overall rolling resistance—can reduce vehicle fuel consumption by 1–2 percent. After examining the fuel-saving technologies and designs that are being developed for original-equipment tires (those supplied with new vehicles) to assist in meeting U.S. CAFE standards, the NRC report also concludes that such a 10 percent reduction in the average rolling resistance of passenger tires is possible over the next decade because many of these technologies can be introduced, not only into the new-vehicle market, but also into the much larger market of replacement tires. The incremental cost of such lower rolling resistance tires is expected to be small.

Aerodynamic Drag Reduction

In the EPA highway driving cycle[13] with an average speed of 48 miles per hour, approximately half of the energy required to propel the vehicle is used to over-

---

[13]The Environmental Protection Agency has developed standard driving cycles that represent urban and highway driving. Fuel-economy ratings and CAFE standards are based on fuel-consumption measurements over these cycles.

come aerodynamic drag. Thus, body designs that reduce aerodynamic drag can achieve meaningful reductions in fuel consumption. The aerodynamic drag on a vehicle is the product of a drag coefficient (C_D), the vehicle frontal area, the vehicle velocity squared, and the air density (divided by 2). Thus drag increases significantly as vehicle speeds increase, especially above 60 miles per hour. A 10 percent reduction in the drag coefficient can lower average vehicle fuel consumption by up to 2 percent. Demonstration vehicles built during the U.S. Department of Energy's Partnership for a New Generation of Vehicles achieved a coefficient of drag as low as 0.22—a 35 percent reduction from the then-current vehicle (NRC, 2000). The cost of reducing the vehicle's drag coefficient, since it would occur when the vehicle is redesigned or a new vehicle is developed, is assumed to be small.

Lubricants

Engine friction has a substantial negative impact on engine efficiency. Friction can be and is being reduced through improvements in engine design and the use of new materials and surface coatings. It can also be influenced by engine lubricant properties, and lower viscosity oils are increasingly being used. The most commonly used engine oils or lubricants are mineral oils that contain additives to improve viscosity, inhibit engine oxidation and corrosion, act as dispersants and detergents, and reduce surface friction. There are strong pressures to reduce both the consumption of engine oil and the additive components that produce ash, in order to minimize the degradation of the exhaust system's emission-control technologies, such as catalysts and particulate traps. Effective and low-cost diesel emission control technologies are critical to any major expansion of diesel engine vehicles in the U.S. LDV market. The use of synthetic engine oils rather than mineral oils is growing: although their cost is higher, they can reduce engine friction and thus improve fuel economy by a few percent. Improvements in mineral oil properties could increase vehicle fuel economy by about 1 percent (NRC, 2008b).

3.3.2.4 Summary of Potential LDV Efficiency Improvements: Performance and Environmental Impacts

Table 3.2 shows the panel's estimates for the potential reductions in petroleum consumption and greenhouse gas emissions that could result over the next 25 years from the adoption of both the evolutionary and the new-vehicle technologies discussed above. These estimates assume that vehicle size and performance are held constant.

TABLE 3.2 Potential Relative Vehicle Petroleum Use and Greenhouse Gas Emissions from Vehicle Efficiency Improvements Through 2035

	Petroleum Consumption (gasoline equivalent)		Greenhouse Gas Emissions[a]	
Propulsion System	Relative to Current Gasoline ICE	Relative to 2035 Gasoline ICE	Relative to Current Gasoline ICE	Relative to 2035 Gasoline ICE
Current gasoline	1.00		1.00	
Current turbocharged gasoline	0.90		0.90	
Current diesel	0.80		0.80	
Current hybrid	0.75		0.75	
2035 gasoline	0.65	1.00	0.65	1.00
2035 turbocharged gasoline	0.60	0.90	0.60	0.90
2035 diesel	0.55	0.85	0.55	0.85
2035 HEV	0.40	0.60	0.40	0.60
2035 PHEV	0.20	0.30	0.35–0.45	0.55–0.70
2035 BEV	None		0.35–0.50	0.55–0.80
2035 HFCV	None		0.30–0.40	0.45–0.60

Note: These estimates assume that vehicle performance (maximum acceleration and power-to-weight ratio) and size remain the same as today's average new-vehicle values. That is, the improvements in propulsion efficiency are used solely to decrease fuel consumption rather than to offset increases in vehicle performance and size. Estimates have been rounded to the nearest 0.05. BEVs and HFCVs are expected to have shorter driving ranges than PHEVs between rechargings or refuelings.
BEV, battery-electric vehicle; HEV, hybrid-electric vehicle; HFCV, hydrogen fuel-cell vehicle; ICE, internal combustion engine; PHEV, plug-in hybrid vehicle.

[a]Greenhouse gas emissions from the electricity used in 2035 PHEVs, 2035 BEVs, and 2035 HFCVs are estimated from the projected U.S. average electricity grid mix in 2035. Greenhouse gas emissions from hydrogen production are estimated for hydrogen produced from natural gas.
Source: Bandivadekar et al., 2008. Estimates based on assessments by An and Santini, 2004; Wohlecker et al., 2007; Cheah et al., 2007; NPC, 2007; and NRC, 2004.

These estimates are based on studies that have evaluated the fuel-consumption reduction potential of many plausible improvements in power train and vehicle technology. The studies have aggregated these improvements through vehicle simulations and drive-cycle analysis, or by appropriately compounding realizable combinations of these improvements. Each entry in Table 3.2 is the fuel consumption of each technology (gasoline equivalent) relative to that of the average vehicle in either the current or the 2035 new-vehicle sales mix, and thus reflects an attempt to incorporate the extent to which these improvements have been deployed across the sales mix. Relative fuel consumption for cars and light trucks is comparable. These numbers are for vehicles with performance levels and interior size essentially the same as those of today's new vehicles, and with a 20 percent vehicle weight reduction, a 25 percent reduction in vehicle drag coefficient, and a 33 percent reduction in the tire rolling-friction coefficient. These reductions in relative fuel consumption are indicative of what could be achieved on average in vehicles by these improvements and changes in power train and vehicle technologies.

Taken together, these engine and transmission improvements, reductions in weight, and other nonpropulsion system improvements could reduce the fuel consumption of a gasoline ICE vehicle by up to 35 percent by about 2035.

Although current diesel-engine vehicles have a 20 percent gasoline-equivalent fuel-consumption advantage over current ICE gasoline-engine vehicles, this gap is likely to narrow (e.g., to 15 percent by 2035), as there is greater improvement potential in the gasoline engine.

Because their technology is relatively new and thus can deliver deeper cuts in vehicle fuel consumption, HEVs and PHEVs have greater potential for improved fuel consumption (e.g., 47 percent and 73 percent, respectively) than do ICE power trains. Note, however, that they continue to depend on petroleum or other liquid fuels.

Reductions in greenhouse gas emissions from gasoline and diesel ICEs, HEVs, and PHEVs are proportional to the reductions achieved in petroleum consumption. Further reductions in greenhouse gas emissions could be achieved by motor vehicles if the effective carbon content of fuels were lowered through the addition of biofuels having low net carbon emissions (NAS-NAE-NRC, 2009b).

BEVs and HFCVs are two longer-term technologies that need not depend on petroleum or alternative hydrocarbon fuels and could have zero tailpipe emissions of criteria pollutants and CO_2.

For PHEVs, BEVs and HFCVs, the well-to-tank emissions produced during the generation of electricity and hydrogen determine the full potential for these vehicle technologies to reduce greenhouse gas emissions. Efficiency improvements in the vehicles themselves, together with low- or zero-emissions generation of the electricity and hydrogen that they require, offer the potential for dramatic reductions in total greenhouse gas emissions. If implemented, these improvements could give the PHEV the edge over the HEV in terms of reducing both petroleum consumption and greenhouse gas emissions.

The panel judges that the estimates shown in Table 3.2 can be realized if manufacturers devote all future improvements in vehicle efficiency to reducing actual fuel consumption such that vehicle performance and size (acceleration and power-to-weight ratio) are kept essentially constant at today's levels. Also, the electricity used to recharge PHEVs and BEVs is assumed to come from the anticipated average U.S. electricity supply mix (Kromer and Heywood, 2008). There are significant regional variations in the greenhouse gas emissions from the electricity supply system, as well as uncertainty about how large a fraction of the future supply will come from nuclear or renewable-energy sources, or will employ effective carbon capture and storage technology. The greenhouse gas emissions for HFCVs are based on the assumption that, in this transition timeframe (through 2035), hydrogen is produced by steam reforming of natural gas, currently the most economic and developed hydrogen production process.

Because vehicle manufacturers compete on—and consumers expect—ever-better performance, resolving the performance, size, and fuel-consumption trade-off is a critical policy challenge. As noted earlier, due to relatively low fuel prices and static fuel-economy standards between the mid-1980s and the early 2000s, vehicle size and performance increases have offset energy efficiency gains in vehicles (Lutsey and Sperling, 2005; An and DeCicco, 2007). Sales of more fuel-efficient vehicles have fluctuated with recent increases in fuel prices (see Figure 3.2). It is unlikely that future energy efficiency improvements will be realized in decreased fuel consumption unless appropriate fiscal, regulatory, or other policies are implemented to promote or require reduced fuel consumption over increased power or performance.

3.3.3 Light-Duty Vehicles—Costs

The estimation of technology costs is more uncertain than is the estimation of the relative benefit of individual technologies. This is particularly the case under the volatile economic conditions of 2008–2009.

The results of the panel's evaluation, carried out under the economic conditions prevailing in mid-2008, are given in Table 3.3. The price increments given are relative to a 2005 gasoline vehicle. The cost estimates shown represent the approximate incremental retail price of future vehicle types (including the cost of emission-control systems), compared with current gasoline-fueled ICE vehicles (EEA, 2007; Bandivadekar et al., 2008).

These future (2035) vehicles incorporate improved engines of the type indicated, more efficient transmissions, a 20 percent reduction in vehicle weight (two-

TABLE 3.3 Estimated Additional Cost to Purchaser of Advanced Vehicles Relative to Baseline 2005 Average Gasoline Vehicle

	Additional Retail Price (2007 dollars)	
Propulsion System	Car	Light Truck
Current gasoline	0	0
Current diesel	1,700	2,100
Current hybrid	4,900	6,300
2035 gasoline	2,000	2,400
2035 diesel	3,600	4,500
2035 hybrid	4,500	5,500
2035 PHEV	7,800	10,500
2035 BEV	16,000	24,000
2035 HFCV	7,300	10,000

Note: Cost and price estimates depend on many assumptions and are subject to great uncertainty. For example, different companies may subsidize new vehicles and technologies with different strategies in mind. Costs listed are additional costs only, relative to baseline average new car and light truck purchase prices (in 2007 dollars) that were calculated as follows:

—Average new car: \$14,000 production cost × 1.4 (a representative retail price equivalent factor) = an average purchase price of \$19,600.

—Average new light truck: \$15,000 × 1.4 = \$21,000.

These are not meant to represent current average costs. Rather, they are the costs used in this analysis. See Box 3.3 for more information on the assumptions underlying the estimates.

For the purpose of these estimates, the PHEV all-electric driving range is 30 miles; the BEV driving range is 200 miles. Advanced battery and fuel-cell system prices are based on target battery and fuel-cell costs from current development programs.

Source: Bandivadekar et al., 2008.

thirds from materials substitution at a cost of \$3/kg), and moderate reductions in tire rolling resistance and aerodynamic drag. They have the same size and performance as those of today's vehicles.

These prices are based on the costs associated with producing a vehicle at the manufacturing plant gate. Note that if the demand for new materials in these vehicles raises material costs significantly, then manufacturing costs would increase accordingly. To account for distribution costs and manufacturer and dealer profit margins, production costs were multiplied by a representative factor of 1.4 to provide representative retail price estimates (Evans, 2008).

More information on the assumptions behind the estimates in Table 3.3 is provided in Box 3.3.

Note that the timescales indicated for these future-technology vehicles are not precise. The rate of price reduction will depend on the rate at which these

BOX 3.3 **Estimating the Cost of Advanced Vehicles**

The cost estimates in Table 3.3 in this chapter are from Bandivadekar et al. (2008). The estimates are based on an extensive review of existing studies assessing the costs and fuel-consumption benefits of future vehicle technologies. The cost estimates assume high-volume production of 500,000 to 1 million vehicles per year. For conventional internal combustion engines, future technology costs were estimated on a component-by-component basis in proportion to the improvement in fuel consumption, based on a comparison of the costs and fuel-consumption benefits of 11 studies. Alternative power train costs—for hybrid, plug-in hybrid, battery-electric, and fuel-cell vehicles—were estimated by aggregating the results of several studies based on a component-by-component assessment.

The costs for pre-commercial and emerging technologies (in particular, those for fuel-cell systems and batteries) assume additional cost reductions resulting from continued technology development. These cost reductions are over and above those realized as a result of high production volumes. Battery costs assume that the material costs decrease by 30 percent relative to a present-day, high-volume estimate; this assumption is consistent with technology-development assumptions in Anderman et al. (2000). Costs for fuel-cell systems were estimated using the cost models developed in Carlson et al. (2005) and the assumptions summarized in Kromer and Heywood (2008).[1]

In addition, since the more recent Massachusetts Institute of Technology report (Bandivadekar et al., 2008) was released, updated fuel-cell and battery data have been made publicly available and provide a more recent assessment. See the following:

- *Batteries:* Kalhammer, F.R., B.M. Kopf, D.H. Swan, V.P. Roan, and M.P. Walsh, 2007, Status and Prospects for Zero Emissions Vehicle Technology: Report of the ARB Independent Expert Panel 2007. Prepared for the State of California Air Resources Board, Sacramento, California, April 13, 2007.
- *Fuel cells:* Sinha et al., 2008, *Direct Hydrogen PEMFC Manufacturing Cost Estimation for Automotive Applications.* Presentation by TIAX, LLC, to the U.S. DOE Fuel Cell Annual Merit Review.
- *Hydrogen storage:* Lasher et al., 2008, *Analyses of Hydrogen Storage Materials and On-Board Systems.* Presentation by TIAX, LLC, to the U.S. DOE Fuel Cell Annual Merit Review.

[1] Kromer and Heywood (2008) assume that continued development enables reduced platinum loadings from those used in TIAX's estimated system costs of $57/kW.

technologies are taken up by the market (Bandivadekar et al., 2008; Evans, 2008). Wide and rapid deployment could bring costs down more quickly.

The results in Table 3.3 show that power trains entering the fleet today, such as improved diesel engines and hybrid vehicles, cost the purchaser from 10 percent to 30 percent more than a current gasoline vehicle costs. This price difference (in constant dollars) is expected to drop to 5–15 percent in the midterm (by 2035) because the increase comes from incorporating new technology whose cost is expected to drop more rapidly, owing to in-use experience, than will the costs of well-established technologies. Longer-term options such as plug-in hybrid and fuel-cell vehicles are estimated to cost between 25 percent and 30 percent more than this 2035 gasoline vehicle. Battery-electric vehicles with standard vehicle performance and size remain costly, approaching double the cost of a future gasoline vehicle. As noted earlier, a more plausible market opportunity for BEVs is small, city cars with reduced range. However, these also will need significantly improved battery performance and reduced battery costs to become competitive.

Retail price increases from technologies that reduce fuel consumption are largely offset by fuel savings over a vehicle's lifetime, but not in all cases. The extent to which this is the case depends on how the more efficient technology is used. As noted earlier, efficiency improvements can be directed toward reducing actual fuel consumption or toward moderating the increase in fuel consumption that would otherwise accompany increased vehicle size and power. An estimate of the full cost of reducing fuel consumption would account for how changes in vehicle attributes such as fuel efficiency, power, and size affect the value that consumers derive from these products.

The net economic benefit of reduced fuel consumption derived by vehicle purchasers obviously depends on the fuel price, the discount rate, and the time period over which the benefit is assessed. It also depends, of course, on the vehicle price increment and the amount of fuel saved. With full emphasis on reducing actual fuel consumption rather than on increasing vehicle performance and size, at a fuel price of $2.50 per gallon and a 7 percent discount rate over 15 years (the average vehicle lifetime) and 150,000 miles, improved gasoline engines fully pay back the retail price increase (relative to a current vehicle) (Bandivadekar et al., 2008). Hybrid and diesel power trains pay back 60 percent and 90 percent of the up-front retail price increase, respectively, under similar discounting assumptions. Longer-term options such as PHEVs and HFCVs are estimated to pay back 50 percent to 70 percent of the increase in retail price at a fuel price of $2.50 per gallon. At a higher fuel price of $5.00 per gallon, the discounted fuel savings of all the technologies listed in Table 3.3, except diesel vehicles and standard-sized BEVs, fully pay back the initial retail price increase. (Battery packs in these future hybrid and electric vehicles are assumed to last for a vehicle's lifetime, so battery replacements are not included in these costs.)

Note that it is widely accepted that consumers discount their fuel savings over a much shorter period—typically 3 to 4 years. This reduces the benefit of the fuel savings significantly. Reductions in the price of future hybrid systems could allow these vehicles to break even. Diesel engines could lose ground relative to future gasoline vehicles owing to the greater potential for increased efficiency of the gasoline ICE and also resulting from the anticipated higher diesel fuel cost over gasoline due to rising diesel fuel demand for freight transportation.

The estimates in Table 3.3, when combined with the estimated fuel-consumption reductions in Table 3.2, indicate that evolutionary improvements in gasoline ICE vehicles are likely to prove the most cost-effective choice for reducing petroleum consumption and greenhouse gas emissions. These vehicles will be sold in large quantities in the near term, so if the overall cost of reducing fuel consumption and greenhouse gas emissions from motor vehicles is to be kept as low as possible, it is critical that efficiency improvements in these vehicles be used primarily to reduce on-the-road fuel consumption.

While current HEVs appear to be less competitive than improved gasoline- and diesel-fueled ICE vehicles, over time they are likely to become a more cost-effective choice in many applications as a consequence of their substantial and increasing fuel efficiency advantage and the anticipated reduction in their price premium.

PHEVs, BEVs, and HFCVs appear to be more costly alternatives for reducing petroleum consumption and greenhouse gas emissions. Among these three technologies, PHEVs are likely to become more widely available in the near term to midterm, whereas BEVs and HFCVs are midterm to long-term alternatives for high-volume production.

3.3.4 Light-Duty Vehicles—Deployment

To have a significant effect on the petroleum use and greenhouse gas emissions of the entire vehicle fleet, the market share of vehicles that are significantly more fuel efficient must become sizable. Common barriers to achieving such a market share include but are not limited to higher initial cost, safety concerns, fuel availability (or lack thereof), reliability and durability concerns, and a lack of consumer awareness. Because the advanced-technology (non-ICE) vehicles are competing against steadily improving gasoline ICE vehicles, the market penetration of such vehicles may be slow unless aided by high fuel prices or by fiscal, regulatory, or other policies.

Even if sufficient market demand exists for certain technologies, time and capital constraints affect how quickly the demand can be satisfied. Typical development times for automotive products are 3–5 years. Then to deploy these new products, vehicle manufacturers and their suppliers must be able to make adequate capital investments to bring new production capacity online, and the supply of critical components (such as advanced batteries) must be assured. Generally, a decade or more is required between the development of a technology to a stage at which it can be deployed, its introduction on a commercial vehicle, and then the achievement of significant sales. For example, it has taken 10–15 years or more for major new technologies such as automatic transmissions to reach significant deployment levels; the same has been the case for the spread of small, high-speed diesel engines in cars in Europe (now some 50 percent of the market) (Bandivadekar et al., 2008).

Currently, there are no quantitative methods that can estimate possible vehicle market shares based on the constraints outlined above. Moreover, there are the technical constraints discussed earlier. For example, both PHEVs and HFCVs will need significant technical breakthroughs to become competitive in the market.

If high energy-storage battery technology progresses sufficiently (see Box 3.2), PHEVs could be deployed more rapidly than could HFCVs and the hydrogen distribution infrastructure, with production volumes building over the 2020–2035 timeframe. The infrastructure issues associated with supplying electricity to PHEVs can be dealt with incrementally as production volumes start to increase over the next decade. In the midterm, beyond about 2020—when PHEV sales volumes could increase to significant levels—the impact on the electrical supply and distribution system would then become more substantial. In contrast, a new infrastructure is needed to supply hydrogen to HFCVs. Thus, PHEVs are increasingly being viewed as a promising midterm to long-term option.

A recent NRC report (NRC, 2008c) concludes that, although "the maximum practicable number of HFCVs that could be on the road by 2020 is around two million," it would take decades—e.g., until 2050—for this technology to have a major impact on oil use and greenhouse gas emissions.

Table 3.4 shows the panel's judgment, based on all these constraints, of the extent to which these advanced-technology vehicles could penetrate the *new*

TABLE 3.4 Plausible Share of Advanced Light-Duty Vehicles in the New-Vehicle Market by 2020 and 2035 (percent)

Propulsion System	2020	2035
Turbocharged gasoline SI vehicles	10–15	25–35
Diesel vehicles	6–12	10–20
Gasoline hybrid vehicles	10–15	15–40
PHEV	1–3	7–15
HFCV	0–1	3–6
BEV	0–2	3–10

Note: The percentage of hydrogen fuel-cell vehicles considered "plausible" is in contrast to the percentages reported in NRC (2008c), which represent "maximum practical" shares.

light-duty vehicle (LDV) market in the United States. The estimates are intended as illustrations of achievable deployment levels, based on historical case studies of comparable technology changes which suggest that relative annual increases of 8–10 percent in the deployment rate are plausible. With changes in the factors that affect vehicle attributes or purchases, such as stricter fuel-economy standards or high fuel prices, the timeline for reaching these market shares could be shortened.

Note that the panel's estimates are not meant to imply that all of these technologies would necessarily be deployed together. It may turn out that some technologies do not prove to be marketable. Others that are more appealing could then capture a higher fraction of new-vehicle sales.

Vehicles with major changes in technology, or with new technology, face many hurdles on their way to market acceptance. Of course, they must be more appealing to a significant fraction of the market than the vehicles that they are intended to replace. That attractiveness has many attributes: for example, performance, capacity, utility, style, fuel consumption, and especially price. The panel's judgment is that none of the alternatives to steadily improving mainstream technology vehicles currently appears attractive enough to guarantee market acceptance.

3.3.5 Total Light-Duty Vehicle Fleet Fuel Consumption—Estimates

As stated above, the Energy Independence and Security Act of 2007 requires that the corporate average fuel economy standard be 35 mpg in 2020.

The panel examined two scenarios to explore how the deployment of the advanced technologies listed in Table 3.2, together with vehicle efficiency improvements (such as reductions in vehicle weight, aerodynamic drag, and tire rolling resistance), could reduce the petroleum consumption of the U.S. in-use vehicle fleet. The methodology used for the analysis follows that described in Cheah et al. (2007) and Bandivadekar et al. (2008). The two scenarios—termed "optimistic" and "conservative"—are described in Box 3.4.

Note that these scenarios are not predictions or forecasts of what the future vehicle fleet would be like, but instead are intended as illustrative examples of the degree of change to the vehicle fleet required to improve fleet average fuel economy. In these scenarios, the panel examined the effects on fleet fuel consumption of the fuel-economy improvements that may be achieved by 2020. It then extrapolated the associated improvement rates (over 2006–2020) out through 2035 (see Box 3.4). The scenarios reflect the relative petroleum consumption of vehicle technologies as shown in Figure 3.3. These values are closely comparable to the rounded numbers on consumption based on more than one source in Table 3.2. The values were estimated assuming that vehicle performance and size are held constant and that all power train and vehicle efficiency improvements are used to reduce fuel consumption rather than to offset increases in performance and size.

Based on the estimated fuel-consumption characteristics of individual vehicle types shown in Figure 3.3 and the fleet efficiency improvements represented in the scenarios, Table 3.5 shows examples of the sales mixes and weight reduction that would be required to meet the CAFE targets and to extend that rate of improvement beyond 2020. Achieving the "optimistic" targets would require that the efficiency improvements provided by these technology changes be used largely to decrease actual fuel consumption. In this case, the emphasis on reducing fuel consumption, or ERFC, parameter would have to be 75 percent, which allows only a modest increase in average vehicle performance (a reduction in the 0-to-60 mph acceleration time of about 1 second from its current average value of about 9 seconds). For the conservative scenario, only half of the efficiency gains that could be made by 2035 are realized in decreased fuel consumption—the rest of the efficiency improvement is used to offset the fuel-consumption impacts of additional increases in vehicle power, weight, and size.

The relative proportions of the various power trains are based on the panel's judgments as to their relative attractiveness (including cost), the degree of change from the baseline technology, and the historically observed limit of about 10 percent in the annual increase in production volumes when attractive changes in power train technology occur (such as the transition from manual to automatic transmissions in the United States, and the buildup of diesel engines in passenger cars in Europe). Note that the emphasis on reducing fuel consumption rather than

BOX 3.4 Future Vehicle Scenarios

Optimistic Scenario

The optimistic scenario assumes that the new vehicle sales mix in 2020 meets the Energy Independence and Security Act (EISA; Public Law 110-140) corporate average fuel economy (CAFE) target of 35 mpg (a 40 percent increase from today's value). It then assumes that fuel efficiency continues to improve at the same rate through 2035. A full 75 percent of this improvement potential is assumed to be devoted to decreasing actual fuel consumption; the rest is assumed to be offset by increased vehicle performance, size, and weight.[1] This assumption is represented by introducing a factor called "emphasis on reducing fuel consumption," or ERFC. In this case, the value of ERFC is 75 percent. The result is that, by 2035, average new-vehicle fuel economy would reach 50 mpg—double today's value.

Conservative Scenario

The conservative scenario assumes that the 2020 CAFE target is met 5 years later, in 2025. It then assumes that fuel efficiency continues to improve at this rate (a lower rate than in the optimistic scenario). However, it also assumes that only half of this improvement is used to decrease actual fuel consumption (an ERFC value of 50 percent), and the rest is assumed to be offset by gains in vehicle performance, size, and weight. The result is that, by 2035, average new-vehicle fuel economy has increased to only 40 mpg—about 60 percent above today's values.

No-Change Baseline

The two scenarios above are compared with a no-change baseline. This baseline extrapolates the history of the past 20 years, during which power train efficiency improvements essentially offset the negative impacts on fuel consumption of increasing vehicle performance, size, and weight (i.e., the no-change baseline assumes an ERFC value of zero).

[1]This assumption reflects the panel's judgment that it is unlikely that there will be *no* increases in vehicle performance, size, and weight.

on increasing performance and size assumes a significant shift in U.S. vehicle purchasers' choices. ERFC is currently low (about 10 percent) in the United States, whereas it averages 50 percent in Europe. The weight-reduction estimates come from a shift from light trucks to cars (the mix in the United States is about 50 percent light trucks); more extensive use of lighter-weight materials such as aluminum; the redesign of components and vehicles; and some reduction in vehicle size, in both light trucks and cars. These weight-reduction estimates (of 700–1050 lb)

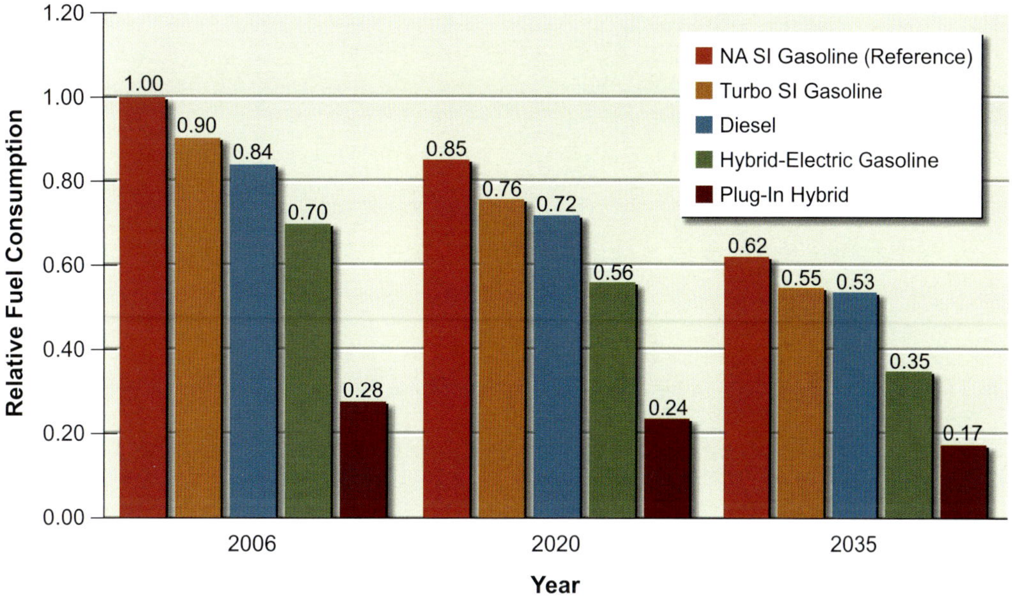

FIGURE 3.3 *Relative fuel consumption (tank to wheels) of future cars by power train, assuming that all efficiency improvements go to raising fuel economy.*
Source: Cheah and Heywood, 2008.

are based on assessing the specific weight-reduction opportunities and aggregating plausible combinations of these other improvements that would also meet these fuel-economy objectives. These sales mix illustrations were selected so as to make comparable the degree of challenge in all the areas where improvements are needed. See Bandivadekar et al. (2008) and Cheah and Heywood (2008) for additional details.

This analysis indicates that achieving the CAFE target and continuing that rate of improvement beyond 2020 will require substantial changes in engine and vehicle technology, as well as significant weight reduction (part of which could result from size reduction) and changes in consumer preferences and purchasing behavior.

Figure 3.4 shows, for the conservative and optimistic scenarios, the corresponding annual gasoline consumption of the U.S. in-use LDV fleet from the present out to 2035. A no-change baseline assumes that all of the efficiency improvements go to vehicle size, weight, and power, as has occurred since 1982. The cumulative fuel savings under each scenario compared with this no-change base-

TABLE 3.5 Illustrative Vehicle Sales Mix Scenarios

	Percent Emphasis on Reducing Fuel Consumption[a]	Percent Light Trucks vs. Cars	Percent Vehicle Weight Reduction	Market Share by Power Train (percent)						Percent Fuel Efficiency Increase from Today
				Naturally Aspirated SI	Turbo SI	Diesel	Hybrid	Plug-in Hybrid	Total Advanced Power Train	
Optimistic[b]										
2020	75	40	17	52	26	7	15	0	48	+38
2035	75	30	25	36	26	9	20	9	64	+100
Conservative[c]										
2025	50	40	17	55	24	7	14	0	45	+38
2035	50	40	20	49	21	7	16	7	51	+62

Note: Assumed average new-vehicle weight (cars and light trucks) currently is 1900 kg (4180 lb). Thus, average weight reductions of 700–1050 lb per vehicle would be required. Neither of these scenarios includes BEVs or FCVs.

[a]The amount of the efficiency improvement that is dedicated to reducing fuel consumption (i.e., that is not offset by increases in vehicle power, size, and weight).

[b]The optimistic scenario meets the new CAFE target of 35 mpg in 2020, and then extrapolates this rate of improvement through 2035. In this case, the average fuel economy in 2035 reaches 52 mpg, roughly double today's value.

[c]The conservative scenario achieves the new CAFE target of 35 mpg only in 2025 (5 years later) and extrapolates this rate of improvement through 2035, when the average fuel economy reaches only 60 percent above today's value.

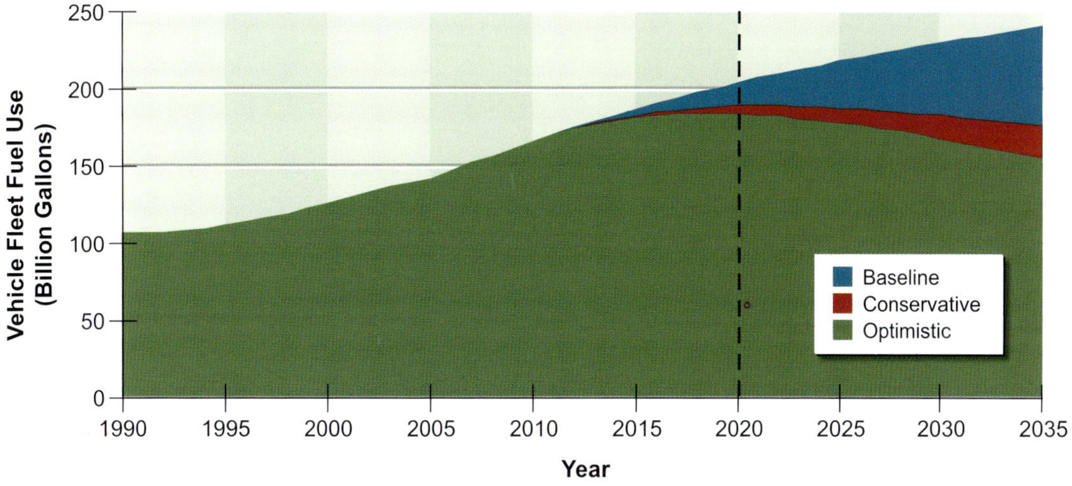

FIGURE 3.4 *Fuel use for the U.S. in-use light-duty vehicle fleet out to 2035. Source: Cheah and Heywood, 2008.*

line are indicated. Note that this no-change baseline includes some growth in overall fleet size and miles driven, but no resulting change in vehicle fuel consumption.

Table 3.6 shows the corresponding cumulative fuel savings of the U.S. in-use LDV fleet through 2035. The cumulative, fleetwide fuel savings can be substantial, so long as the proposed fuel-economy standards are met and the rate of improvement is sustained.

Table 3.7 gives the corresponding annual fuel savings from the no-change baseline in 2020 and 2035.

These illustrative scenarios show that substantial changes in vehicle weight and size, significant improvements in the efficiency of ICE power trains, and the increasing production over time of hybrid systems will all be needed to reduce the in-use fuel consumption of the U.S. LDV fleet. The market will need to respond by purchasing these improved vehicles in steadily growing volumes despite their higher price, and it will need to forgo expectations of ever-increasing vehicle performance. If the trends indicated by these scenarios are to occur, the assumed production-vehicle changes (or their equivalents) will need to start soon. If all this does happen, then in-use U.S. LDV gasoline consumption would level off by about 2020, offsetting the fuel-consumption growth path that the United States has been following over the past few decades. Fuel consumption could then decline back to 2007–2008 levels by 2035.

TABLE 3.6 Cumulative Fuel Savings from the Baseline Shown in Figure 3.4

	Today through 2020 (billion gallons)	2020 through 2035 (billion gallons)
Optimistic scenario	86	834
Conservative scenario	64	631

Note: The no-change baseline assumes constant sales mix by power train, constant ratio of light trucks versus cars, 0.8 percent compounded annual growth in new-vehicle sales, and 0.1–0.5 percent increase in vehicle travel.

TABLE 3.7 Annual Fuel Savings in 2020 and 2035 from the No-Change Baseline Shown in Figure 3.4

	2020 (billion gallons/year)	2035 (billion gallons/year)
Optimistic scenario	21	86
Conservative scenario	16	66

Note: The no-change baseline assumes no change in average new-vehicle fuel consumption, a constant ratio of light trucks versus cars, and a 0.8 percent compounded annual growth in new-vehicle sales. It also assumes that growth in vehicle travel slows from 0.5 percent to 0.1 percent per year over 25 years, and that any efficiency improvements are fully offset by increases in vehicle performance, size, and weight.

3.3.6 Environmental Impacts of Light-Duty Vehicles—Life-Cycle Context

A full assessment of the effects on the environment of an LDV would cover energy consumption and all environmental effects, including greenhouse gas emissions, over the entire vehicle lifetime, which includes the vehicle manufacturing and disposal stages as well as vehicle use. Currently, the energy use and greenhouse gas emissions associated with manufacturing are each some 10 percent of the total fuel use and emissions over the vehicle life cycle. This fraction rises as vehicles become more fuel efficient: for hybrid and fuel-cell vehicles, the fraction is 15–20 percent.[14]

For a full life-cycle assessment, the energy and greenhouse gas emissions involved in fuel supply would also be included. The energy required to produce gasoline or diesel fuel ranges from 20 to 25 percent of the fuel energy delivered to the vehicle fuel tank, depending on the petroleum source and the refining details

[14]Values from the Argonne National Laboratory's GREET 2.7 model.

(Bandivadekar et al., 2008). For biofuels, electricity, and hydrogen, this component is more complex and depends strongly on how these other sources of energy are generated.

When the energy consumption and emissions from all four life-cycle stages—manufacturing, use, disposal, and fuel supply—are added together, the relative benefits (of one vehicle type over another) are diminished, because the energy and greenhouse gas emissions associated with supplying petroleum-based fuels are proportional to the amount of fuel used, whereas the manufacturing effects are not.

The need discussed above to assess the environmental impacts of transportation in a full life-cycle context goes beyond the scope of this chapter. The World Business Council for Sustainable Development reviewed these broader issues in its study *Mobility 2030,* which describes the major challenges that future transportation systems must address. These involve reducing emissions of greenhouse gases and other air pollutants, mitigating ecological damage, lowering traffic-related deaths and injuries, reducing noise, easing congestion, and enhancing mobility opportunities. This broader set of challenges is the context in which an assessment of transportation's energy consumption and greenhouse gas emissions must be grounded (WBCSD, 2004).

3.3.7 Passenger Aircraft for Air Transportation

As shown in Figure 3.1, air transportation represents almost half of nonhighway transportation energy use, or about 10 percent of total transportation energy use in the United States. Several studies have examined opportunities for increasing energy efficiency in commercial passenger aircraft (see Kahn Ribeiro et al., 2007). Airline investment decisions are driven by fuel efficiency, since fuel expenditures are the largest operating cost for most airlines. For example, Boeing's and Airbus's newest generation of airliners, the Boeing 787 Dreamliner and 747-8 and the Airbus A350-XWB, employ weight-reducing carbon composite structural materials and less energy-intensive electrical systems. These aircraft represent a 15–20 percent improvement in fuel efficiency over the aircraft that they replace.

As shown in Figure 3.5, there have been many energy-saving technological improvements in commercial aircraft since the introduction of jet airliners, spanning the 1960s-era Boeing 707 to the Boeing 777. The new 787 Dreamliner and Airbus's forthcoming A350 are extending these improvements. Business jets are likewise becoming more energy-efficient. Fuel performance has become a major selling point for makers of these aircraft. For example, Honda Motor Company is

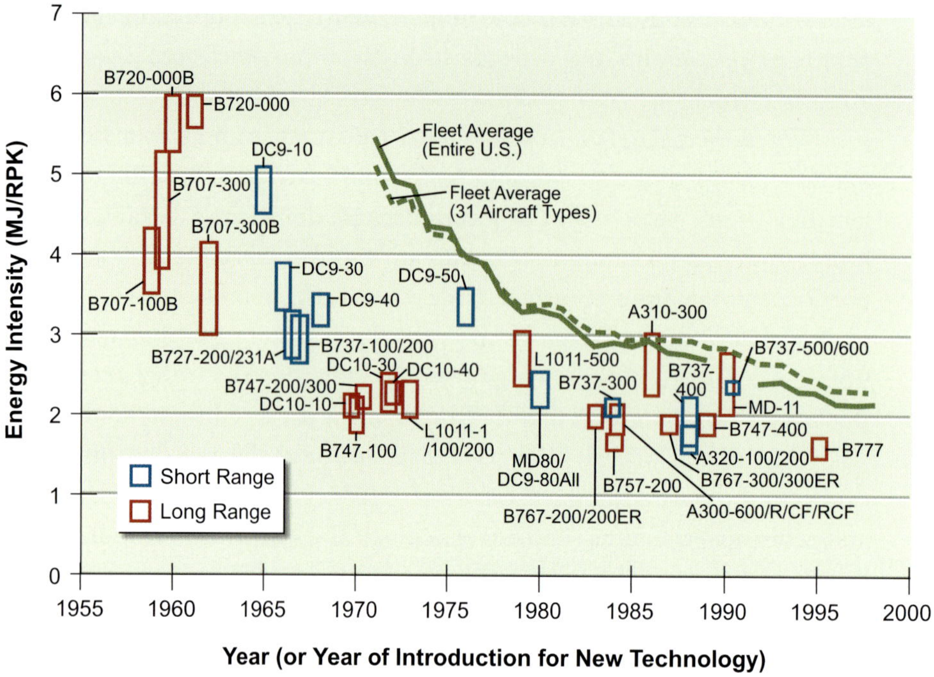

FIGURE 3.5 *Commercial aircraft efficiency trends. The dotted line is the fleet average for only the 31 aircraft types shown.*
Note: RPK, revenue passenger-kilometer; 1 million joules is about 0.95 thousand Btu.
Source: Lee et al., 2004, adapted from Lee et al., 2001. Reprinted with permission from Elsevier.

developing a six-seat jet aircraft that the automaker plans to market for business aviation. This aircraft has a number of features aimed at reducing weight and drag (lightweight engines, all-composite fuselage, and over-the-wing engine mount) and thus fuel burn.

Commercial aircraft must satisfy a number of demands and constraints, including performance with respect to safety, noise, passenger comfort, and emissions of air pollutants. As noted in Lee et al. (2001), as the fuel efficiency of commercial airliners has increased, some of these gains have been used to provide additional passenger amenities (e.g., more luxurious first-class seating, sophisticated entertainment systems) as well as to reduce noise. This fractional ERFC parallels the experience with automobiles discussed above.

In addition to the design of the aircraft themselves, the systems in which they operate have a major influence on energy efficiency. The efficient use of aircraft, along with choosing the most suitable aircraft to fulfill market service requirements, is critical to improving system energy efficiency. Figure 3.6 shows the historical trends in aircraft seating capacity and load factors (passengers per available seat). As noted in Lee et al. (2001), "Load factor gains have been attributed to deregulation in the U.S. and global air travel liberalization, both of which contributed to the advent of hub-and-spoke transportation systems" (p. 185). While hub-and-spoke services may lead to more circuitous trips than are required with point-to-point service (necessitating additional miles traveled and operations to and from the connecting hub airport), on balance they can boost energy efficiency because they enable more intense utilization of aircraft. Likewise, air-traffic-control and management procedures influence energy efficiency. Air-traffic management that leads to more precise and less circuitous flight paths, reduced taxiing and idling, and more efficient climbs and descents will reduce fuel burn. Next-generation, satellite-based air navigation systems and new air-traffic-control procedures, such as continuous descent approaches, promise to shorten flights and yield further gains in operational efficiency. These gains will complement those from advances in aircraft engines, materials, and wing designs (such as raked wing tips that reduce cruise drag).

Thus, energy efficiency in air transportation must be viewed on a comprehensive, systems basis that considers the energy performance of aircraft designs as well as how they are used. According to Lee et al. (2001), the energy intensity

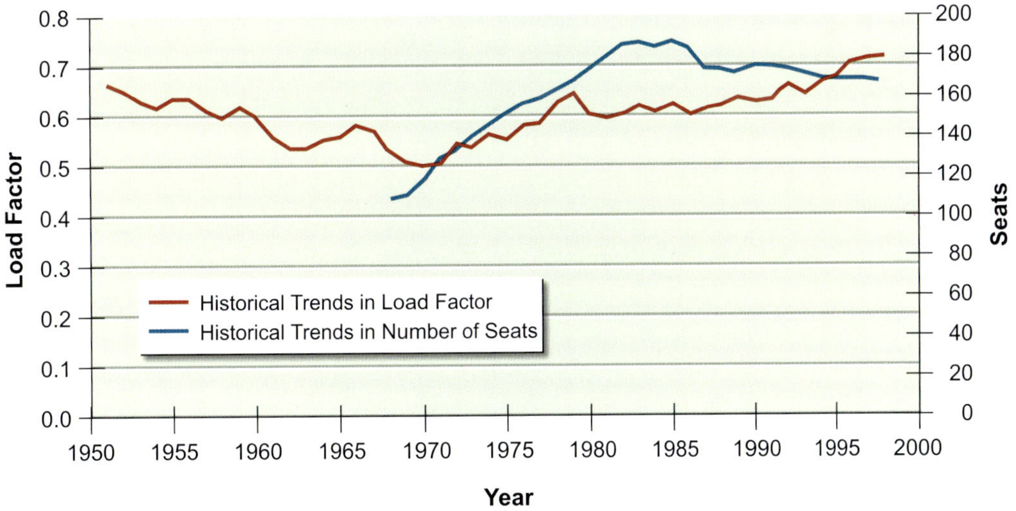

FIGURE 3.6 *Historical trends in aircraft seating capacity and load factors for flights operated by U.S. carriers.*
Source: Lee et al., 2001. © 2001. Reprinted with permission from Elsevier.

of new aircraft (measured by energy consumed per seat-mile flown) declined by 60 percent during the first 40 years of jet travel. The authors estimate that 57 percent of this decline stemmed from increases in engine efficiency, 22 percent from increases in aerodynamic performance, 17 percent from increased load factors, and 4 percent from operational changes such as flight time efficiency (that is, reduced time on the ground or in noncruise portions of the flight). They anticipate energy efficiency improvements of 1–2 percent per year for the next two decades, yielding a total improvement of more than 30 percent over this period.

The Federal Aviation Administration expects air travel demand (in passenger emplanements) to grow about 3 percent per year over the next several decades.[15] This presents a major challenge to efforts to reduce fuel consumption in this sector, because energy efficiency per passenger emplanement is expected to improve by only 1–2 percent per year (Lee et al., 2004). This energy efficiency improvement will not be enough to counter the expected growth in demand.

3.4 FREIGHT TRANSPORTATION

The United States spends about 6–7 percent of its gross domestic product (GDP) on the movement of freight. According to the Federal Highway Administration (FHWA, 2007), about 21 billion tons of freight were moved in 2006 (including 4 billion tons in pipeline movements).[16] The FHWA expects U.S. freight transport to continue to grow by 2 percent per year over the next two to three decades as the economy grows and domestic and international trade increases, resulting in an 85 percent increase in freight tonnage by 2035 (to 37 billion tons). Factoring in 0.5 percent annual growth in energy efficiency in the freight sector means that total energy use for freight movement will grow by 40 percent or more.

Table 3.8 shows projections of freight tonnage by mode for 2035. These projections are based in large part on assumptions for GDP growth, as freight volumes have historically tracked economic growth.

Trucking dominates freight shipment in the United States in terms of both tonnage and shipment value (on the latter measure, it accounts for 95 percent of shipments). The dominance of the truck mode is not expected to change during

---

[15]See http://www.faa.gov/data_statistics/aviation/aerospace_forecasts/2008-2025/.

[16]See http://ops.fhwa.dot.gov/freight/freight_analysis/nat_freight_stats/docs/07factsfighres/table2_1.htm.

TABLE 3.8 Weight of Freight Shipments by Mode: 2007 and 2035 (millions of tons)

	2007				2035			
	Total	Domestic	Exports[a]	Imports[a]	Total	Domestic	Exports[a]	Imports[a]
Total	21,225	19,268	619	1,338	37,210	33,666	1,112	2,432
Truck	12,896	12,691	107	97	22,813	22,230	262	320
Rail	2,030	1,872	65	92	3,525	3,292	57	176
Water	682	575	57	57	1,041	874	114	54
Air	14	4	4	6	61	10	13	38
Intermodal[b]	1,505	191	379	935	2,598	334	660	1,604
Pipeline and unknown[c]	4,091	3,934	6	153	7,172	6,926	5	240

[a]Data do not include imports and exports that pass through the United States from a foreign origin to a foreign destination by any mode.

[b]Mail and courier shipments and all intermodal combinations except air and truck.

[c]Pipeline and unknown shipments are combined because data on region-to-region flows by pipeline are statistically uncertain.

Source: U.S. Department of Transportation, Federal Highway Administration, Office of Freight Management and Operations, Freight Analysis Framework, Version 2.2, 2007. Available at http://www.ops.fhwa.dot.gov/freight/freight_analysis/nat_freight_stats/docs/08factsfigures/table2_1.htm.
See also http://ops.fhwa.dot.gov/freight/freight_analysis/nat_freight_stats/docs/07factsfigures/table 2_1.htm.

the next 25 years. It is important to note, however, that this forecast was made before the run-up in diesel prices to more than $5 per gallon in 2008. Sustained higher diesel prices may lead to some marginal shifts in traffic to other modes and perhaps to lower overall growth in freight traffic.

3.4.1 Heavy-Duty Vehicles

The trucking sector is the main user of heavy-duty vehicles, defined as trucks and buses having gross vehicle weights exceeding 10,000 lb. HDVs consume about 25 percent of the fuel used in the highway sector, the vast majority diesel (Figure 3.7). The largest HDVs used in transportation—those having gross vehicle weights in excess of 33,000 lb—account for half of the energy used by the HDV fleet. These vehicles include the tractor-trailer combinations that are dominant for long-haul freight transportation. It merits noting that HDVs are used for construction, mining, agriculture, and other nontransportation purposes. Although these off-road vehicles are not examined further in this section, they do use a large portion of total HDV energy.

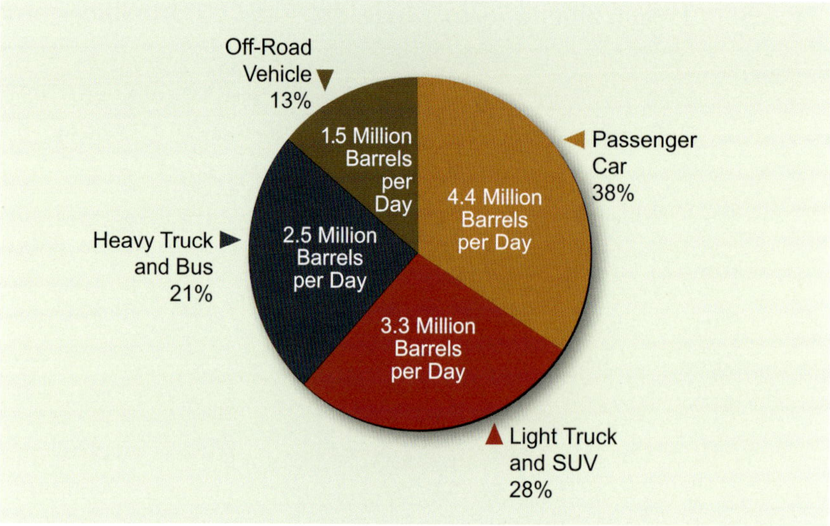

FIGURE 3.7 *Total U.S. highway and off-road vehicle fuel use in 2003 (diesel and gasoline only), in million barrels per day.*
Source: Transportation Energy Data Book, Edition 25 (Davis and Diegel, 2006).

Fuel efficiency is an important factor in diesel-engine and truck design because fuel costs account for a major portion of HDV operating costs. It is not uncommon for a tractor used intensely for long-distance freight transportation to travel more than 800,000 miles in its service life. Some tractors get about 5 mpg with diesel fuel; maintaining the engine, improving tires and aerodynamics, and limiting speed could boost their performance by 1–2 mpg. Over an 800,000-mile life, a tractor getting 5 or 7 mpg would use 160,000 and 114,000 gallons of diesel fuel, respectively. The more efficient tractor would save 46,000 gallons of diesel fuel, or $230,000 at $5 per gallon. This example shows how high diesel fuel prices create an environment that compels carriers to focus on vehicle efficiency, both in their vehicle purchase decisions and in their fleet maintenance and operations.

HDV energy efficiency is a complex issue, however, because trucks perform a wide variety of duties and operate in many environments. Measuring energy efficiency across this sector is therefore complicated. For example, one truck moving a 30-ton payload 500 miles may average only 5 mpg, while another operating over the same distance carrying a 10-ton payload may average 7 mpg. The former truck will be more energy-efficient on a ton-mile basis, whereas the latter will appear to be more energy-efficient when considered on a vehicle-mile basis. The nature of the payload (e.g., whether it consists of weight-limited, high-density freight or space-limited, low-density freight) is therefore an important factor in measuring energy efficiency—and the expected payload is a factor in the design of the vehicle. Likewise, HDVs used mainly for local deliveries and services, such as tanker trucks and refuse and dump trucks, will appear to have low energy efficiency because they operate in congested, stop-and-go environments that are inherently fuel intensive.

Many factors influence HDV energy use. Vehicle design factors include the energy consumed by driveline friction, air resistance, the use of auxiliaries, and tire rolling resistance. Operating variables such as speed, road type, idling, and weather conditions also influence energy use, in addition to payload characteristics and whether the truck is fully or partially loaded. Trailer characteristics are also important factors in HDV efficiency, affecting aerodynamic drag and rolling resistance. Truck operators often do not own the trailers—shippers frequently own them—and these fleets can have long service lives. Trailer efficiency improvements, therefore, tend to lag tractor improvements.

The pressure to reduce fuel costs has led truck manufacturers to make continuous improvements in engine efficiency through various technological improvements, including more sophisticated fuel injection systems, improved combustion, and higher cylinder pressures due to increased turbocharging. Automated manual systems are an example of transmission improvements that yield energy savings. Technologies that are on the horizon include CVT and power-shift transmissions, as well as hybrid-electric systems that can be used to modulate auxiliaries (pumping, fans, compressors, air-conditioning, and power steering) and reduce idling. Reducing idling can be especially important in urban duty cycles and for sleeper cabs, where idling alone can account for 5–10 percent of vehicle fuel use (Davis et al., 2008). More efficient auxiliary power units could increase fuel economy, as could the use of utility-supplied electricity when an HDV is parked at a truck stop.

The aerodynamic designs of the tractors in operation today are far better than those of a decade or more ago. Many more tractors are equipped with side skirts, roof fairings, and aerodynamic fronts. At common highway speeds (60–70 mph), overcoming aerodynamic drag represents about 65 percent of the total energy expenditure for a modern Class 8 combination truck. The drag coefficient is defined as the drag/(dynamic pressure × projected area). The EPA estimates that this coefficient has been reduced from 0.8 to 0.65 during the past two decades (EPA, 2004). The EPA believes that the implementation of known technologies and techniques to improve aerodynamics can lead to a further 20 percent reduction in the drag coefficient. As an example, a Kenworth T2000 tractor, designed with a built-in aerodynamic shield, small radiator, rounded corners, and recessed lamps and tanks, is approximately 15 percent more fuel efficient than the "classic" Kenworth W900L tractor, designed without an aero shield and having a large radiator, many corners, and protruding lamps, tanks, and pipes (Jensen, 2006).

The U.S. Department of Energy's (DOE's) Project on Heavy Vehicle Aerodynamic Drag (McCallen et al., 2003) anticipates that continued research on truck aerodynamics, aided by wind tunnels and computer simulations, can lead to even further reductions in today's drag coefficients, perhaps by as much as 50 percent. Although the authors do not specify a timeframe for this outcome, such a reduction would reduce HDV fuel use by approximately 25 percent when traveling at highway speeds.

Rolling resistance has been reduced through rigorous tire-maintenance programs by carriers, automated tire-pressure monitoring and refilling systems, and the advent of "super single" tires. Further reductions in the weight of tractors represent an area of opportunity, and more gains can be made through improvements in trailer aerodynamics, mass, and rolling resistance characteristics.

With regard to truck and carrier operations, there are numerous areas where energy savings can be achieved. Many, however, involve factors outside the direct control of truck operators, such as government size and weight limits for trucks, highway congestion, and road speed limits. Operational areas that are under operator control include route optimization (congestion avoidance and distance minimization), the reduction of empty mileage, more aggressive fleet management and maintenance, travel speed and acceleration control, "smart" gearing, and cruise management through global positioning systems (anticipating grade and speed-limit changes). Speed governors are designed into most new tractors but often go unused or are disabled by drivers. Greater use of these existing controls could save fuel.

Table 3.9 summarizes the potential for fuel efficiency gains in long-haul trucking from various near-term options discussed above, as estimated by the vice president of advanced engineering for Volvo Powertrain.[17]

Looking farther out in time, the U.S. DOE's 21st Century Truck Program examined the prospects for alternative fuels in HDV applications, including trucks that fall into lighter classes than the long-haul tractor-trailer combinations of

[17]A. Greszler, presentation to the NRC Transportation Research Board, May 1, 2008.

TABLE 3.9 Largest Near-Term Opportunities for Improving the Fuel Efficiency of Long-Haul Trucks

Opportunity	Estimated Fuel Efficiency Gain[a]	Technology Readiness	Issues/Obstacles
Low-rolling-resistance tires (super singles) on tractors and trailers	3%	Available for high-volume use. Increasingly deployed.	Cost and life factors Skepticism by operators Trailer ownership split Road damage concerns[b]
Turbo compound	3%–5%	Concept proven with some production, but outside the United States.	Cost and reliability Package space
Trailer side skirts	4%	Commercially available.	Trailer/truck ratio >3 Trailer ownership split Skirt damage Knowledge/incentives
Mandatory limit of road speed to 65 mph (controlled via truck software)	5% average	Available in all Class 8 trucks since the mid-1990s.	Drivers paid by mile Car traffic meshing/safety Congressional action needed
Elimination of idling in sleeper mode	5%–7%	Available: APU, battery, storage systems, shore power in some stops, engine stop-start systems, IdleAire system.	Storage system performance Shore power availability IdleAire system availability and cost Cost and weight for onboard systems California APU DPF requirement Stop/start cycle disturbs sleep
Increase in weight, length, and trailer combination limits	Fewer trucks needed on road	None required.	Safety concerns Road damage concerns State variations
Optimization of power train and engine to duty cycle	2%–5%	Available.	Customer awareness Adequate sales engineering support Variation in duty cycle
Trailer gap reduction	3%	Commercially available. Deployed in some fleets.	Mix of trailers hauled Turning-radius reduction DPF size

[a]The percentage reductions are not intended to be additive because some of the changes, if made, would reduce the impact of those that follow.

[b]"Super single" tires show promise but may have drawbacks, such as pavement wear, under some circumstances.

Source: Estimates from A. Grezler, vice president of advanced engineering for Volvo Powertrain. Presentation to the Transportation Research Board, May 1, 2008.

TABLE 3.10 Summary of Commercial Truck Greenhouse Gas Reduction Measures

Measure	Description	Phase-in Scenario	Primary Studies Referenced[a]
Class 2b efficiency	25% CO_2 g/mi reduction	Logistical S-curve for new truck deployment from 2010 to 2020	Austin et al., 1999; DeCicco et al., 2001; EEA, 2001; NRC, 2002; Plotkin et al., 2002; Weiss et al., 2000
Class 3–6 efficiency	40% CO_2 g/mi reduction	Logistical S-curve for new truck deployment from 2010 to 2020	Vyas et al., 2002; An et al., 2000; Lovins et al., 2004; Langer, 2004
Class 7–8 efficiency	34% CO_2 g/mi reduction	Logistical S-curve for new truck deployment from 2010 to 2020	Vyas et al., 2002 ; Muster, 2001; Lovins et al., 2004; Schaefer and Jacoby, 2006; Langer, 2004
Ethanol fuel substitution	Increase mix of ethanol to 15% by volume of gasoline by 2020	Phased in linearly from 8% in 2010 to 15% in 2020 (all new additions above baseline are from cellulosic feedstock)	Bowman and Leiby, 1998; Wang et al., 1999
Biodiesel fuel substitution	Increase mix of biodiesel to 5% by volume of diesel by 2020	Phased in linearly from ~0% in 2010 to 5% in 2020	Sheehan et al., 1998; Hill et al., 2006; Farrell and Sperling, 2007; EPA, 2007

[a] See Kasseris and Heywood, 2007, for references.

Classes 7 and 8 (DOE, 2006). The measures examined for reducing greenhouse gas emissions by trucks are summarized in Table 3.10. Their projected influence on new truck fuel economy through 2030 is shown in Figure 3.8.

3.4.2 Air Freight

The NRC report *Potential Impacts of Climate Change on U.S. Transportation* (NRC, 2008a) notes that commercial aircraft account for 12 percent of transport energy use worldwide and 8 percent of that in the United States. The vehicle-centric energy efficiency technology for air freight is essentially the same as that for passenger-based airliners; such technologies are discussed in Section 3.3.7.

3.4.3 Railroads

Railroads account for about 2.5 percent of transport energy use in the United States (Davis et al., 2008). Freight railroads in this country are nearly all diesel powered, compared with Japanese and European systems, which are electrified

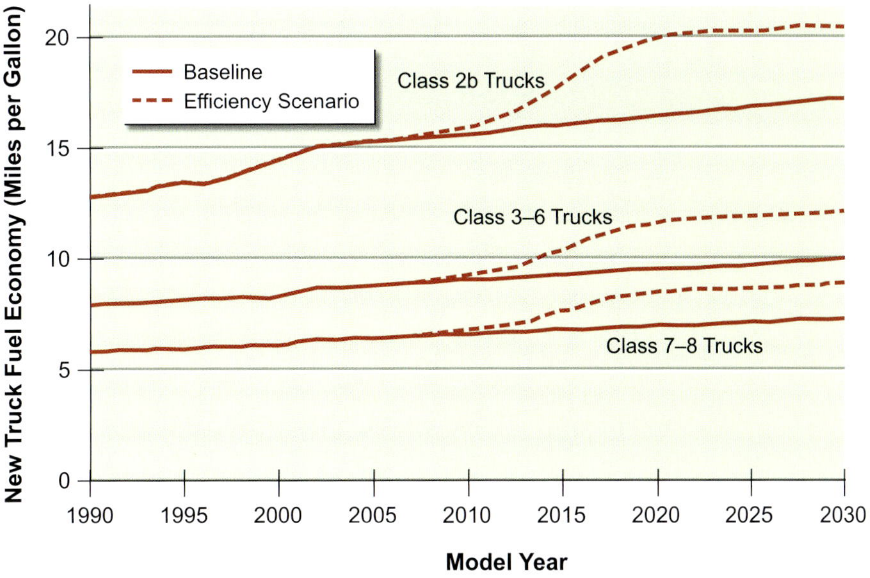

FIGURE 3.8 *Trends in the fuel economy of new commercial trucks. Historical and projected trends in the fuel economy of Class 2b (light) trucks; Classes 3–6 (medium) trucks; and Classes 7 and 8 (heavy) trucks. Class 2b trucks are those with gross vehicle weights (GVWs) of 8,500–10,000 lb; trucks in Classes 3–6 have GVWs of 10,000–26,000 lb; and trucks in Classes 7 and 8 have GVWs of more than 26,000 lb.*
Source: Lutsey, 2008.

through much of their systems. Improvements in railroad technology offer modest opportunities for gains in U.S. transportation energy efficiency. Areas of opportunity include advanced high-efficiency locomotive engines, reductions in aerodynamic drag, track lubricants, lower train weight, regenerative braking, hybrids for switching engines in yards, and higher-efficiency propulsion systems.[18]

Railroad operations represent another area in which energy efficiency gains can be achieved. Opportunities include increased railcar capacity (from 286,000 to 315,000 lb) and train length (e.g., 8,500-foot-long intermodal trains), optimized line-haul speeds enabled by technologies such as positive train control, and further structural changes in railroad economics such as a continued shift to larger and more uniform shipments (e.g., "retail" to "wholesale" railroading such as unit trains) that permit more efficient operations generally.

---

[18]James J. Winebrake, "Scenarios for Reducing the Greenhouse Gas Intensity of Fuels Used in Goods Movement," presentation to the NRC Transportation Research Board, May 1, 2008.

Examining freight from an intermodal perspective suggests further opportunities to save fuel by shifting some freight from truck to rail. The candidates for diversion to rail include truckload trailers of commodities that are not time sensitive and that are traveling more than 500 miles, as well as less-than-truckload long-haul freight. By and large, however, railroads have difficulty competing with trucks for freight whose delivery is time sensitive, such as overnight mail. Because trucks consume 10 times more energy than rail (and waterborne) per ton of freight moved, a 10 percent diversion of freight from truck to rail could produce a 9 percent energy savings. This is a substantial amount of traffic diverted, however. Even a small percentage shift in freight from truck to rail would require a considerable increase in railroad capacity. Increasing miles of track would be expensive and would likely be opposed by nearby residents, especially if new routes are needed.

3.4.4 Waterborne Shipping

Waterborne shipping makes use of oceans, inland and coastal waterways, and the Great Lakes. These routes use different vessels, require different infrastructure, and transport different commodities. The main fuels used are diesel oil (about 70 percent) and heavy fuel oil (about 30 percent).

Measured in tonnage, the oceangoing segment of this sector accounts for about half of the freight moved on water into or within the United States. Oil tanker traffic is, of course, one important reason for this share. Another is the increase in manufactured goods shipped in international trade by container ships. More than 75 percent of the U.S. international trade (in dollar value) is with five countries: Canada, Mexico, Japan, China, and Germany (NRC, 2004b). Most of the trade with Canada and Mexico is by truck and rail, whereas most of the goods traded with Japan, China, and Germany are transported by container ships and other oceangoing vessels. In March 2007, the International Council on Clean Transportation concluded that "carbon dioxide emissions from shipping are double those of aviation and increasing at an alarming rate, which will have a serious impact on global warming, according to research by the industry and European academics" (ICCT, 2007a). However, measured in terms of the CO_2 emitted or energy consumed per ton-mile or per value of freight moved, ocean shipping is highly efficient, since the vessels carry very large payloads over long distances.

On the domestic inland rivers, the Great Lakes, and coastal waterways, tugboats, barges, and self-propelled vessels are used to move (mostly) bulk commodities, such as petrochemicals, coal, grain, lumber, and minerals. They are important modes of transportation for these commodities in the specific regions in which the water routes are available. The energy efficiency of domestic marine shipping is comparable to that of railroads on a ton-mile basis (Davis et al., 2008).

Various opportunities exist in the near term to improve the energy efficiency of waterborne transportation. On the technology side, they include better shore power management and electrification, high-efficiency propulsion technology, improved hull design, and the use of alternative fuels. The potential for technical measures to reduce CO_2 emissions from diesel fuel has been estimated at 5–30 percent in new vessels and 4–20 percent in older ones. On the operations side, near-term opportunities include improved terminal operations to reduce idling, queues, and delays; improved vessel-loading and -unloading operations; better hull maintenance; and speed reduction or optimization. Increasing vessel size will also reduce energy use per ton of freight shipped, especially for container ships. Marintek (2000) estimated that these operational measures could provide up to a 40 percent increase in energy efficiency.

In analyzing measures that can be taken to improve energy efficiency, Kromer and Heywood (2008) estimated the potential gains in energy efficiency in marine shipping to be 20–30 percent by 2020. Speed reduction was found to offer the greatest potential, followed by implementation of new and improved technology. Speed reduction, however, would require strong incentives to achieve, in view of the incentives to move shipments rapidly. Moreover, because of continued growth in commerce and waterborne traffic, it is likely that total energy use will continue to rise (Marintek, 2000).

3.5 FUELS OLD AND NEW

Current U.S. transportation systems—land, water, and air—overwhelmingly use petroleum-based hydrocarbon fuels. These fuels dominate because they are liquid at ambient temperature, have very high energy density, and fit well with today's engine technologies: spark-ignition engines, diesels, and gas turbines. As an illustration of their attractiveness, when refueling a car today, the fuel's chemical energy flows through the nozzle in one's hand at the rate of 570,000 Btu per minute, providing another 400 miles of driving with a 5-minute refueling time.

Current U.S. fuels and engine technologies have evolved together over many decades. Thus, U.S. petroleum extraction, delivery, refining, and distribution systems are cost-effective, and these fuels are well matched to what end-users—

vehicle owners and operators—need. The current U.S. petroleum-based fuel-supply system is vast in scale and does its job well. A major problem, of course, is that these established fuels—gasoline, diesel fuel, aviation kerosene—are some 86 percent carbon by weight and when burned in engines emit almost all this carbon as CO_2. The discussion below summarizes the current status of and anticipated developments in mainstream petroleum-based and alternative fuels. It also discusses whether these fuels offer any useful opportunities for augmenting the energy efficiency of the fuel-supply system as well as the tank-to-wheels efficiency of the vehicle. A more complete discussion can be found in the parallel effort of the America's Energy Future Panel on Alternative Liquid Transportation Fuels (NAS-NAE-NRC, 2009b).

Transportation fuels affect internal-combustion-engine performance and efficiency directly through their combustion characteristics (knock resistance, or octane rating, for gasolines; self-ignition, or cetane rating, for diesel). They have indirect effects through their energy density and therefore weight for a given vehicle's driving range, and through constraints imposed on engine operation because fuel composition affects vehicle air pollutant emissions control. The issue of fuel energy density is especially critical in the longer term for jet aircraft. Also, in a broader, life-cycle context, the energy consumed and the greenhouse gas emissions released during fuel production affect the overall energy and emissions impacts of the total vehicle-plus-fuels system. Lubricants affect engine energy efficiency through their role in engine friction.

3.5.1 Petroleum-Based Fuels

The characteristics of petroleum-based fuels have developed to match the needs of today's land-based spark-ignition and diesel engines, marine use in boats and large ships, and the requirements of aviation's jet engines. For reliable and efficient end use, a broadly based set of fuel-property requirements must be met. Current challenges are the rising cost of these fuels as demand grows rapidly; concerns about their long-term availability in ever-growing (and very large) volume; the need for cleaner fuels with decreasing levels of contaminants owing largely to ever-more stringent air pollutant emissions requirements; and much tighter fuel specifications (also for emissions control reasons). An exacerbating factor is that a growing proportion of the crude oils used is heavier (more dense), and these oils have higher inherent levels of contaminants.

Changes in gasoline and diesel fuel specifications are currently being explored that could reduce refinery energy requirements and permit useful—

though modest—improvements in energy performance and efficiency. Fuel characteristics that would aid the development of new engine combustion concepts such as HCCI are also being explored.

Since petroleum-based fuels dominate the transportation sector, they have developed very large-scale refining and distribution systems. More than 300 billion gallons of refinery products are distributed across the nation each year. The ability of alternative fuel streams to be compatible with and integrated into these refining and distribution systems is an important aspect of their attractiveness.

3.5.2 Natural Gas

The use of natural gas (methane) in road transportation varies around the world, but it is typically about 1 percent of the amount of petroleum-based fuel use. In a few countries (for example, Argentina and Italy) where tax policies make it an economically attractive fuel, its use is about 10 percent of transportation fuel consumption (Yeh, 2007). In the 1990s, natural gas made inroads into U.S. municipal bus fleets in order to achieve lower air pollutant emissions. However, diesel engines with effective exhaust cleanup technology are now proving to be a cheaper option in that market (Cohen, 2005; Cohen et al., 2003). Natural gas has a higher octane rating than that of typical gasolines, and it has good combustion characteristics, and thus could improve spark-ignition engine efficiency. Despite these advantages, methane is a gaseous fuel that must be compressed and stored on the vehicle in high-pressure tanks that are bulky, heavy, and costly. There are additional concerns over safety issues associated with compressed natural gas (which also constrain vehicle use), and uncertainty as to whether a secondary market for reselling natural gas vehicles used by fleet operators would develop.

Based on the available evidence, the panel's overall assessment of natural gas as a transportation fuel is that the drawbacks of a gaseous fuel (e.g., lower specific engine power, reduced driving range, a significant energy penalty for compression in vehicle fueling, the loss of vehicle interior space owing to fuel-storage tanks, extra cost, and methane emissions) currently more than offset the attraction of the lower carbon-to-hydrogen ratio of this fuel and its potential for improving efficiency. Moreover, demand for natural gas in other applications is rising rapidly, threatening to increase its price and make it less attractive as a vehicle fuel. Recently, technologies for extracting natural gas from shales have raised the prospect of significant increases in domestic production at moderate prices. If domestic natural gas supplies expand significantly and the cost of natural gas remains significantly lower than the cost of liquid petroleum-based fuels, its overall attractiveness as a transportation fuel relative to other applications will need to be reconsidered.

3.5.3 Nonpetroleum Hydrocarbon Fuels

Oil sands (e.g., in Canada) and heavy oils (from Venezuela) are already contributing a growing fraction (about 5 percent) to liquid transportation fuels. Over time, other nonpetroleum sources of hydrocarbon fuels, such as gas-to-liquids, oil shale, and coal, are likely developments. These pathways can either produce high-quality transportation fuels directly or provide an input stream to appropriately modified refineries. These high-quality fuels can be blended with petroleum products to improve overall fuel quality and thus petroleum refinery efficiency. Such sources of transportation fuels are expected to steadily increase in volume. However, with current technology, the energy used in the production of nonpetroleum-based fuels is higher, and the amount of greenhouse gases emitted during their production is also higher, than is the case for petroleum-based fuels (NAS-NAE-NRC, 2009a).

3.5.4 Biomass-Based Fuels

Liquid transportation fuels derived from biomass have the potential to contribute significantly to supplying energy for vehicles. Sources of biomass include corn grain, corn stover, switchgrass, miscanthus, forest wastes, and other dedicated fuel crops. End-products include ethanol (and possibly other alcohols), biodiesel and, potentially, gasoline- and diesel-like fuels. Also important are the life-cycle greenhouse gas emissions that result from growing and harvesting the biomass and producing and distributing the specific biofuels. Critical questions that still need to be resolved are the availability of suitable land for these crops, fertilizer and water requirements, land degradation over time, water pollution issues, and the net energy requirements during production. These issues are discussed extensively in the NRC report on alternative liquid transportation fuels (NAS-NAE-NRC, 2009b). Biofuels would, of course, displace petroleum-based fuels.

Biofuels—currently about 3 percent of land transportation fuel supply—could potentially grow in volume to some 10 percent on an energy-equivalent basis over the next 10 or so years (NAS-NAE-NRC, 2009b). The current biofuels are ethanol, about 80 percent, and the rest biodiesel. Integrating these new fuels into the petroleum fuel production and supply system creates logistical problems that need to be resolved (e.g., because of its water-absorbing and solvent characteristics, ethanol cannot be transported in existing pipelines used for oil). Currently, the basic issues with biofuels are their delivered costs relative to those of petroleum fuels and their compatibility with the existing petroleum-based fuel production and distribution system.

From a broader perspective, the critical question is the life-cycle greenhouse gas emissions that result from growing and harvesting the biomass and producing and distributing the specific biofuels produced, and whether the advantageous characteristics of the fuel (e.g., ethanol with its greater knock resistance, which could be used to increase the engine compression ratio) can be used to improve efficiency. The substantial potential of biofuel as an important contributor to greater efficiency in the U.S. transportation sector still needs extensive evaluation.

3.5.5 Electricity

Plug-in hybrids and battery-electric vehicles draw electricity from the electric grid. Their impact on the electric grid obviously depends on the number of these vehicles, on how much they are driven each day, and on when and where they recharge their batteries and how rapidly they recharge. Limited numbers of these vehicles will be available over the next few years. It is plausible that, as has happened with today's conventional hybrids, production volumes will slowly expand over the next decade as these technologies are tested and improved and the significant cost premium is reduced. This introduction and initial growth phase can likely be accommodated by the electric grid with modest adjustments, although consumers will likely be restricted in their recharging options.

It is useful to consider how the impacts on the electric grid would evolve if sales volumes became increasingly larger. (See, for example, Samaras and Meisterling, 2008.) The electrical energy that must be supplied per mile traveled in a future PHEV or BEV is about one-third of the gasoline energy that would be supplied in an equivalent ICE vehicle. In the extreme case in which all LDVs in the United States were electric and the annual VMT were the same as that of standard vehicles today, the annual electricity demand would be about 30 percent of the total amount of electricity currently generated each year. If all this recharging were done overnight (which is unlikely), then current U.S. generating capacity would be able to recharge about half of these vehicles, although the variation from state to state would be substantial (6 percent to 63 percent) (Samaras, 2008). Growth in travel and electricity demand would also have to be factored in. However, the power requirements for recharging are significant; for example, a 30-mile battery recharge requires about 15 kWh, which is 3 kW for 5 hours, and at 110 volts would require close to 30 amps for this time period.

There are many longer-term system issues with PHEVs or electric vehicles. These include the plausible fraction of total vehicles that would satisfy the recharging location, time of day, and charging-power-level constraints; and how the consumer vehicle purchase and use patterns would be affected by the range and recharging limitations of vehicles having different operating characteristics. It is too early in the development of electric-vehicle technology to be able to project how large a fraction of the vehicle market might eventually be met by such vehicles. Note that the impact on greenhouse gas emissions of using electricity as an energy source in transportation will depend on how much electricity is produced, distributed, and used for that purpose and how that energy is generated (i.e., what the primary energy source is, and—if it is fossil fuels—whether carbon capture and storage technology is effectively deployed).

3.5.6 Hydrogen

Hydrogen fuel presents an especially challenging set of issues, since there is currently no hydrogen distribution system. A recent NRC study, *Transition to Alternative Transportation Technologies: A Focus on Hydrogen* (NRC, 2008c), examines what would be needed to implement such a transition and the timescales involved. It concludes that reductions in petroleum use and greenhouse gas emissions could grow steadily over the 2020–2050 timeframe but that substantial government actions and assistance would be needed for this to happen. Establishing a hydrogen production, distribution, and refueling system that provides a sufficiently widespread availability of the fuel so as not to impede the growth in fuel-cell vehicle deployment is a challenging (but doable) task.

3.6 SYSTEM-LEVEL ISSUES

The history of transportation is one of continuous innovation. Most innovations are small and incremental. Some innovations accumulate and lead to a restructuring and reorganization of activities. Energy costs and supply often play a role in motivating innovation—for instance, the transition from sailing ships to steamships—but usually system innovations and changes are motivated by other factors. Major changes in transportation systems are costly and are complicated to bring about for many reasons, including the difficulty of coordinating the interests of businesses (such as vehicle and energy companies); the difficulty of overcoming the inertia of business practices; the existence of government rules dealing with safety, interstate commerce, and so on that were promulgated for previous practices and products; the nonuniformity of government rules across jurisdictions; and the need to blend new and old infrastructures and technologies. However, when major system transitions do occur, they may provide opportunities to boost the overall energy efficiency of the transportation sector.

The freight sector offers examples. The development of standard, 20- and 40-ft (6.1- and 12.2-meter) shipping containers, for example, has stimulated intermodal transfers among trucks, rail, ships, and even cargo airplanes. These containers can be collected by truck at factories and transported to rail terminals where they are carried for the long-haul portion of a trip, often to a seaport for loading onto specially designed container ships. By facilitating the transfer of cargo among modes, the "container revolution" has led to dramatic changes in logistics and in patterns of trade. Because of the greater energy efficiency of rail and water transport compared with trucking, containerization has presumably led to significantly lower energy use per ton-mile of freight. The impact on total energy use, however, is unclear, as the utility of containers has enabled far more extensive, international logistics systems.

In passenger transport, opportunities for increasing fuel efficiency through system-level changes may be greater, if only because of the current pattern of largely single-occupant vehicle usage. One catalyst is the use of information and communication technologies, referred to in the transportation community as intelligent transportation systems (ITSs). The preponderant ITS effort has been incremental in nature. Local governments have learned to use information to manage the use of roads better, and travelers have gained access to navigation devices and information services that ease driving tension, reduce destination search times, and can be used for emergency services. The net effect is a small reduction in driving and energy use—from smoother vehicle flow and reduced vehicle-miles traveled. More ambitious initiatives include using wireless and advanced information technologies to offer new mobility services such as demand-responsive jitney (inexpensive small bus) services, dynamic ridesharing ("smart carpooling"), smart car sharing, smart parking, and so on. These services have the potential for significant reductions in vehicle use and therefore in energy use and greenhouse gas emissions. But there are many barriers to their successful adoption, including consumer resistance, the difficulty of competing against subsidized conventional transit services, opposition from entrenched interests such as taxis and transit operators, and insurance costs. Companies are emerging to offer these services, but their market share is very small.

Still more innovative are automated highway lanes for cars and trucks, using advanced control technologies, sensors, and wireless communication technologies. But these efforts have faltered in the face of litigation and safety concerns. And even as these automation technologies enter the market, initially as "smart" cruise control, automated vehicle parking, and automated emergency braking, they will have minimal effects on energy use. Furthermore, as they are fully implemented, they might even increase vehicle travel and energy use, as the ease of "driving" induces people to live farther from work and to drive longer distances.

Bus rapid transit service, which makes use of dedicated lanes and fare collection before bus entry (Levinson et al., 2002), combines the speed of subways with the flexibility of buses. For more personal service, smart paratransit, real-time carpools, and car-sharing services could reduce VMT.[19]

ITS and other advanced technologies may be used to create broader system changes with potentially much larger energy and greenhouse gas emission benefits. When transportation and land use are considered together, it is possible to imagine how new transportation systems could be developed that bring about improvements in energy efficiency. While a shift toward dense urban corridors would be at odds with long-term trends, changes in individual preferences (e.g., interest in urban amenities) and values (e.g., environmental concerns) may foster such a movement.

For such a diversified system to evolve, numerous changes would need to occur, not only in people's preferences but also in policies and institutions that govern land-use management and the provision of transportation services. The panel cannot delve into these broader topics in this report. When taking a longer-range view of options, however, if the goal is to reduce overall energy use and greenhouse gas emissions, the interconnections among land use, transportation, and life styles should not be neglected. For instance, while some new transit services might by themselves consume more energy per passenger-mile traveled than single-occupant vehicles, the net effect of greater mobility and locational choices could be an overall reduction in energy use and greenhouse gas emissions.

---

[19]See Chapter 2 of Sperling and Gordon (2009).

It is thus important to explore, at this still early stage of ITS development and implementation, how ITS can be used to address the multiple needs and problems associated with surface transportation in a synergistic manner.

Note that for any of the above-described changes to have a significant impact on U.S. transportation's fuel consumption and greenhouse gas emissions, they would have to be implemented on a substantial scale.

This brief overview identifies many opportunities for reducing vehicle energy use and greenhouse gas emissions through changes in the ways that the U.S. transportation infrastructure is managed and used. However, major insights and improvements will come from a broader, deeper understanding of transportation system issues for all transportation modes. Developing better data and tools that can be used to further that understanding is an important task.

3.7 CHALLENGES AND BARRIERS

- In the United States, many factors, including a century of falling energy prices and rising incomes, together with personal preferences and various government policies, have contributed to decentralized land-use patterns and a transportation-intensive economy.
- Low-priced energy led to consumer purchasing behavior, vehicle designs, and operating decisions that emphasized convenience, style, and speed over fuel economy in automobiles and light trucks, and with added emphasis on cost-effectiveness in medium- and heavy-duty trucks, ocean shipping, and the air transport of passengers and freight.
- The primary barriers to realizing greater energy efficiency in the transportation sector are the expectations of individuals and companies about future energy prices, fuel availability, and government policies. Although an extensive menu of technologies exists for saving energy in transportation, before decision makers choose to invest in these technologies, they must be convinced that energy price increases (or other factors that stimulate market demand) will persist.
- A barrier to rapid changes in the mix of LDV annual sales is the capacity of the automotive industry to change both power trains and platforms rapidly, across all models, and its ability to set up a high-volume supplier base in high-risk items such as high-energy-storage batteries.

The vehicle design cycle can be 3–5 years if the change involves major new technologies or materials.

- Even when new or improved vehicle technologies are available on the market, barriers to purchase include high initial cost, safety concerns, reliability and durability concerns, and lack of awareness. For new technologies to reach a substantial fraction of vehicle sales usually takes more than a decade unless mandated by law or consumers clearly demand the new or improved technology.

3.8 FINDINGS

T.1 In the transportation sector, the potential for energy savings and petroleum displacement resides both in increasing the efficiency with which liquid fuels (especially petroleum) are used and in shifting some of the vehicle fleet's energy demand to electricity (including hydrogen fuel-cell vehicles). The overall energy use and greenhouse gas emissions (and other environmental effects) associated with such a shift depend on how the electricity or hydrogen is generated.

T.2 An extensive menu of technologies exists today for increasing energy efficiency in transportation. Achieving the average new-vehicle fuel economy targets for 2020 set by the Energy Independence and Security Act of 2007 (EISA; P.L. 110-140), which represent a 40 percent increase over today's value (and a 30 percent reduction in average fuel consumption), is thus a feasible, although challenging, objective. Reaching the EISA targets, and continuing to decrease fuel consumption, will require a shift from the historic U.S. emphasis on ever-increasing vehicle power and size to an emphasis on using efficiency improvements to improve vehicle fuel consumption.

T.3 In the near term, fuel-consumption reductions will come predominantly from improved gasoline and diesel engines, improved transmissions, and reduced vehicle weight and drag. Through at least 2020, evolutionary improvements in vehicles with gasoline internal-combustion engines are likely to prove the most cost-effective approach to reducing petroleum consumption. Gasoline-electric hybrids will likely play an increasingly important role as their production volume increases and their cost, relative to that of conventional vehicles, decreases. Meeting the EISA

standards is likely to require that, over the next decade or two, an ever-larger fraction of the new vehicle fleet be hybrids or plug-in hybrids.

T.4 Beyond 2020, continuing reductions in fuel consumption are possible. Plausible efficiency improvements in light-duty vehicles, alongside weight reduction and more extensive use of hybrid and plug-in hybrid (and possibly battery-electric) vehicles, could reduce transportation fuel consumption to below the levels implied by the higher 2020 fuel-economy standards mandated by the EISA. An especially important R&D focus is developing marketable vehicles that use electricity, which will require improving the performance and reducing the cost of high-energy-storage batteries.

T.5 A parallel longer-term prospect is fuel cells with hydrogen as the energy carrier. To be attractive, major improvements, especially in reducing costs, are needed. Widespread implementation requires significant investment in efficient, low-greenhouse-gas-emissions hydrogen supply and distribution systems. Onboard hydrogen storage is a key R&D issue. Establishing a new propulsion system technology and new fuel infrastructure on a large scale is a formidable task, and significant deployment of fuel-cell vehicles is unlikely before 2035.

T.6 There are opportunities to reduce energy use in freight transportation by improving both vehicle efficiency and freight system logistics and infrastructure. Reductions of 10–20 percent in the fuel economy of heavy- and medium-duty vehicles appear feasible over a decade or so. A broad examination is needed of the potential for improving the effectiveness of the freight system to reduce energy consumption further.

T.7 Air transport and waterborne shipping have become more energy-efficient in response to higher fuel prices. Jet engine and aircraft technology has the potential to improve the efficiency of new aircraft by up to 35 percent over the next two decades. However, improvements in aviation efficiency for passenger transport are unlikely to fully offset projected growth in air travel. Major additional issues are the full greenhouse gas and other environmental impacts of aviation fuel use at high altitude and of growing airline travel; the potential for using biomass-based fuels in jets; and whether the use of low-grade residual fuel in oceangoing vessels will continue.

T.8 Most transportation efficiency studies and proposals have focused on the considerable energy efficiency gains that could be achieved with

improved vehicles rather than in the transportation system as a whole. This emphasis is appropriate given the potential for and impact of such gains. However, major insights and improvements can result from a broader and deeper understanding of transportation system issues. The potential overall impact of such broader, system-based changes, such as densifying and reorganizing land use and collective modes of travel, needs further exploration and quantification. Developing better data and tools that can be used to analyze and forecast how different policies and investments might affect vehicle use and travel is thus an important task.

3.9 REFERENCES

Alamgir, M., and A.M. Sastry. 2008. Efficient batteries for transportation applications. SAE Paper 2008-21-0017. SAE Convergence, Detroit, Mich. October.

An, F., and D. Santini. 2004. Mass impacts on fuel economies of conventional vs. hybrid electric vehicles. SAE Technical Paper 2004-01-0572. Warrendale, Pa.: SAE International.

An, F., and J. DeCicco. 2007. Trends in technical efficiency trade-offs for the U.S. light vehicle fleet. Technical Paper Series No. 2007-01-1325. Warrendale, Pa.: SAE International. April.

Anderman, M., F.R. Kalhammer, and D. MacArthur. 2000. Advanced Batteries for Electric Vehicles: An Assessment of Performance, Cost, and Availability. Sacramento, Calif.: California Air Resources Board, Year 2000 Battery Technology Advisory Panel.

Anderman, M. 2007. Status and Prospects of Battery Technology for Hybrid Electric Vehicles, Including Plug-in Hybrid Electric Vehicles. Briefing to the U.S. Senate Committee on Energy and Natural Resources, January 26.

APTA (American Public Transportation Association). 2008. 2008 Fact Book. Washington, D.C.: APTA.

Bandivadekar, A.P. 2008. Evaluating the Impact of Advanced Vehicle and Fuel Technologies in the U.S. Light-Duty Vehicle Fleet. Ph.D. Thesis, MIT Engineering Systems Division. Cambridge, Mass.: Massachusetts Institute of Technology.

Bandivadekar, A.P., K. Bodek, L. Cheah, C. Evans, T. Groode, J. Heywood, E. Kasseris, K. Kromer, and M. Weiss. 2008. On the Road in 2035: Reducing Transportation's Petroleum Consumption and GHG Emissions. Laboratory for Energy and the Environment Report No. LFEE 2008-05 RP. Cambridge, Mass: Massachusetts Institute of Technology. July.

BTS (Bureau of Transportation Statistics). 2006. Long distance transportation patterns: Mode choice. America on the Go. Findings from the National Household Travel Survey. May. Washington, D.C.: BTS. May. Available at http://www.bts.gov/publications/america_on_the_go/. May.

Carlson, E.J., P. Kopf, J. Sinha, S. Sriramulu, and Y. Yang. 2005. Cost Analysis of PEM Fuel Cell Systems for Transportation. NREL/SR-560-39104. Golden, Colo.: National Renewable Energy Laboratory.

Cheah, L., and J. Heywood. 2008. Meeting CAFE Standards by 2020 and Beyond. Research Brief, Sloan Automotive Laboratory. Cambridge, Mass.: Massachusetts Institute of Technology. November. Available at http://web.mit.edu/sloan-auto-lab/research/beforeh2/reports.htm. November.

Cheah L., C. Evans, A. Bandivadekar, and J. Heywood. 2007. Factor of Two: Halving the Fuel Consumption of New U.S. Automobiles by 2035. LFEE Report 2007-04 RP. MIT Laboratory for Energy and the Environment. Cambridge, Mass: Massachusetts Institute of Technology. Available at http://web.mit.edu/sloan-auto-lab/research/beforeh2/files/cheah_factorTwo.pdf.

Cohen, J.T. 2005. Diesel vs. compressed natural gas for school buses: A cost-effectiveness evaluation of alternative fuels. Energy Policy 33(13):1709-1722.

Cohen, J.T., J.K. Hammitt, and J.I. Levy. 2003. Fuels for urban transit buses: A cost-effectiveness analysis. Environmental Science and Technology 37:1477-1484.

Crabtree, G., M. Dresselhaus, and M. Buchanan. 2004. The hydrogen economy. Physics Today 57(12):39-44.

Davis, S.C., and S.W. Diegel. 2006. Transportation Energy Data Book: Edition 25. ORNL-6974. Oak Ridge, Tenn.: Oak Ridge National Laboratory.

Davis, S.C., S.W. Diegel, and R.G. Boundy. 2008. Transportation Energy Data Book: Edition 27. ONRL-6981. Oak Ridge, Tenn.: Oak Ridge National Laboratory.

DOE (Department of Energy). 2006. 21st Century Truck Partnership: Roadmap and Technical White Papers. Report 21CTP-0003. Washington, D.C.: DOE. Available at http://www1.eere.energy.gov/vehiclesandfuels/pdfs/program/21ctp_roadmap_2007.pdf.

EEA (Energy and Environmental Analysis, Inc.). 2007. Update for Advanced Technologies to Improve Fuel Economy of Light Duty Vehicles. Draft Final Report. Submitted to U.S. Department of Energy. Arlington, Va.: EEA. August.

EIA (Energy Information Administration). 2006. Emissions of Greenhouse Gases in the United States 2005. DOE/EIA-0573(2005). Washington, D.C.: Department of Energy, Energy Information Administration.

EIA. 2008. Annual Energy Review 2007. DOE/EIA-0384(2007). Washington, D.C.: Department of Energy, Energy Information Administration.

EIA. 2009. Monthly Energy Review (February). Washington, D.C.: Department of Energy, Energy Information Administration.

EPA (Environmental Protection Agency). 2004. A Glance at Clean Freight Strategies: Improved Aerodynamics. Report EPA420-F-04-012. Washington, D.C.: Office of Transportation Air Quality. February. Available at http://www.epa.gov/smartway/documents/aerodynamics.pdf.

EPA. 2007. Light-Duty Automotive Technology and Fuel Economy Trends: 1975 Through 2007. EPA420-R-07-008. Washington, D.C. September. Available at http://www.epa.gov/OTAQ/cert/mpg/fetrends/420r07008.pdf.

Evans, C. 2008. Putting Policy in Drive: Coordinating Measures to Reduce Fuel Use and Greenhouse Gas Emissions from U.S. Light-Duty Vehicles. M.S. Thesis, Technology and Policy Program. June. Cambridge, Mass.: Massachusetts Institute of Technology.

FHWA (Federal Highway Administration). 2007. Freight Facts and Figures 2007. Washington, D.C. Available at http://ops.fhwa.dot.gov/freight/freight_analysis/nat_freight_stats/docs/07factsfigures.

Frost & Sullivan. 2008. World Hybrid/Electric Vehicle Battery Markets. MC1812769. June 26.

Hughes, J., C. Knittel, and D. Sperling. 2008. Evidence of a shift in the short-run price elasticity of gasoline demand. Energy Journal 29(1):113-134.

ICCT (International Council on Clean Transportation). 2007a. Air Pollution and Greenhouse Gas Emissions from Ocean-going Ships: Impacts, Mitigation Options and Opportunities for Managing Growth. Washington, D.C.: ICCT. March.

ICCT. 2007b. Passenger Vehicle Greenhouse Gas and Fuel Economy Standards: A Global Update. Washington, D.C.: ICCT. July. Available at http://www.lowcvp.org.uk/assets/reports/ICCT_GlobalStandards_2007.pdf.

Jensen, J. 2006. Go with the Flow—Truck Aerodynamics. ExpeditersOnline.com. September. Available at http://www.expeditersonline.com/artman/publish/truck-aerodynamics.html.

Johnson, T. 2008. Diesel Emission Control Technology in Review. SAE Technical Paper 2008-01-0069. Warrendale, Pa.: Society of Automotive Engineers. April.

Johnson, T. 2009. Diesel Emission Control in Review. SAE Technical Paper 2009-01-0121. Warrendale, Pa.: Society of Automotive Engineers. April.

Kahn Ribeiro, S., S. Kobayashi, M. Beuthe, J. Gasca, D. Greene, D.S. Lee, Y. Muromachi, P.J. Newton, S. Plotkin, D. Sperling, R. Wit, and P.J. Zhou. 2007. Transport and its infrastructure. In Climate Change 2007: Mitigation. Contribution of Working Group III to the Fourth Assessment Report of the Intergovernmental Panel on Climate Change (B. Metz, O.R. Davidson, P.R. Bosch, R. Dave, and L.A. Meyer, eds.). Cambridge, U.K., and New York, N.Y.: Cambridge University Press.

Kasseris, E., and J. Heywood. 2007. Comparative Analysis of Automotive Powertrain Choices for the Next 25 Years. SAE Technical Paper 2007-01-1605 SP-2091. Warrendale, Pa.: SAE International.

Kromer, M., and J. Heywood. 2008. A comparative assessment of electric propulsion systems in the 2030 U.S. light-duty vehicle fleet. SAE Technical Paper 2008-01-0459. Warrendale, Pa.: SAE International.

Lee, J., S. Lukachko, I. Waitz, and A. Schafer. 2001. Historical and future trends in aircraft performance, cost, and emissions. Annual Review of Energy and the Environment 26:167-200.

Lee, J.J., S.P. Lukachko, and I.A. Waitz. 2004. Aircraft and energy use. In Encyclopedia of Energy, Volume 1. Amsterdam, The Netherlands: Elsevier.

Levinson, H., S. Zimmerman, J. Clinger, and S. Rutherford. 2002. Bus rapid transit: An overview. Journal of Public Transportation 5(2):1-30.

Lutsey, Nicholas P. 2008. Prioritizing Climate Change Mitigation Alternatives: Comparing Transportation Technologies to Options in Other Sectors. Research Report UCD-ITS-RR-08-15. Institute of Transportation Studies, University of California, Davis. Available at http://pubs.its.ucdavis.edu/publication_detail.php?id=1175.

Lutsey, N., and D. Sperling. 2005. Energy efficiency, fuel economy, and policy implications. Transportation Research Record 1941:8-25.

Marintek. 2000. Study of Greenhouse Gas Emissions from Ships. Final Report to the International Maritime Organization. Prepared by a consortium run by the Norwegian Marine Technology Institute—Marintek. Available at http://unfccc.int/files/ methods_and_science/emissions_from_intl_transport/application/pdf/imoghgmain.pdf.

McCallen, R., K. Salari, J. Ortega, L. DeChant, C. Roy, J. Payne, B. Hassan, W. Pointer, F. Browand, M. Hammache, T.-Y. Hsu, J. Ross, D. Satran, J. Heineck, S. Walker, D. Yaste, R. Englar, A. Leonard, M. Rubel, and P. Chatelain. 2003. FY2003 Annual Report: DOE Project on Heavy Vehicle Aerodynamic Drag. November. UCRL-TR-200618. Berkeley, Calif.: Lawrence Livermore National Laboratory. Available at https://e-reports-ext.llnl.gov/pdf/301858.pdf.

NAS-NAE-NRC (National Academy of Sciences-National Academy of Engineering-National Research Council). 2009a. America's Energy Future: Technology and Transformation. Washington, D.C.: The National Academies Press.

NAS-NAE-NRC. 2009b. Liquid Transportation Fuels: Technological Status, Costs, and Environmental Impacts. Washington, D.C.: The National Academies Press.

NPC (National Petroleum Council). 2007. Technologies for Transportation Efficiency. Topic paper of the Transportation Efficiency Subgroup. Council Committee on Global Oil and Gas. Washington, D.C.: NPC. July.

NRC (National Research Council). 1992. Automotive Fuel Economy: How Far Should We Go? Washington, D.C.: National Academy Press.

NRC. 2000. Review of the Research Program of the Partnership for a New Generation of Vehicles: Sixth Report. Washington, D.C.: National Academy Press.

NRC. 2002. Effectiveness and Impact of Corporate Average Fuel Economy (CAFE) Standards. Washington, D.C.: National Academy Press.

NRC. 2004a. The Hydrogen Economy: Opportunities, Costs, Barriers, and R&D Needs. Washington, D.C.: The National Academies Press.

NRC. 2004b. The Marine Transportation System and the Federal Role. Transportation Research Board Special Report 279. Washington, D.C.: The National Academies Press.

NRC. 2006. Tires and Passenger Vehicle Fuel Economy. Transportation Research Board Special Report 286. Washington, D.C.: The National Academies Press.

NRC. 2008a. Potential Impacts of Climate Change on U.S. Transportation. Transportation Research Board Special Report 290. Washington, D.C.: The National Academies Press.

NRC. 2008b. Assessment of Technologies for Improving Light Duty Vehicle Fuel Economy: Letter Report. Washington, D.C.: The National Academies Press.

NRC. 2008c. Transitions to Alternative Transportation Technologies—A Focus on Hydrogen. Washington, D.C.: The National Academies Press.

NRC. 2008d. Review of the Research Program of the FreedomCar and Fuel Partnership: Second Report. Washington, D.C.: The National Academies Press.

Ricardo, Inc. 2008. A Study of Potential Effectiveness of Carbon Dioxide Reducing Vehicle Technologies. Prepared for EPA Office of Transportation and Air Quality. Washington, D.C. January. Available at http://www.epa.gov/OMS/technology/420r08004a.pdf.

Samaras, C. 2008. PHEVs, electricity demand and infrastructure. Memorandum to Lester Lave, Carnegie Mellon University, November.

Samaras, C., and K. Meisterling. 2008. Life cycle assessment of greenhouse gas emissions from plug-in hybrid vehicles: Implications for policy. Environmental Science and Technology 42:3170-3176.

Schipper, L. 2006. Automobile fuel economy and CO_2 emissions in industrialized countries: Troubling trends through 2005/6. Washington, DC: EMBARQ, The World Resources Institute Center for Sustainable Transport. Available at http://pdf.wri.org/automobile-fuel-economy-co2-industrialized-countries.pdf.

Schuring, D., and S. Futamura. 1990. Rolling loss of pneumatic highway tires in the eighties. Rubber Chemistry and Technology 62(3):315-367.

Small, K., and K. Van Dender. 2007. Fuel Efficiency and Motor Vehicle Travel: The Declining Rebound Effect. Working Paper 05-06-03. Irvine, Calif.: Department of Economics, University of California.

Sperling, D., and D. Gordon. 2009. Two Billion Cars: Driving Towards Sustainability. Cambridge, U.K.: Oxford University Press.

Vidal, J. 2007. CO_2 output from shipping twice as much as airlines. The Guardian. March 3. Available at http://www.guardian.co.uk/transport/Story/0,,2025723,00.html.

Wang, M., C. Saricks, and D. Santini. 1999. Effects of fuel ethanol use on fuel-cycle energy and greenhouse gas emissions. ANL-38. Argonne, Ill.: Department of Energy, Argonne National Laboratory, Center for Transportation Research.

WBCSD (World Business Council on Sustainable Development). 2004. Mobility 2030: Meeting the Challenges to Sustainability. Sustainable Mobility Project Report. 180 pages. Available at http://www.wbcsd.org/DocRoot/fI3llMAvneJpUcnLgSeN/mobilityfull.Pdf.

Winebrake, J.J., J.J. Corbett, and P.E. Meyer. 2007. A total fuel life-cycle analysis of energy use and emissions from marine vessels. Transportation Research Board 2007 Annual Meeting, Washington, D.C.

Wohlecker, R., M. Johannaber, and M. Espig. 2007. Determination of weight elasticity of fuel economy for ICE, hybrid and fuel cell vehicles. SAE Technical Paper 2007-01-0343. Warrendale, Pa.: SAE International.

Yeh, S. 2007. An empirical analysis on the adoption of alternative fuel vehicles: The case of natural gas vehicles. Energy Policy 35(11):5865-5875.

4 Energy Efficiency in Industry

Building on improvements in energy efficiency in U.S. industrial manufacturing that have occurred over the past several decades in response to volatile fossil-fuel prices, fuel shortages, and technological advances is essential to maintaining U.S. industry's viability in an increasingly competitive world. The fact is that many opportunities remain to incorporate cost-effective, energy-efficient technologies, processes, and practices into U.S. manufacturing. This chapter describes the progress made to date and the magnitude of the untapped opportunities, which stem both from broader use of current best practices and from a range of possible advances enabled by future innovations. It focuses on the potential for improving energy efficiency cost-effectively in four major energy-consuming industries—chemical manufacturing and petroleum refining, pulp and paper, iron and steel, and cement—and discusses the role of several crosscutting technologies as examples. In addition, this chapter identifies major barriers to the deployment of energy-efficient technologies, outlines the business case for taking action to improve the energy efficiency of U.S. manufacturing, and presents the associated findings of the Panel on Energy Efficiency Technologies.

4.1 ENERGY USE IN U.S. INDUSTRY IN A GLOBAL CONTEXT

As shown in Chapter 1, Figure 1.1, industry is responsible for 31 percent of primary energy use in the United States. Figure 4.1 illustrates how this energy use was distributed among industries, particularly the most energy-intensive ones, in 2004.

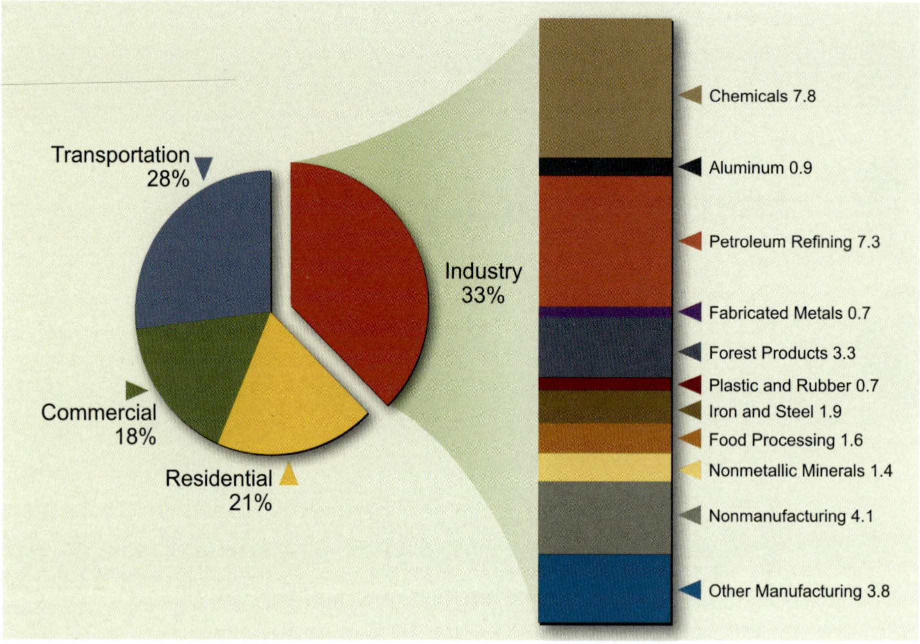

FIGURE 4.1 *Total energy use in the U.S. industrial sector in 2004, quadrillion Btu (quads). Values include electricity-related losses. Total U.S. energy use in 2004 was 100.4 quads; total U.S. industrial energy use in 2004 was 33.6 quads.*
Source: Craig Blue, Oak Ridge National Laboratory, based on EIA (2004) (preliminary) and estimates extrapolated from EIA (2002).

Globally, industry is the largest consumer of energy—the energy that it consumes exceeds that devoted to transportation, the residential sector, and commercial buildings combined. According to the *International Energy Outlook 2009,* the industrial sector worldwide used 51 percent of the total delivered energy (or 50 percent of the primary energy) in the year 2006, and its demand was projected to grow by an annual rate of 1.4 percent between 2006 and 2030 (EIA, 2009a).[1] Before 1973, manufacturing was the largest energy consumer in most member countries of the Organisation for Economic Cooperation and Development (OECD), but in recent years its dominance has subsided as industrial output has slowed, energy efficiency has increased, and other sectors have surged ahead (Schipper, 2004). As a result, industrial energy demand in OECD countries was anticipated to grow only 0.6 percent annually. In contrast, industrial-sector energy

[1]See http://www.eia.doe.gov/oiaf/ieo/excel/ieoendusetab_1.xls.

consumption in non-OECD countries was projected to increase by 2.1 percent per year over the same period, with the most rapid growth occurring in China and India.

As of 2006, industry accounted for 33 percent of the primary energy consumed in the United States and 28 percent of carbon dioxide (CO_2) emissions (EIA, 2008). Overall, the quantity of energy used by U.S. industries is huge, estimated at 32.6 quadrillion British thermal units (quads) of primary energy in 2006 at a cost of $205 billion. About 5 quads, or 21 percent of this total, was for non-fuel uses of coal, gas, and oil—for example, the use of oil refining by-products in asphalt, natural gas employed as a feedstock for petrochemicals, and petroleum coke used in the production of steel (EIA, 2009b). U.S. industries use more energy than the total energy used by any other Group of Eight (G8) nation and about half of the total energy used by China (DOE, 2007b).

The average annual rate of growth of energy in the U.S. industrial sector is projected to be 0.3 percent out to 2030, while CO_2 emissions from U.S. industry are projected to increase more slowly, at 0.2 percent annually (EIA, 2008). These low rates are due partly to the presumed introduction of energy-efficient technologies and practices in industry. They also reflect the projected restructuring of the economy away from energy-intensive manufacturing and toward service and information-based activities. Many of the commodities that were once produced in the United States are now manufactured offshore and imported into the country. The energy embodied in these imported products is not included in the standard energy metrics published by the Energy Information Administration (EIA) of the Department of Energy (DOE). According to an analysis by Weber (2008), products imported into the United States in 2002 had an embodied energy content of about 14 quads, far surpassing the embodied energy of exports from the United States (about 9 quads).

The most energy-intensive manufacturing industries are those producing metals (iron, steel, and aluminum); refined petroleum products; chemicals (basic chemicals and intermediate products); wood and glass products; mineral products such as cement, lime, limestone, and soda ash; and food products. As shown in Figure 4.1, these industries are responsible for more than 70 percent of industrial energy consumption. Industries that are less energy-intensive include the manufacture or assembly of automobiles, appliances, electronics, textiles, and other products.

4.1.1 Recent Trends in Industrial Energy Use

Primary-energy use in the industrial sector declined in the 1970s following the run-up of energy prices. Energy consumption bottomed out in the mid-1980s and then increased steadily through the turn of the century, exceeding its previous peak. Table 4.1 shows energy use for selected years within this period (excluding nonfuel uses). In recent years, industrial energy use has declined partly as a result of the economic restructuring noted above. Energy use in the manufacturing sector continues to be significantly higher than in the nonmanufacturing sectors, which include agriculture, forestry and fisheries, mining, and construction. Energy-use trends in some sectors have been relatively stable, such as in chemical manufactur-

TABLE 4.1 Total U.S. Industrial Energy Use (Excluding Nonfuel Uses of Coal, Oil, and Natural Gas), in Selected Years from 1978 to 2004 (in quadrillion Btu)

Use[a]	1978	1985	1990	1995	2002
Food Manufacturing, Beverage, and Tobacco (311/312)	1.36	1.4	1.35	1.72	1.77
Textile Mills, Textile Mill Products (313/314)	0.53	0.43	0.46	0.54	0.44
Apparel, Leather and Allied Products (315/316)	0.19	0.10	0.11	0.16	0.66
Wood Product Manufacturing (321)	0.64	0.52	0.59	0.67	0.70
Paper Manufacturing (322)	2.38	2.66	3.16	3.17	3.14
Printing and Related Support Activities (323)	0.16	0.15	0.20	0.22	0.23
Petroleum and Coal Products Manufacturing (324)	3.09	2.01	3.37	3.37	3.92
Chemical Manufacturing (325)	4.20	3.05	4.22	4.22	4.06
Plastic and Rubber Products Manufacturing (326)	0.45	0.44	0.52	0.67	0.86
Nonmetallic Mineral Product Manufacturing (327)	1.62	1.16	1.29	1.23	1.32
Primary Metal Manufacturing (331)	5.01	2.43	2.73	2.74	2.70
Fabricated Metal Product Manufacturing (332)	0.66	0.58	0.65	0.75	0.72
Machinery Manufacturing (333)	0.50	0.38	0.43	0.44	0.39
Computer and Electronic Product Manufacturing (334)	0.29	0.39	0.47	0.47	0.38
Electrical Equipment, Appliance, and Component Manufacturing (335)	0.24	0.23	0.29	0.34	0.27
Transportation Equipment (336)	0.73	0.66	0.70	0.77	0.82
Furniture and Related Product Manufacturing (337)	0.12	0.09	0.12	0.13	0.14
Miscellaneous Manufacturing (339)	0.14	0.11	0.12	0.14	0.17
Total (Manufacturing)	22.3	16.8	20.8	21.7	22.1
Total (Non-Manufacturing)	Not available	6.0	4.8	5.5	3.3

[a]North American Industry Classification System codes are given in parentheses. Totals may not equal sum of components due to independent rounding.

Source: U.S. Department of Energy, U.S. Energy Intensity Indicators, Trend Data, Industrial Sector. Available at http://www1.eere.energy.gov/ba/pba/intensityindicators/.

TABLE 4.2 Primary Energy Consumption by Type of Fuel in the U.S. Industrial Sector (quadrillion Btu, or quads)

	1978	1985	1990	1995	2002
Petroleum	9.87	7.74	8.28	8.61	9.57
Natural gas	8.54	7.08	8.50	9.64	8.67
Coal	3.31	2.76	2.76	2.49	2.03
Renewable energy	1.43	1.91	1.67	1.91	1.68

Source: U.S. Department of Energy, U.S. Energy Intensity Indicators, Trend Data, Industrial Sector. Available at http://www1.eere.energy.gov/ba/pba/intensityindicators/.

ing and wood product manufacturing. In other sectors, however, energy use has increased significantly. For example, energy use in the plastic and rubber products manufacturing sector almost doubled between 1978 and 2002.

Petroleum and natural gas are the two most common fuels consumed by the industrial sector (Table 4.2). While the use of petroleum and natural gas increased by 24 and 22 percent, respectively, from 1985 to 2002, coal consumption dropped by approximately 27 percent. The use of renewable energy has fluctuated over the years, totaling 1.43 quads in 1978, rising to 1.91 quads in 1985, and then retreating to 1.68 quads in 2002.

4.1.2 Energy-Intensity Trends and Comparisons

Between 1985 and 2003, industrial-sector gross domestic product (GDP) increased by 64 percent, while industrial energy use increased by only 12 percent (Figure 4.2), resulting in a significant decline in the energy intensity of the industrial sector (DOE/EERE, 2008). As previously noted, over the past decade structural factors (the change in manufacturing output relative to industrial output and the shift among manufacturing sectors to less energy-intensive industries) have caused a decline in energy intensity and in total industrial energy use.

By comparing the energy intensity of manufacturing across 13 countries that are members of the International Energy Agency (IEA), Schipper (2004, p. 18) provides a glimpse into the relative efficiency of U.S. manufacturing. A simple comparison of manufacturing energy use per dollar of output suggests that the United States has a slightly higher than average manufacturing energy intensity. This is corroborated by statistics from the IEA (2004, p. 69) on energy use per unit of manufacturing value added in countries that are members of the OECD.

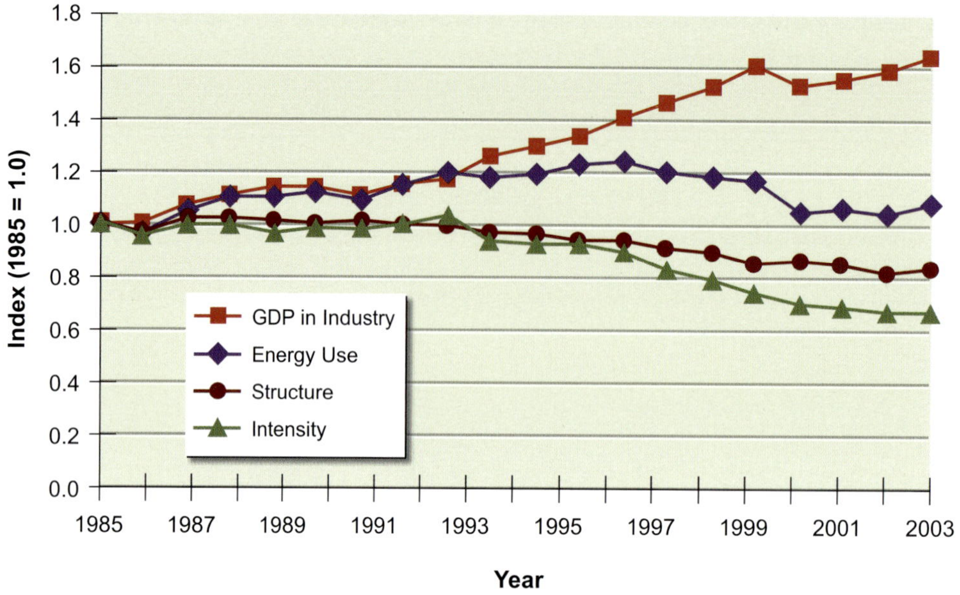

FIGURE 4.2 *Trends in U.S. industrial sector gross domestic product (GDP), energy use, structure, and energy intensity, 1985–2003. Industrial GDP increased 64 percent between 1985 and 2003; energy intensity (energy use per dollar of GDP) declined by 19 percent over the same period, with most of the decline occurring since 1993. "Structure" represents the change in manufacturing as a fraction of total industrial output, and the changes that have occurred within manufacturing.*
Manufacturing, which is more energy-intensive than nonmanufacturing, has seen a growth in GDP relative to total industrial GDP, with most of that change occurring since 1995. This factor has added about 6 percent to energy use, most of this effect occurring after the recession in the early 1990s. Manufacturing industries that are less energy-intensive have grown relative to those manufacturing industries that are highly energy-intensive, thus reducing the energy intensity of manufacturing as a whole.
Source: DOE/EERE, 2008.

The United States is considered a country with medium energy intensity country along with Finland, Sweden, and the Netherlands. High-energy-intensity countries include Norway, Australia, and Canada. At the same time, the United States has a less energy-intensive manufacturing sectoral structure relative to the other 12 IEA member countries, many of which are big producers of raw materials (e.g., Australia, Canada, the Netherlands, Norway, and Finland).[2] Correcting for this difference raises the U.S. energy-intensity index compared with that of other IEA coun-

[2]Taking into account the activity of multinational corporations headquartered in each country.

TABLE 4.3 "Business as Usual" Forecast of U.S. Industrial Energy Consumption (quadrillion Btu, or quads)

Industry	2006	2020	2030
Refining	3.94	6.07	7.27
Aluminum	0.39	0.36	0.33
Iron and steel	1.44	1.36	1.29
Cement	0.45	0.43	0.41
Bulk chemical	6.83	6.08	5.60
Paper	2.18	2.31	2.49
Total	32.6	34.3	35

Source: EIA, 2008a.

tries. While the analysis by Schipper is based on somewhat dated statistics (focusing on 1994), the panel's assessment is that its fundamental conclusion regarding the relative energy inefficiency of U.S. manufacturing remains valid.

The EIA's *Annual Energy Outlook 2007* forecasted that U.S. industrial energy consumption would increase from approximately 34.1 quads in 2006 to 35.8 quads in 2020 and 38.7 in 2030 (EIA, 2007). This baseline forecast assumed the continuation of current policies and some autonomous, or naturally occurring, efficiency improvement (see Section 4.2.1.4).

The EIA's *Annual Energy Outlook 2008* reduced the 2007 forecast's projected increase in U.S. industrial energy consumption substantially to reflect the nation's economic slowdown, rising energy prices, and the passage of the Energy Independence and Security Act of 2007 (P.L. 110-140) (EIA, 2008). With rising prices and more policy levers encouraging energy efficiency, greater energy efficiency improvement is anticipated to occur naturally as part of the 2008 baseline forecast. Specifically, the 2008 EIA estimate of U.S. industrial energy consumption for 2006 is 32.6 quads, 34.3 quads for 2020, and 35.0 quads for 2030 (Table 4.3). With a lower anticipated rate of growth in energy consumption, the potential for further cost-effective efficiency improvements must be recalibrated. This has been done by scaling the percentage savings for 2007 to the 2008 projections (see Section 4.2.1.1).

4.2 POTENTIAL FOR ENERGY SAVINGS

4.2.1 Review of Studies of Energy Efficiency Potential

Two major studies that have attempted to assess the potential for cost-effective energy efficiency improvements across the U.S. industrial sector—*Scenarios for a Clean Energy Future* (IWG, 2000) and *The Untapped Energy Efficiency Opportunity in the U.S. Industrial Sector* (McKinsey and Company, 2007)—are described below. In addition, many studies have examined the potential for energy efficiency in individual manufacturing industries such as aluminum, chemicals, and paper; others have focused on the potential impact of specific technologies (such as membranes or combined heat and power [CHP]) or families of technologies (e.g., sensors and controls, fabrication and materials). Such cross-sectional studies are the subject of Section 4.3 (focusing on major energy-consuming industries) and Section 4.4 (focusing on crosscutting technologies and processes). Because they do not treat the industrial sector comprehensively, these studies do not enable a sector-wide estimation of economic energy efficiency potential. However, they provide valuable benchmarking of the two comprehensive studies discussed below. In addition, there are state-level and international assessments of industrial energy efficiency potential, which are also drawn on below.

4.2.1.1 U.S. Industrial-Sector Assessments

In the DOE-sponsored study *Scenarios for a Clean Energy Future* (CEF), prepared by the Interlaboratory Working Group on Energy-Efficient and Clean Energy Technologies (IWG), a portfolio of advanced policies[3] was estimated to reduce energy consumption in the industrial sector by 16.6 percent relative to a business-as-usual (BAU) forecast, at no net cost to the economy (IWG, 2000; see also Brown et al., 2001, and Worrell and Price, 2001). The assumptions made in the study regarding cost-effectiveness are detailed in Box 4.1. The policies were assumed to be implemented in the year 2000; the 16.6 percent reduction was the difference between the BAU forecast for 2020 and the scenario trajectory

---

[3]The effects of many policies for reducing energy use and greenhouse gas emissions from industry are modeled in *Scenarios for a Clean Energy Future* (IWG, 2000). These include industry-wide agreements to reduce greenhouse gas emissions, the expanded deployment and marketing of ENERGY STAR® buildings, the rapid expansion of industrial energy assessment programs, and a carbon cap-and-trade system.

BOX 4.1 **Cost-Effectiveness of Industrial Energy Efficiency Investments**

Investment decisions can be characterized by the internal rate of return (IRR), also called the hurdle rate, used to trigger an expenditure. The IRR involves a discounted cash flow analysis that is based on a firm's cost of capital plus or minus a risk premium to reflect the project's particular risk profile. McKinsey and Company (2007, 2008) assumes that investments with an IRR greater than 10 percent are cost-effective. In their studies, each investment opportunity is treated individually; no integrated analysis is conducted to determine whether investments in one technology might impact the economics of other investment options.

The CEF study, *Scenarios for a Clean Energy Future* (IWG, 2000), does not use a single hurdle rate. Rather, it draws on a variety of best-in-class modeling approaches that employ economic metrics seen as appropriate to particular sectors and technologies.

For example, in the buildings sector, the business-as-usual hurdle rate is assumed to be about 15 percent (in real terms). In the advanced scenario, the potential impact of individual policies on energy demand was assessed in detailed spreadsheets using lower discount rates (typically about 7 percent), reflecting the influence of supporting policies that remove barriers to the adoption of energy-efficient technologies. The hurdle rates and other parameters inside the buildings-sector modules of the National Energy Modeling System (NEMS; the energy modeling system used by the U.S. Department of Energy's Energy Information Administration) were then changed so that the model replicates the energy savings calculated from the CEF spreadsheets (IWG, 2000).

In the industrial sector, the business-as-usual hurdle rate was generally assumed to be approximately 30 percent. In the advanced scenario, industrial subsectors were assessed using a hurdle rate of 15 percent to reflect the impact of the policy instruments that reduce transaction costs and financial risks. Combined heat and power (CHP) was modeled separately using Resource Dynamics Corporation's DISPERSE model because of limitations of the NEMS model (IWG, 2000).

As a final step, the NEMS integration model was used to assess the full range of effects of the economy-wide technology and policy scenarios. The integration step allows technology trade-offs and allows the effects of changes in energy use in each sector to be taken into account in the energy-use patterns of other sectors (IWG, 2000).

in 2020 as defined by advanced policies. The annual energy cost savings from the advanced scenario was estimated to exceed the sum of the annualized policy implementation costs and the incremental technology investments. (See Box 4.2 for further description of the CEF study.)

BOX 4.2 The *Scenarios for a Clean Energy Future* Study

The study *Scenarios for a Clean Energy Future* (CEF; IWG, 2000) was conducted by scientists at five U.S. Department of Energy (DOE) national laboratories with more than $1 million in funding from the DOE and the U.S. Environmental Protection Agency. Published in November 2000, it involved a comprehensive analysis of U.S. technology and policy opportunities, using a combination of engineering-economic analysis and a modified version of the DOE Energy Information Administration's National Energy Modeling System (CEF-NEMS). In the study the major sectors of the economy (buildings, industry, transportation, and electricity) were analyzed separately to identify the most cost-effective energy policy and technology alternatives for addressing multiple energy-related challenges facing the nation. Using CEF-NEMS, an integrated assessment of technology and policy options was produced. Seven supplemental studies are published in the CEF report's 600-page appendix (e.g., an assessment of combined heat and power opportunities). The appendix also contains details of the engineering-economic analysis so as to enable full public disclosure and replication by others. The report had extensive peer review, including that of a blue-ribbon advisory committee, and the results were the subject of a special issue of *Energy Policy* published in 2001 (see Brown et al., 2001).

Taken from the *Annual Energy Outlook 1999*, the BAU forecast used in the CEF study (IWG, 2000) estimated that the U.S. industrial sector would require 41.2 quads of energy in 2020. In contrast, the advanced portfolio of policies (defined earlier in the CEF study and assumed implemented by 2020) produced a scenario with industry requiring only 34.3 quads of energy (saving 6.9 quads of energy, a 16.6 percent reduction). The 2008 EIA projection (EIA, 2008) forecasts a BAU industrial-sector consumption of only 34.3 quads of energy in 2020. Scaling the 16.6 percent savings estimate to this lower level of future baseline industrial energy consumption suggests a savings of 5.7 quads, or a possible policy-induced reduction in industrial energy use to 28.4 quads.[4] These sector-wide

[4]When the panel applies older estimates of percentage improvement to newer (and lower) BAU estimates of energy to estimate the absolute energy savings, it is possible to create double-counting even if there was no double-counting in the original estimate of percentage improvement. That is, some of the energy efficiency improvements in the original estimate may have become a part of the BAU forecast (partially explaining the reduction in the BAU). The panel expects this problem to be negligible or nonexistent, because new energy efficiency opportunities

savings estimates do not account for the possible efficiencies available from CHP systems, because at the time of the *Annual Energy Outlook 1999*, the model used by the EIA—the National Energy Modeling System—was unable to model CHP technology in an integrated manner.

The *Scenarios for a Clean Energy Future* study commissioned an off-line analysis of the economic energy-savings potential of new CHP under "advanced" policies. This assessment concluded that CHP could reduce the energy requirements of the industrial sector by 2.4 quads in 2020 (IWG, 2000; Lemar, 2001). Scaling this estimate to reflect the downward forecast of future industrial energy consumption suggests an economic savings potential of 2 quads.[5] In combination with the sector's other energy efficiency opportunities identified in the CEF study, this brings the total estimate of economic energy-savings potential to 7.7 quads, or 22.4 percent of the *Annual Energy Outlook 2008* (EIA, 2008) forecasted consumption of 34.3 quads in 2020.

Building on the CEF study, on other assessments, and on original research, a more recent publication by McKinsey and Company (2007)[6] concurred that U.S. industries have a significant opportunity for energy efficiency gains (Figure 4.3). Financially attractive investments (defined as those with internal rates of return [IRRs] of 10 percent or greater) are estimated to offer 3.9 quads in energy-usage reduction in 2020, compared with the business-as-usual forecast based on the reference case of the *Annual Energy Outlook 2007* (EIA, 2007). These investments are estimated by McKinsey and Company (2007) to generate $30–$55 billion in increased earnings, before interest and taxes, by 2020; this earnings growth would, in turn, generate a $210–$385 billion increase in the market value of industrial companies. As shown in Figure 4.3, an additional 1.0 quad is identified by McKinsey and Company (2007) as "additional opportunities through driving R&D," bringing the estimated energy efficiency potential in the industrial sector to 4.9

---

arise each year as infrastructure and equipment age and as new and improved technologies are introduced into the marketplace.

[5]The EIA 1998 forecast of industrial energy consumption in 2020 was 41.2 quads, and the EIA 2008 forecast is 34.3. Multiplying 2.4 quads times the ratio of these two forecasts (0.83) results in the estimated 2.0 quad savings from the use of CHP.

[6]The McKinsey and Company study has been widely criticized for ignoring adoption and transaction costs and the potential impacts on product attributes. For example, it does not include the cost of policy or program implementation, as is done in great detail in the CEF study (IWG, 2000; see Appendix E-1 of that study).

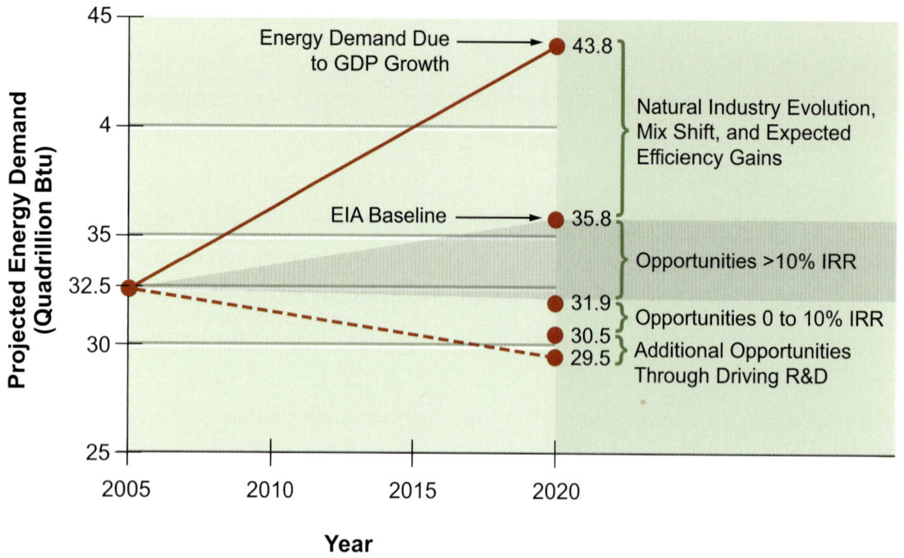

FIGURE 4.3 *Summary of industrial energy efficiency opportunities through 2020 identified by McKinsey and Company.*
Note: GDP = gross domestic product; IRR = internal rate of return; R&D = research and development.
Source: McKinsey and Company, 2007.

quads. Lower-returning projects with positive IRRs below 10 percent are also acknowledged by McKinsey and Company.

Contained within the 3.9 quads of energy efficiency potential are several crosscutting energy-saving opportunities totaling 1.5 quads. CHP represents 46 percent of this opportunity (or 0.7 quad) and is characterized by McKinsey and Company (2007, p. 3) as "the leading cross-segment opportunity." This estimate for CHP is considerably less than the 2.0 quad estimate from *Scenarios for a Clean Energy Future* (IWG, 2000). Because McKinsey and Company (2007) does not publish its background data, it is not possible to reconcile these two results. A recent National Research Council (NRC) study concluded that CHP economics are likely to improve in the near term through technology advancements and new niche applications for which CHP offers an economic advantage (NRC, 2007). Perhaps some of this future potential for CHP is included in the McKinsey and Company estimate of savings from new research and development (R&D) investments. There may also be differences in the more limited potential assigned to small (<5 MW) projects by McKinsey and Company (2007, p. 47) compared with the CEF study and others (Pace Energy Project, 2002).[7] An additional study by Bailey and Worrell (2005) provides estimates of the opportunity for "non-traditional" CHP technologies. It identifies 7.4 quads of potential savings relative to 2002 U.S. energy consumption. However, only 5 of the 19 technologies identified are related to CHP technologies (i.e., advanced cogeneration, steam-injected gas turbine, gas turbine process heater, gas turbine drying, and fuel cells), resulting in an estimated technical energy efficiency potential in industry of about 4.4 quads. This is comparable to the proposition examined by Shipley et al. (2008) that the United States could create a 20 percent generating capacity from CHP by the year 2030, which would lead to a fuel savings of 5.3 quads, or approximately half of the total energy currently consumed by U.S. households. The report also estimates that such an investment in CHP would create 1 million new green-collar jobs and $234 billion in new investments across the United States.

Table 4.4 summarizes the two studies' estimates of energy-savings potential in various industrial subsectors. It also shows estimates from other U.S. studies and global estimates from the IEA (see Section 4.2.1.2 below). The CEF study estimates a large potential for economic energy savings in pulp and paper manufacturing (6.3 percent), iron and steel (15.4 percent), and cement (19.1 percent) (IWG, 2000, Table 5.8; Worrell and Price, 2001). Applying savings at these percentages to the latest BAU forecast of energy consumption in 2020 (based on the EIA, 2008) results in savings estimates of 0.14, 0.21, and 0.08 quad, respectively.

On a segment-by-segment basis, McKinsey and Company (2007) concluded that the largest untapped opportunities for U.S. industrial energy efficiency savings reside in pulp and paper and in iron and steel. Because of the limited documentation underpinning these estimates, the panel treats them as qualitatively instructive. Relative to the McKinsey and Company study, the CEF study estimates for the iron and steel and cement industries are similar, but the estimate for the pulp and paper industry is significantly lower.

Table 4.5 summarizes the savings estimates in a different way, showing the overall range of savings identified for each energy-intensive industry, and for industry as a whole, and the baseline for the analysis.

---

[7]McKinsey and Company (2007, p. 47) postulates that "the economics of smaller facilities (<5 MW) are less attractive and offer diminished potential for additional energy savings beyond business as usual gains. Other issues with smaller CHP projects include: a) higher operating costs and lower heat rates; and b) high fixed costs (e.g., engineering, design, legal)."

TABLE 4.4 Economic Potential for Energy Efficiency Improvements in Industry in the Year 2020: Sector-wide and by Selected Subsectors and Technologies

	Estimates for U.S. Industry			
	CEF Study (IWG, 2000) Scaled to *AEO 2008* (quads)	McKinsey and Company (2008) (quads)	Other U.S. Studies (quads)	Global Estimates from IEA (2007) (%)
Petroleum refining	n.a.	0.3	0.61–1.21 to 1.40–3.28[a]	13–16
Pulp and paper	0.14[b]	0.6	0.37 to 0.85[c]	15–18
Iron and steel	0.21[d]	0.3	0.79[e]	9–18
Cement	0.08[f]	0.1	0.29[g]	28–33
Chemical manufacturing	n.a.	0.3	0.19[h] to 1.1[i]	13–16
Combined heat and power	2.0	0.7	4.4–6.8[j]	
Total, industrial sector	7.7 (22.4%)	4.9 (14.3%)		18–26

Note: This table appeared in Lave (2009) before this report was completed. The data in Table 4.4 have been updated since the Lave (2009) article was published. CEF study, *Scenarios for a Clean Energy Future* (IWG, 2000); *AEO 2008, Annual Energy Outlook 2008, with Projections to 2030* (EIA, 2008); n.a., not available.

[a]Based on a range of 10–20 percent savings (LBNL, 2005) to 23–54 percent savings (DOE, 2006b) from a baseline forecast of 6.08 quads.

[b]6.1 percent of the 2.31 quads of energy consumption forecast for the paper industry in 2020 by the *Annual Energy Outlook 2008* (EIA, 2008).

[c]Based on 16 percent savings (Martin et al., 2000a) and 37 percent savings (DOE, 2006c) from the baseline forecast of 2.31 quads.

[d]15.4 percent of the 1.36 quads of energy consumption forecast for the iron and steel industry in 2020 by the *Annual Energy Outlook 2008* (EIA, 2008).

[e]Based on 58 percent savings (AISI, 2005) from the baseline forecast of 1.36 quads.

[f]19.1 percent of the 0.43 quads of energy consumption forecast for the cement industry in 2020 by the *Annual Energy Outlook 2008* (EIA, 2008).

[g]Based on 67 percent savings (Worrell and Galitsky, 2004) from the baseline forecast of 0.43 quads.

[h]National Renewable Energy Laboratory, 2002.

[i]DOE, 2007.

[j]Bailey and Worrell, 2005.

4.2.1.2 International Assessments

The Intergovernmental Panel on Climate Change (IPCC) came to conclusions similar to those of the CEF (IWG, 2000) and McKinsey and Company (2007) studies regarding the industries with the largest carbon-mitigation potentials worldwide. Specifically, the IPCC identified the steel, cement, and pulp and paper industries as having the largest potential for energy savings (IPCC, 2007).

TABLE 4.5 Summary of Estimated Cost-Effective Energy Savings in Industry Resulting from Improved Energy Efficiency (quads)

	Energy Use in Industry			
		Business as Usual (BAU) Projection (AEO 2008 reference case)		
Industry	2007	2020	2030	Savings over BAU in 2020[a,b]
Petroleum refining	4.39	6.07	7.27	0.3–3.28
Iron and steel	1.38	1.36	1.29	0.21–0.76
Cement	0.44	0.43	0.41	0.29
Chemical manufacturing	6.85	6.08	5.60	0.19–1.1
Pulp and paper	2.15	2.31	2.49	0.14–0.85
Total savings—all industries (including those not shown)				4.9–7.7[c] 14–22%

[a]Based on Table 4.4, which provides results from a review of studies for specific energy-using industries and for industry as a whole, and for industry-wide combined heat and power (CHP).

[b]Savings shown are for cost-effective technologies, defined as those providing an internal rate of return of at least 10 percent or exceeding a firm's cost of capital by a risk premium.

[c]Includes CHP systems, which contribute an estimated savings in 2020 of 0.7–6.8 quads.

Tracking Industrial Energy Efficiency and CO_2 Emissions (IEA, 2007), which estimates energy and carbon savings from the adoption of best-practice commercial technologies in manufacturing industries, suggests an overall level of energy-savings potential of 18–26 percent globally, with large-percentage savings from petroleum refining, pulp and paper, iron and steel, cement, and chemical manufacturing (see Table 4.4). It concluded that, on the basis of physically produced industrial output, Japan and Korea have the highest levels of manufacturing industry energy efficiency, followed by Europe and North America—levels that reflect differences in "natural resource endowments, national circumstances, energy prices, average age of plant, and energy and environmental policy measures" (IEA, 2007, p. 20).

Since the IEA's estimated energy savings are global percentages, their applicability to the U.S. context is not exact. In particular, care is needed to avoid unrealistic assessments of the savings potential in older industrial plants as compared with new, state-of-the-art facilities. International comparisons, however, underscore the potential for efficiency upgrades by U.S. industry.

4.2.1.3 State Assessments

At least two states—New York and California—have conducted assessments of the economic potential for energy efficiency improvements in the industrial sector. These studies help to set parameters for estimation of economic energy efficiency potential at the national scale.

KEMA, Inc. (2006) provides an assessment of the electric and gas energy efficiency potential in existing industrial facilities in four California utility territories, focusing on the year 2016. With a base use of 32,800 GWh forecast for 2016, the study estimates that 4970 GWh of electricity use (i.e., 15.1 percent) could be eliminated by economic efficiency investments—that is, investments that are cost-competitive with supply-side options. For natural gas, 468 million therms of natural gas are forecast to be the magnitude of economic efficiency opportunity in the industrial sector, representing 13 percent of the base use of 3590 million therms in 2016. Figure 4.4 presents the two supply curves, which identify the least-expensive efficiency measures. The least-cost options are arrayed on the left side of the curve in ascending order based on levelized energy costs. The width of each line is proportional to the amount of energy that can be saved. KEMA (2006) concludes that pumping has the largest electric end-use savings potential, followed by compressed air and lighting. Similarly, boilers represent the largest source of natural gas savings potential, followed by process heating.

A similar potential for energy efficiency improvement is described in a 2003 assessment for New York State. According to Optimal Energy, Inc. (2003), the New York Energy Research and Development Authority (NYSERDA) forecasts that the industrial sector will require 33,100 GWh of electricity in the year 2022. Optimal Energy estimates that 5,000 GWh (15 percent) of this base use could be displaced by economic electricity-efficiency measures. The assessment did not evaluate natural gas or other energy-savings opportunities.

A combined heat and power market-potential study conducted by the NYSERDA identified over 5000 MW of installed CHP capacity at more than 210 sites in New York State. Close to 80 percent of this capacity is at industrial sites, represented by a few facilities that have large CHP systems (Pace Energy Project, 2002).

The New York study identified numerous commercial and emerging technologies that can be used for CHP—including the internal combustion engine, steam turbine, gas turbine, micro-turbine, and fuel cell—which constitute nearly 8500 MW of technical potential for new CHP at 26,000 sites. Near-term market-

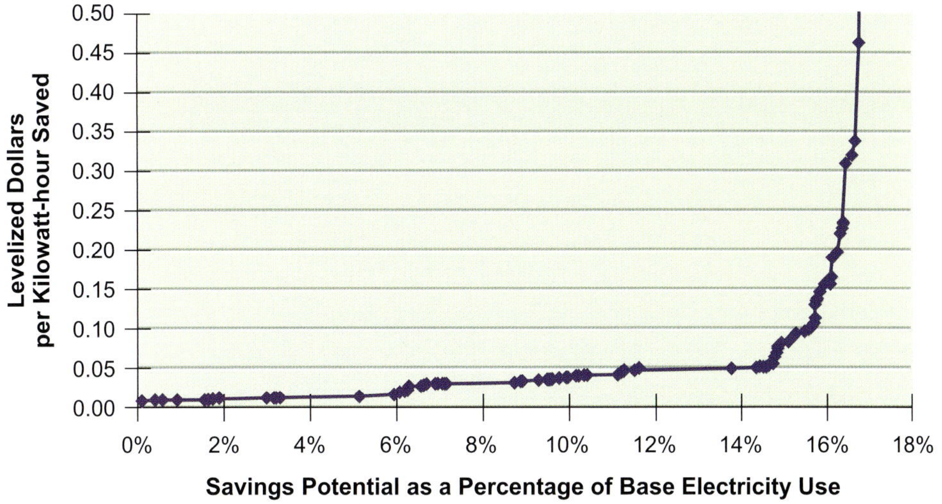

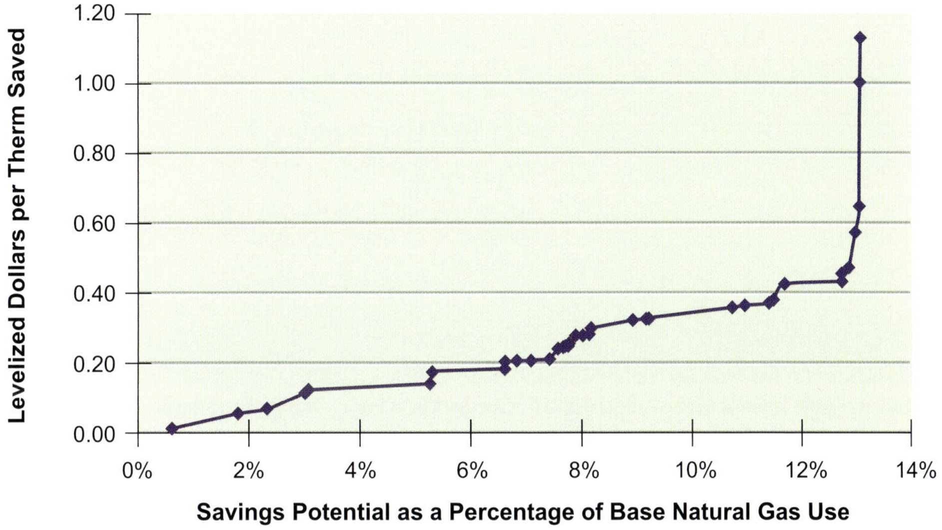

FIGURE 4.4 *Energy efficiency supply curves for California through 2016. The width of each portion of the curves is proportional to the amount of energy that can be saved. The fact that these two curves reach a maximum as the cost of efficiency options rises may simply reflect the limited set of technologies considered that have a marginal return on investment.*
Source: KEMA, 2006.

penetration forecasts range from 764 MW to nearly 2200 MW over the coming decade. Close to 74 percent of remaining capacity is below 5 MW and is located primarily at commercial and institutional facilities. Achieving this remaining potential depends on the degree to which many of the obstacles outline in Section 4.5 can be overcome.

4.2.1.4 Naturally Occurring Efficiency Improvement

The McKinsey and Company (2007) analysis assumes a significant amount of energy efficiency improvement in the BAU forecast, based on EIA modeling (see in Figure 4.3 the difference between the EIA baseline and the energy demand attributable to GDP growth). The naturally occurring improvement results from capital stock turnover of outdated technologies, as well as from cost reductions and performance improvements achieved from economies of scale and advances in science and technology. Thus, the level of energy efficiency improvement anticipated in the year 2020 relative to today could be large. For example, DOE (2004) identified 5.2 quads of cost-effective energy-savings opportunities from a range of end-uses in industrial energy systems, including steam generation, fired heaters, on-site power generation, motor systems, and facility heating, ventilation, and air-conditioning (HVAC), and lighting systems. More than 35 percent of this total opportunity (1.8 quads) was identified in waste-heat recovery, such as from gases and liquids in chemicals, petroleum, and forest products, including hot gas cleanup and the dehydration of liquid-waste streams. The second largest opportunity (1.4 quads) was identified in best practices in energy management and integration. These are the kinds of potential cost savings that EIA assumes will be absorbed in the BAU case. Relative to today's energy efficiency practices, industrial energy efficiency improvements in 2020 could save considerably more energy than the 3.9 quads estimated by McKinsey and Company (2007), if the "naturally occurring" efficiency improvements are taken into account.

Looking to the midterm (2020–2035), a wide array of advanced industrial technologies could make significant contributions to reducing industrial energy consumption and CO_2 emissions. Possible revolutionary changes include novel heat and power sources and systems and innovative concepts for new products and processes that take advantage of developments in nanotechnology and micromanufacturing. Examples include the microwave processing of materials and nanoceramic coatings, which show great potential for boosting the efficiency of industrial processes.[8] In addition, advances in recycling (resource recovery and utilization)—for example, of aluminum—could reduce the energy intensities of U.S. industry. Many of these approaches provide other benefits as well, such as improved productivity and reduced waste streams.

4.2.2 The Role of Innovation

Most of the current dialogue focuses on new technology that lowers industry's energy use. In some cases, more important energy savings come from adapting the new technology for use in other sectors. For example, developing a new generation of fuel cells may lead to greater savings in motor vehicles. Other possibilities include "on-demand" manufacturing that applies ink-jet printing systems to three-dimensional fabrication, or new plastics that double as integrated photovoltaic systems (Laitner and Brown, 2005). This role of industry in the development of emerging technologies highlights even greater energy savings than might be apparent from looking at industry's own energy-use patterns alone. With the growing focus on corporate sustainability, industry is adopting a much broader view of its energy and environmental responsibilities, extending its concern to issues surrounding the sustainability of the products and services that it offers, and including the sustainability of its chain of suppliers. Wal-Mart, for example, has included indicators of energy sustainability in metrics used to select product and service providers.[9] Accordingly, contractors that create minimal environmental impact are preferred.

4.3 OPPORTUNITIES FOR ENERGY EFFICIENCY IMPROVEMENTS IN FOUR MAJOR ENERGY-CONSUMING INDUSTRIES

In the chemical and petroleum refining, pulp and paper, iron and steel, and cement industries, numerous opportunities exist for energy efficiency improvements. These opportunities are characterized below in three timeframes: 5–10 years, 10–25 years, and beyond 25 years. For each industry, the size of the economic energy-savings opportunity is described, along with associated costs, performance

---

[8]See http://cleantech.com/news/3476/ceramic-nanotechnology-delivers-efficiency.
[9]Jim Stanway, Wal-Mart, personal communication, 2007.

improvements, and environmental impacts for each.[10] In addition, the discussion characterizes the opportunity for expanding best practices through reinvigorated deployment programs or by removing governmental interventions that might inhibit private-sector funding. Finally, for each of the four major energy-consuming industries, promising R&D is identified that is likely to be required for these technologies to be ready for launch into the marketplace. In each case, government partnerships that might help prepare the technologies for widespread commercialization are noted.

4.3.1 Chemical Manufacturing and Petroleum Refining

The chemical industry manufactures an extensive array of organic and inorganic chemicals and materials. Raw materials include hydrocarbons from petroleum refining, mined chemicals and minerals, and even such animal and plant products as fats, seed oils, sugars, and timber. For energy sources the industry uses petroleum-based feedstocks, natural gas, coal, and electricity—and, to a lesser but growing extent, biomass.[11] Products include thousands of bulk and fine organic and inorganic chemicals, polymers, agricultural chemicals, and fertilizers. Production levels are often in million- and billion-pound quantities but do extend to such high-value, low-volume products as pharmaceutical intermediates, specialty adhesives, and even perfume ingredients. Most large chemical companies are research intensive because of the continual need to generate new and improved products, to improve quality and yields, and to conform to environmental regulations.

Companies are often concentrated near petroleum refineries, around shipping ports, or in places where cheap hydroelectric power is available. Energy costs are almost always a major part of total costs, so the need for energy efficiency is great. For some energy-intensive products, energy for fuel and power needs and feedstocks account for up to 85 percent of total production costs. Reflecting higher fuel costs during 2007, the industry spent $73 billion on purchases of fuel and power and energy feedstocks. Overall, energy costs (including feedstock costs) represent 20 percent of production costs and 10 percent of the value of industry shipments (American Chemistry Council, 2008; U.S. Census Bureau, 2007).

The petroleum industry is similar to the chemical industry in its use of

---

[10]A major issue for each industry is the input-output of material flows, including the consumption of conventional fuels as feedstocks (i.e., "nonfuel" uses). It is beyond the scope of this analysis to replicate or characterize such inputs and outputs of alternative production processes.

[11]See http://www.chemicalvision2020.org/alt_feedstocks.html.

TABLE 4.6 Petroleum and Chemical Industry Energy Use, Selected Years from 1985 to 2002 (quads)

Year	Fuels	Purchased Electricity	Net Energy for Heat and Power	Feedstocks[a]	Total Net Energy Use	Electricity Losses[b]	Total Primary Energy
Petroleum Refining Energy Use (SIC 2911, NAICS 324110)							
1985	2.46	0.11	2.57	2.45	5.02	0.23	5.25
1988	2.95	0.10	2.90	3.26	6.31	0.21	6.52
1991	2.79	0.10	2.89	2.87	5.76	0.21	5.97
1994	3.87	0.11	3.98	2.39	6.26	0.24	6.50
1998	3.48	0.12	3.48	3.75	7.13	0.21	7.34
2002	3.09	0.12	3.09	3.31	6.39	0.12	6.51
Chemical Industry Energy Use (SIC 28, NAICS 325)							
1985	1.78	0.43		1.35	3.57	0.90	4.46
1988	2.27	0.42		1.68	4.36	0.86	5.22
1991	2.25	0.44		2.36	5.05	0.91	5.97
1994	2.35	0.52		2.46	5.33	1.08	6.41
1998	3.70	0.58		2.77	6.06	0.99	7.05
2002	3.77	0.52		3.75	6.47	1.58	8.04

Note: SIC, Standard Industrial Classification; NIACS, North American Industry Classification System.

[a]Petroleum feedstock used to produce nonenergy products only (e.g., petrochemicals, lubricating oils, asphalt).

[b]Electricity losses incurred during the generation, transmission, and distribution of electricity are based on a conversion factor of 10,500 Btu/kWh.

Source: Based on data in select DOE reports, 1988–2005.

energy sources and process equipment, but it normally produces a limited range of refined hydrocarbon products in high volume for the transportation industry. Many refining companies have a bulk-chemical arm to manufacture a limited spectrum of high-volume organic chemicals and bulk-polymer intermediates that are natural extensions of their refining operations. Petroleum companies vary in research intensiveness, but they are generally less dependent on finding new products and processes than the chemical industry is.

For these industries, energy efficiency and product yield are generally key to profitability and emissions abatement. Table 4.6 shows U.S. energy consumption for these industries from 1985 through 2002 (EIA, 2002). As can be seen, energy use from year to year was somewhat erratic, but it generally increased. These changes reflect varying industrial production levels, changing product mixes, and efficiency gains. For example, ExxonMobil achieved a 35 percent reduction in the energy intensity of its global refining and chemical operations from 1974 to 1999 and has identified a further 10–15 percent cost-effective energy-savings opportunity in all plants around the world (Expert Group on Energy Efficiency, 2007).

Benchmarking data indicate that most petroleum refineries can economically improve energy efficiency by 10–20 percent (LBNL, 2005), and analysis of individual refining processes indicate energy savings ranging from 23–54 percent (DOE, 2006b). Common technologies include high-temperature reactors, distillation columns for liquid-mixture separation, gas-separation technologies, corrosion-resistant metal- and ceramic-lined reactors, sophisticated process-control hardware and software, pumps of all types and sizes, steam generation, and many others. In the DOE (2006b) petroleum bandwidth study, the largest potential bandwidth savings are found in crude distillation, with savings of up to 54 percent of current average energy for atmospheric distillation (39 percent for vacuum distillation). Alkylation follows closely, with a potential bandwidth savings of 38 percent, and the remaining processes also exhibit significant potential for improving energy efficiencies. According to experts working in the field of petroleum refining and energy management, identifying plantwide energy savings of approximately 30 percent would be typical. However, these savings estimates are calculated on a relative basis. The absolute energy consumption of petroleum refineries in the United States must be adjusted to account for increasingly heavy crude slates over the coming years. When one adjusts for the use of heavier crude slates, the energy consumption of a refinery increases per equivalent amount of refined product.

Numerous reports and studies are available that describe near-, intermediate-, and potentially longer-term technologies to increase energy efficiency (and decrease related carbon emissions) (Expert Group on Energy Efficiency, 2007; DOE, 2006b) for both the chemical and the petroleum-refining industries. Three recent studies provide a wide range of efficiency-potential estimates. On the low side is the estimate that 0.014 quad could be saved by five technologies included in the DOE (2006a) *Chemical Bandwidth Study*. These technologies are applicable to the production of ethylene, chlorine, ethylene oxide, ammonia, and terephthalic acid. On the high side is the estimate that 1.1 quads of potential energy could be saved (DOE, 2007a). This assessment is based on 16 DOE Industrial Technologies Program (ITP) portfolio technologies (0.58 quad of savings) and five additional R&D technologies from the *Chemical Bandwidth Study* (0.52 quad of savings). Clearly, the magnitude of energy efficiency improvement will tend to expand or contract depending on the number of technologies that are considered. Interestingly, between these two extremes is an assessment based on a single type of technology and an estimated energy efficiency potential of 0.19 quad in the chemical industry today (DOE, 2002). This assessment is based on 12.4 percent of fuel-savings potential from steam system improvements.

As discussed in Box 4.3, in the chemical industry (as well as in petroleum refining), gaining even the so-called low-hanging fruit for increased energy efficiency faces significant obstacles. However, as is also pointed out, good management practices supported by the top levels of management aim to accomplish the savings over a reasonable time period.

One useful perspective on these risks can be seen in the NRC (2007) report *Prospective Evaluation of Applied Energy Research and Development at DOE*.[12] Based on an examination of 22 high-payoff projects, the NRC panel found that great potential existed for energy and carbon-emissions reductions, but technical and market risks were generally quite high. From an individual company's viewpoint, the decision to pursue any one of these technology developments could be too risky. This risk is often the reason that DOE partners with individual firms or groups of companies for technology development and demonstration.

While both the chemical and petroleum-refining industries are capable of prolonged and expensive R&D efforts for their own process improvements, advances in crosscutting technologies (such as process-control hardware and software, separation processes and equipment, and heat-management equipment) are often best accomplished by, or in collaboration with, vendors. While chemical and petroleum-refining companies typically develop process designs and specify the desired performance of technologies, they then purchase these technologies, thereby saving their R&D organizations for new-product development and specific process innovations.

In summary, the chemical and petroleum-refining industries have many similarities in raw materials, energy sources, process equipment and control, and the opportunity to achieve significant energy efficiency improvements. They differ in key ways centered around the breadth of product lines and the areas in which innovation will gain them a competitive advantage. Both purchase much of their process equipment and controls from specialty providers, which themselves carry out R&D to improve their offerings. Factors that can impede the use of technol-

---

[12]See, in NRC (2007), the subsection, "Report of the Panel on DOE's Chemical Industrial Technologies Program."

BOX 4.3 Barriers to Plucking Low-Hanging Energy Efficiency Opportunities

Many reports and studies have been written about the tremendous energy-savings opportunities that exist in U.S. energy-intensive industries (as outlined in Section 4.3 in this chapter). A large portion of these savings are often described as ready for the taking with little or no technical risk. As discussed throughout Section 4.3, these claims are true for the most part. So why doesn't a given company move quickly to make the necessary investments, which often pay themselves back in a year or less? Why are these "sure thing" projects often spread out over years or even sometimes ignored? The answer is simple: competition for capital within a corporation and, in some situations, a lack of time and/or personnel to install the improvements.

The top management of a company is responsible for the health of the corporation. A critical component of this duty is the allocation of limited capital among marketing, sales, manufacturing, research and development, and other functions. Capital allocations are further split into new plant construction, plant improvements and maintenance, office building expansions, and so on. There is tremendous demand at all times for capital in a successful company. This allocation process, as with all the other allocations, is largely decided on the basis of business need. At any given time, more product may be more important than lower energy consumption. Safety and compliance are always the number one priority. So in a given year, there may not be money for energy efficiency.

But even if there is capital for energy efficiency improvements, other constraints exist. Most manufacturing plants have annual shutdowns for maintenance and other alterations. These are carefully planned, with all activities to be done in the shortest

ogy for energy efficiency improvements include the availability of the necessary capital, which must compete with other corporate needs, and, for specific innovations, the costs and risks associated with the marketplace. Generally speaking, both industries are endowed with strong technology and engineering organizations and are aware of the status of the technologies that they need for future improvements. Both, however, are careful in allocating R&D funds. As discussed above, the criteria for such expenditures are strongly influenced by payback times, potential gains in competitive advantage, and the projected timescales and technical and marketing risks for a given innovation.

As a result of these factors, typical industry practice regarding energy-intensive facilities such as large-scale distillation columns is to maintain and use them for as long as possible, primarily because of the large capital investment that time possible. The object is to get up and running so that no product shortages will be created. This shutdown period is especially constraining in times of high product demand. If an energy efficiency improvement cannot be fit into the shutdown period, either because of a long implementation period or possibly a lack of available personnel, it may be passed over for the time being.

The discussion above in no way implies that energy efficiency, with its positive environmental impacts, always gets short shrift—quite the contrary. In energy-intensive industries such as chemical manufacturing, energy efficiency is often where the "big money" is. Well-run companies are usually aware of this and have implemented numerous "best practices" to ensure that they gain these savings as rapidly as possible.

The Dow Chemical Company management, for example, strives to commit its entire organization to energy efficiency in manufacturing (Fred Moore, Dow Chemical Company, December 2007). Top management sets aggressive, 10-year energy efficiency goals for each process. This adds up to a publicly stated energy efficiency goal for the company. To ensure that these goals are met, the company has put into place an energy management organization that is distributed throughout the business units, with reporting lines to the top of the company. Each business unit must have specific 10-year plans that show how it intends to accomplish its part of the company goal and the schedule for doing so. These individual plans consist of three 10-year subplans covering what will be done, what the unit would like to do, and what the unit needs in terms of innovation. Frequent reporting on progress in all three plans is required. Salary, bonus, and career progression are all linked to the goal. From 1994 to 2005, Dow's programs achieved a 22 percent reduction in energy intensity. The company's goal for the next 10 years is an additional 25 percent.

they represent. The consequence is that the motivation to replace them with more efficient equipment is often very low.

Three studies estimate the potential for energy savings in the chemical manufacturing industry. The highest estimate is 1.1 quads (18 percent) in 2020 (DOE, 2007a). The lowest, 0.19 quad (3 percent), comes from NREL (2002). The McKinsey and Company (2008) estimate falls within this range.

Three studies also provide estimates of energy-saving potential in the petroleum industry. The highest estimate is a range of 1.40 to 3.28 quads in 2020 (23 to 54 percent of projected energy consumption in this industry) published in a DOE (2006b) report. The lowest estimate, 0.3 quad in 2020 (5 percent), comes from McKinsey and Company (2008). An LBNL (2005) study provides an intermediate range of 0.61 to 1.21 quads saved in 2020 (10 to 20 percent).

4.3.2 Pulp and Paper Industry

Pulp and paper production, which constitutes a majority of the forest products industry, consumes about 2.4 quads of energy annually (Table 4.7). Drying and the recovery of chemicals are the most energy-intensive parts of the papermaking process. The pulp and paper sector of the forest products industry is both capital- and energy-intensive. Energy is the third-largest manufacturing cost for the forest and paper products industry (AFPA, 2007). According to the Energy Information Administration (EIA, 2004), the forest products industry consumed 3.3 quads of energy in 2004, placing it third after the chemical and petroleum-refining industries in terms of energy use. Paper and paperboard mills consume the most energy in the pulp and paper sector, and more than half of the energy source is derived from net steam (the sum of purchases, generation from renewables, net transfers, and other energy used to produce heat and power or as feedstock or raw material inputs) (EIA, 2002). Steam is needed mainly for paper drying, but it is also used for pulp digesting and other uses. In papermaking, drying is the largest energy consumer, requiring large amounts of steam and fuel for water evaporation (DOE, 2005a). Electricity is required in increasing quantities to run equipment such as pumps and fans and to light and cool buildings, among other uses.

TABLE 4.7 First Use of Energy for All Purposes in the Pulp and Paper Industry (Fuel and Nonfuel), in Primary Energy, 2002 (trillion Btu)

	Net Electricity[a]	Residual and Distillate Fuel Oil	Natural Gas	LPG and NGL	Coal, Coke and Breeze	Other[b]	Total
Total: Pulp and Paper Industry	223	113	504	6	240	1276	2363
Paper mills, except newsprint	78	51	206	1	143	523	1002
Paperboard mills	56	38	188	*	84	542	908
Pulp mills	5	w	24	*	w	175	224
Newsprint mills	38	w	16	*	w	27	94

Note: LPG = liquefied petroleum gas; NGL = natural gas liquid; "w" = data withheld to avoid disclosing data for individual establishments; * = estimates lower than 0.5 trillion Btu.

[a]"Net electricity" is defined as the sum of purchases, transfers in, and generation from noncombustible renewable resources, minus quantities sold and transferred out. It excludes electricity inputs from on-site cogeneration or generation from combustible fuels since that energy is counted under generating fuel such as coal.

[b]"Other" is defined as net steam (the sum of purchases, generation from renewables, and net transfers), and other energy used to produce heat and power or as feedstock/raw material inputs.

Source: EIA, 2002, Data Table 1.2.

Several energy-efficient methods of drying have been developed, many of which are cost-effective today. One of these, a systems approach, involves using waste heat from heat-generating processes, including from power generation and ethanol production, as the energy source for evaporation (Thorp and Murdoch-Thorp, 2008). These opportunities to recycle waste heat are only practical if the power production does not use condensing turbines (that is, if it is relatively inefficient), or if the ethanol distillation is conducted at relatively high temperature and pressures. Advanced water-removal technologies can also reduce energy use in drying and concentration processes substantially (DOE, 2005b). The Oak Ridge National Laboratory (ORNL) and BCA, Inc. (2005) estimated that membrane and advanced filtration methods could significantly reduce the total energy consumption of the pulp and paper industry. High-efficiency pulping technology that redirects green liquor to pretreat pulp and reduce lime kiln load and digester energy intensity is another energy-saving method for this industry (DOE, 2005b). Modern lime kilns are available with external dryer systems and modern internals, product coolers, and electrostatic precipitators (DOE, 2006c).[13]

In most kraft mills today, the black liquor produced from delignifying wood chips is burned in a large recovery boiler. Because of the high water content of the black liquor, its combustion is inefficient, and the possibility of electricity production from secondary steam production is limited by the steam's low pressures. The gasification of black liquor not only allows efficient combustion but also enables the use of a gas turbine or a combined-cycle process with high electrical efficiency, thereby offering the potential for increasing the production of electricity within pulp mills. The surplus of energy from the pulp process also allows for the possible production of useful heat, fuels, and chemicals (that is, the operation of "biorefineries") (Worrell et al., 2004).

The *Pulp and Paper Industry Energy Bandwidth Study* concluded that applying current design practices for the most modern mills can reduce the energy consumption of the pulp and paper industry by 25.9 percent and that the implementation of advanced technologies could reduce mill energy consumption by even more (41 percent) (DOE, 2006c). Of course, it is unrealistic to assume that long-existing facilities can be easily upgraded to new, state-of-the-art facilities. The

---

[13]Electrostatic precipitators (ESPs) are more energy-efficient than wet scrubbers are, because energy in ESPs is applied only to the particulate matter that is being collected, but in wet scrubbers, energy is applied directly to the fluid medium, thus consuming more energy (ANL, 1990).

BOX 4.4 **Reducing Energy Consumption in the U.S. Pulp and Paper Industry**

McKinsey and Company (2007) identified the following opportunities to reduce by 2020 the energy consumed in the U.S. pulp and paper industry:

	100 percent =	0.60 quadrillion Btu (quads)
Papermaking	30 percent	(0.18)
Multiprocess improvements	17 percent	(0.10)
Steam efficiencies	17 percent	(0.10)
Fiber substitution	14 percent	(0.08)
Pulping	12 percent	(0.07)
Other process steps	11 percent	(0.07)

largest potential energy savings in the industry are estimated to be in paper drying, liquor evaporation, and lime kilns.

Similarly, the McKinsey and Company (2007) study for the DOE Industrial Technology Program indicates that the pulp and paper industry can reduce energy consumption by 25 percent (0.6 quad) by 2020 by accelerating the adoption of proven technologies and process improvements. As shown in Box 4.4, a majority of the savings is expected to come from papermaking, multiprocess improvements, steam efficiencies, and fiber substitution.

Martin et al. (2000a) studied the opportunities to improve energy efficiency in the U.S. pulp and paper industry. Their case study results indicate that the technical potential for primary energy savings amounts to 31 percent, without accounting for an increase in recycling. The cost-effective savings potential is 16 percent. When recycling is included, the technical potential increases to 37 percent and the cost-effective savings potential remains the same.

In sum, the estimates of cost-effective energy efficiency potential in 2020 range from a low of 6.1 percent from the CEF study (IWG, 2000) to a high of 37 percent (DOE, 2006c). This range includes the 16 percent estimate produced by Martin et al. (2000a) and the 26 percent estimate produced by McKinsey and Company (2007). Applying these savings estimates to the pulp and paper industry's current consumption of approximately 2.31 quads annually results in a range of energy savings of 0.14 to 0.85 quad by the year 2020. Additional savings are possible from the use of combined heat and power technologies.

4.3.3 Iron and Steel Industry

Iron and steel manufacturing, the fourth-largest user of energy in the industrial sector, consumes 1.4–1.9 quads per year (RECS, 2004; EIA, 2007). The energy-use breakdown by fuel is as follows: natural gas, 28.7 percent; petroleum, 7.6 percent; coal, 49.3 percent; renewables, 0.5 percent; and purchased electricity, 13.9 percent. Direct-energy costs represent 5–15 percent of the total cost of making steel, with additional energy costs embedded in expenditures for raw materials. The role of the U.S. steel industry in world markets has been eroding over the past three decades with manufacturing moving offshore, particularly to Asia. Table 4.8 indicates that between 1997 and 2006, imports of iron and steel products increased from 41 million tons per year to 65 million tons per year. During the same time period, exports of iron and steel products increased from 7.8 million tons per year to 12.7 million tons per year.

Between 2002 and 2006, U.S. production of raw steel increased from 101 million to 109 million tons per year, while China's annual production increased from 201 million to 462 million tons (WSA, 2008). Over the 10-year period from 1996 to 2005, steel production declined in the United States at an average annual rate of 0.1 percent, while growing at an annual rate of 1.2 percent in Japan and 12.1 percent in China.

U.S. industry consumes approximately 120 million tons of metallics to produce 100 million tons of steel. There are two basic methods for producing crude steel: the blast furnace and the basic oxygen furnace (BOF), which use mainly iron ore; and the electric arc furnace (EAF), which uses mainly reduced iron and pig iron. In 2006, integrated steelmakers produced roughly 43 percent of raw steel, while EAF operations produced the remaining 57 percent. For a detailed discussion of these processes, see recent reports by the IEA (2007) and Worrell and Neelis (2006).

Energy intensities for the two steel production methods vary substantially, reflecting the fact that the BOF produces new steel, whereas the EAF uses recycled steel. In 2003, BOFs required 19.55 million Btu/ton while EAFs required 5.26 million Btu/ton. In 2002, the same uses required 21.23 million and 5.23 million Btu/ton, respectively. The calculated minimum energy requirement for ore-based steelmaking is 8.5 million Btu/ton (Fruehan et al., 2000). In 2006, yield losses totaled 8 million tons. The losses occur in many different operations and appear as "home" scrap and waste oxides; integrated producers also lose a small percentage of coal and coke.

TABLE 4.8 Change in Imports and Exports of Iron and Steel Products Between 1997 and 2006

Products	1997 (thousand tons)	2006 (thousand tons)	Percent Change
Imports			
Steel mill products			
Ingots, blooms, billets, slabs, etc.	6,358	9,317	46.5
Wire rods	2,237	3,046	36.2
Structural shapes and pilings	1,141	1,146	0.5
Plates	2,939	3,416	16.2
Rails and accessories	238	352	47.9
Bars and tool steel	2,627	5,111	94.5
Pipe and tubing	3,030	7,545	149.0
Wire drawn	655	903	38.0
Tin mill products	638	749	17.4
Sheets and strips	11,294	13,686	21.2
Total steel mill products	31,157	45,273	45.3
Other steel products	3,233	6,941	114.7
Total steel products	34,389	52,214	51.8
Iron products and ferroalloys	6,659	13,110	96.9
Grand Total, Imports	41,048	65,324	59.1
Exports			
Steel mill products			
Ingots, blooms, billets, slabs, etc.	210	219	4.3
Wire rods	85	1,501	77.3
Structural shapes and pilings	481	892	85.7
Plates	780	1,806	131.7
Rails and accessories	92	164	77.6
Bars and tool steel	835	1,104	32.2
Pipe and tubing	1,352	1,489	10.1
Wire drawn	137	182	33.4
Tin mill products	410	240	-41.3
Sheets and strips	1,654	3,480	110.4
Total steel mill products	6,036	9,728	61.2
Other steel products	**1,333**	**1,702**	**27.7**
Total Steel Products	7,369	11,430	55.1
Iron products and ferroalloys	458	1,260	175.1
Grand Total, Exports	7,827	12,689	62.1

Source: Adapted from AISI, 2006.

Yield losses reduce the overall energy efficiency of steelmaking. The industry consumes about 18.1 million Btu/ton of product, including electricity generation and transmission and distribution losses—22 percent more than the practical minimum energy consumption of about 14 million Btu/ton. These energy losses, about 4 million Btu/ton, are a result of process efficiencies and the production energy embedded in the yield losses. The BOF process itself is not a major energy user. It is the inherent energy of the charge materials that impact the overall energy intensity of the steelmaking process. To produce hot rolled steel from iron ore takes almost five times as much energy per ton as making the same product from scrap steel, as the above energy intensities show. Scrap yields roughly half the steel made and consumed (but these two are not identical).

Prior to the 1990s, steelmaking in the United States was more energy-intensive than that in Germany, Japan, and Korea. It appears that, although the energy intensity of the U.S. steel industry improved significantly between 1995 and 2005, it was still higher in 2005 than that in those three countries (Table 4.9): in 1995, the steel industry in the United States was 57 percent more energy-intensive than that in Korea and about 22 percent more energy-intensive than that in Japan and in Germany. In 2005, the U.S. steel industry was still more energy-intensive than Korea's and about 6 percent more energy-intensive than Japan's and Germany's. The report *Saving One Barrel of Oil per Ton* states, for example, that "energy consumption in blast furnace ironmaking has decreased by more than 50 percent since 1950" (AISI, 2005). Yet blast furnace operation uses nearly 40 percent of all the energy consumed by the iron and steel industry. One means of improving the efficiency of blast furnace ironmaking has been by recovering blast

TABLE 4.9 International Comparison—Energy Intensity of the Iron and Steel Industry of Selected Countries

	Energy Intensity, 1995 (Btu/tonne)	Energy Intensity, 2005 (Btu/tonne)	Percent Difference Compared with the U.S., 1995	Percent Difference Compared with the U.S., 2005
Germany	8,114	7,660	–22	–7
Japan	8,059	7,743	–23	–6
Korea	4,463	7,438	–57	–10
United States	10,418	8,246	0	0

Source: Data from International Energy Agency, online statistical database; World Steel Association, online database.

furnace gas and using it elsewhere in the overall production process. If this recovery is taken into the evaluation, the energy intensity of the blast furnace operation is said, by a representative of the steel industry, to be 12–14 million Btu/ton, and a total of 18–21 million Btu/ton total for finished goods. Over the past 20 years, the efficiency of iron and steel manufacturing in all countries has improved substantially, but the worldwide average has not improved much. This is due to the growth of iron and steel manufacturing in China, where the overall efficiency has not changed much. In China, there is a difference of about 20 percent between the average and the best plant due to the blast furnace size and the amount of heat recovery.

It is important to use caution in comparing countries because differences can be caused by the actual efficiency of production, the amount of recycled material used, the process (BOF versus EAF) employed, and the type of final product (Schipper, 2004). Efficiency depends on the size and age of the plant—larger and newer facilities are often more energy-efficient than older ones. Savings can occur over time as a result of changes within plants or in processes and from shifts to plants and processes that are more energy-efficient; differences in resources, prices, and other factors also matter. Schipper (2004) points out as an important caveat that processes that are efficient in one country could be significantly less so in another country.

To remain competitive, the U.S. iron and steel industry must become more resource efficient and less capital intensive. Figure 4.5 shows the energy con-

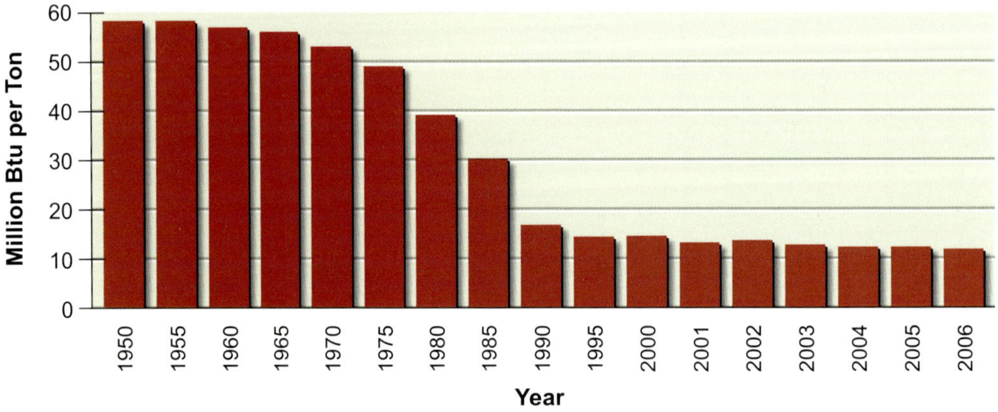

FIGURE 4.5 *Energy consumption in the U.S. steel industry per ton of steel shipped, 1950–2006.*
Source: DOE, 2008.

sumption per ton of steel shipped by U.S. industry from 1950 to 2006. Energy consumption per ton of steel has decreased 27 percent since 1990, while CO_2 emissions decreased by 16 percent. For 2002–2005, energy intensity per ton of steel decreased by 12 percent. In 2005, the American Iron and Steel Institute (AISI) announced a goal of using 40 percent less energy per ton of steel in 2025 compared to what was being used in 2003 (AISI, 2005). According to AISI, only a small portion of this reduction could be obtained by the implementation of best practices; instead, major advances would require the development and implementation of "transformational technologies." Some of the best opportunities (in terms of cost/benefit) include EAF melting advances, BOF slag heat recovery, the integration of refining functions, heat capture from EAF waste gas, and increased direct carbon injection. The majority of these technologies would be available before 2020 assuming continued technological R&D. With standard rates of stock turnover, one could expect these technological changes to be implemented in the midterm timeframe (2020–2035).

Fruehan (2008), in a study for DOE in partnership with the industry, analyzed various combinations of technologies—including the rotary hearth furnace (RHF); the CIRCOFER process, in which coal is charred and ore is partly metallized in a single first step, and then completed in a bubbling second step; and the RHF with a submerged arc furnace (SAF)—to determine whether combinations of proven technologies could enhance the overall process. Several revolutionary new steelmaking technologies—such as the use of hydrogen as an iron ore reductant or furnace fuel (under development), or electrolytic and/or biometallurgical-based iron and steel production (in the concept definition and development stage)—could be ready in the midterm (that is, between 2020 and 2035).

McKinsey and Company (2008) identified the iron and steel industry as one of the two (pulp and paper being the other) largest opportunities to reduce energy use in the industrial sector. Box 4.5 indicates that the iron and steel industry can reduce energy consumption by 0.3 quad (22 percent) by 2020 by accelerating the adoption of proven technologies and process improvements. These technologies are consistent with those mentioned above. Many of them have IRRs greater than 20 percent. The AISI (2005) study provides a higher estimate of energy efficiency potential—0.79 quad or 58 percent of the projected energy consumption in the iron and steel industry in 2020. The CEF study (IWG, 2000) estimates a potential of only 0.21 quad (15 percent) in 2020.

The barriers to implementing energy efficiency in the iron and steel industry are similar to those of the other energy-intensive industries: lack of sustained cor-

BOX 4.5 Reducing Energy Consumption in the U.S. Iron and Steel Industry

McKinsey and Company (2007) identified the following opportunities to reduce by 2020 the energy consumed in the U.S. iron and steel industry:

	100 percent =	0.30 quadrillion Btu (quads)
Secondary casting	39 percent	(0.12)
Arc furnace processes	19 percent	(0.06)
Blast furnace processes	11 percent	(0.03)
Integrated casting	8 percent	(0.02)
Multiprocess improvements	10 percent	(0.03)
Other process steps	13 percent	(0.04)

porate interest, reduced levels of engineering research, low energy prices, and competition for capital, which are discussed in Section 4.5.

The steel industry has been in a challenging financial position for several decades. In the past 30 years, new alloys using thinner steel have led to 25 percent less steel use in cars. Steel has also lost market share to aluminum, composites, and plastics. Steel has become a commodity product, and profit margins continue to shrink. Over the next 20 to 30 years, steel may be replaced by carbon-fiber-reinforced materials, especially as carbon nanotube materials develop. These may be at least as strong as steel and much lighter. Some have recommended that the steel industry focus on specialty steels for tools, stainless steel, or high-silicon steels, which might provide value-added exports and more economic benefits. This would require a focus on new technologies instead of incremental process improvements.[14]

4.3.4 Cement Industry

The cement industry is among the largest industrial energy consumers in the United States and the world, accounting for 5 percent of the energy used in the U.S. manufacturing sector, or 1.3 quads (EIA, 2002, Table 2a), and about 9 percent of global industrial energy use (IEA, 2007, p. 9). The industry also accounts

[14]F. Harnack, United States Steel Corporation, personal communication, June 2008.

for about 5 percent of global anthropogenic CO_2 emissions and 2 percent of anthropogenic CO_2 emissions in the United States (Worrell et al., 2001, 2004).

The cement process involves three components: first, the mining and preparation of inputs, most importantly limestone (kiln feed preparation); second, the chemical reactions that produce clinker (clinker production); and third, the grinding of clinker with other additives to produce cement (finish grinding). The feed for older kilns is a slurry of inputs (the wet kiln process), while large new plants mix dry materials for introduction to the kiln. Energy use varies with the process and characteristics of the plant, but in general about 90 percent of the energy use and all of the fuel use occur in the manufacture of clinker in the kiln. The chemical process that converts limestone to lime, the key ingredient in clinker, produces roughly the same amount of CO_2 gas as that generated by the energy use in its production for coal-fired kilns. Technologies that allow production of cement with a lower per ton share of clinker thus yield multiple benefits: savings in fuel consumption and reductions in greenhouse gas emissions of a factor of two or more above that associated with the energy efficiency alone.

In a wet kiln, the burners introduce heat at one end while at the other, "cool" end, the slurry is introduced and then dried (zone 1) and heated (zone 2). Calcining, or the conversion of calcium carbonate to lime, occurs in the third zone, at temperature of 750°C to 1000°C, followed by the sintering zone where a mix of chemicals is reacted to create clinker. The clinker is then cooled. In modern dry kilns the first zone is omitted, and the heating is done in a preheater tower, followed by a second calciner tower, so that a shorter kiln is employed only for the sintering stage (Figure 4.6).

Larger plants are more energy-efficient per ton, and the advanced processes are substantially more efficient. In the United States, energy efficiency varied from 6.2 million Btu/ton of clinker for smaller wet-kiln plants to 3.8 million Btu/ton of clinker for dry preheater-precalciner kilns (Table 4.10). Coal dominates fuel consumption in U.S. plants, although they utilize an increasing proportion of waste materials, used tires, and petroleum coke (Table 4.11).

Energy efficiency in the U.S. cement industry improved steadily during the 1970s and 1980s as wet kilns were replaced by more modern facilities (Figure 4.7). Energy efficiency deteriorated during the 1990s, rising by approximately 10 percent per ton, as both the fall in energy prices and the increased demand for cement caused some of the less efficient manufacturing units to be redeployed. As of 2000, the U.S. cement industry was among the least energy-efficient of the cement industries in the world, using nearly 80 percent more

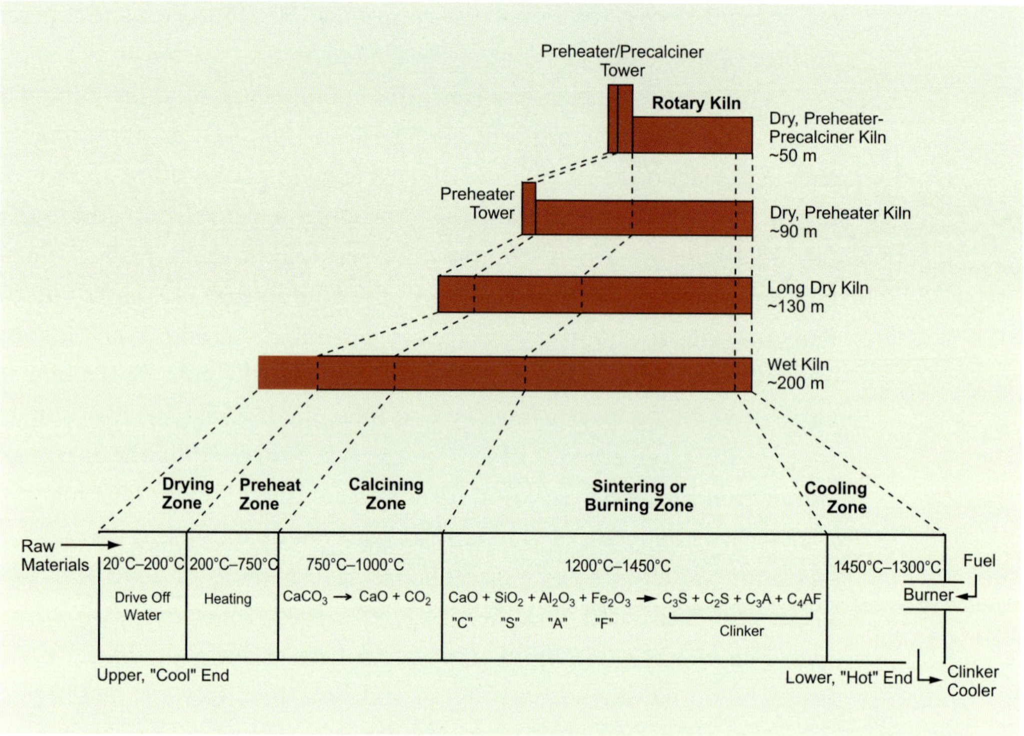

FIGURE 4.6 *Diagram of functional zones for different cement kiln technologies. The feed for older (wet) kiln processes must first be dried. Modern dry-kiln feeds require little or no drying.*
Source: van Oss, 2005.

TABLE 4.10 Energy Intensity, Major Cement Processes in the United States

Process	Primary Energy Use (million Btu/ton clinker)
Wet kilns <0.5 million tons/yr capacity	6.2
Wet kilns ≥0.5 million tons/yr capacity	5.6
Dry kilns <0.5 million tons/yr capacity	4.9
Dry kilns ≥0.5 million tons/yr capacity	4.1
Long dry kilns	5.1
Dry preheater kilns	4.1
Dry preheater-precalciner kilns	3.8

Source: van Oss, 2005.

energy in the production of clinker—the most energy-intensive ingredient in the final product—than was used in Japan, the world leader in energy efficiency, while also using 10 percent more clinker in the production of cement than was used in Japan (Table 4.12). Partly because Japan has severely limited domestic energy supplies, it has made huge investments in R&D to improve energy efficiency in order to reduce its dependency on foreign energy resources. The United States has not faced a comparable sense of urgency in the industrial sector.

As the apparent energy performance of other countries suggests, considerable technological opportunities exist to improve the energy efficiency of the U.S. cement industry. The Clean Energy Future study (IWG, 2000) estimated that a 19 percent improvement in energy efficiency was economically attractive, and the McKinsey and Company (2007) study concluded that a 21 percent improvement in energy use was feasible based on commercially available and commercially attractive technologies—with an estimated IRR of at least 10 percent. The details of the McKinsey and Company calculations are not available, but other sources suggest that while the McKinsey estimate may be optimistic, improvements are clearly feasible (Worrell et al., 2004). The panel first reviews the categories of potential improvements and then considers some issues that may arise in their adoption. This discussion is based on the estimates provided in Worrell et al. (2004).

Upgrading a kiln from wet to dry and from a long dry kiln to a preheater, precalciner kiln results in major energy efficiency gains but for a price that requires a payback period of at least 10 years. Worrell et al. (2004) conclude that these upgrades are attractive only when an old kiln needs to be replaced. However, they discuss a wide range of less drastic upgrades that yield commercially attractive benefits at each stage of the process.

At the first stage, Worrell et al. (2004) identify technologies with short-term payback periods (less than 3 years) that yield only modest energy efficiency improvements for the dry kiln process, controls (savings up to 3 percent in the electricity used), and possibly blending and roller mills (payback periods not provided; total savings of up to 10 percent). Over half of the energy used at this stage, however, can be eliminated through available technologies with payback periods of more than 10 years. Key technologies are efficient conveyer systems of the dry kilns and high-efficiency classifiers for the wet kilns.

Short-payback options in clinker production include advanced control systems, combustion improvements, indirect firing, and the optimization of components such as the heat shell. Although opportunities vary with specific

TABLE 4.11 Source of Fuel and Electricity Consumption by the U.S. Cement Industry in 2000

		Fraction of Contributed Heat and Total Energy	
Fuel	Quantity	Heat (%)	Total Energy (%)
Coal	10.1 (million metric tons)	67	60
Coke, petcoke	1.8 (million metric tons)	14	13
Natural gas	338.3 (million cubic meters)	3	3
Fuel oils	123.7 (million liters)	1	1
Used tires	0.4 (million metric tons)	3	3
Solid wastes	1.0 (million metric tons)	6	5
Liquid wastes	929.1 (million liters)	6	5
Electricity	12.6 (billion kilowatt-hours)	nil	10

Note: Average unit consumption of energy (fuel and energy consumed reflect the U.S. mix of wet and dry kilns; values likely would differ in countries operating a different mix of technologies):

- Electricity—143.9 kilowatt-hours per metric ton of cement,
- Heat—4.7 million Btu (1 million Btu = 1.055056 gigajoules) per metric ton of clinker,
- Total energy (includes electricity)—4.9 million Btu per metric ton of cement.

Source: van Oss, 2005, p. 29; data from U.S. Geological Survey's annual survey of U.S. plants.

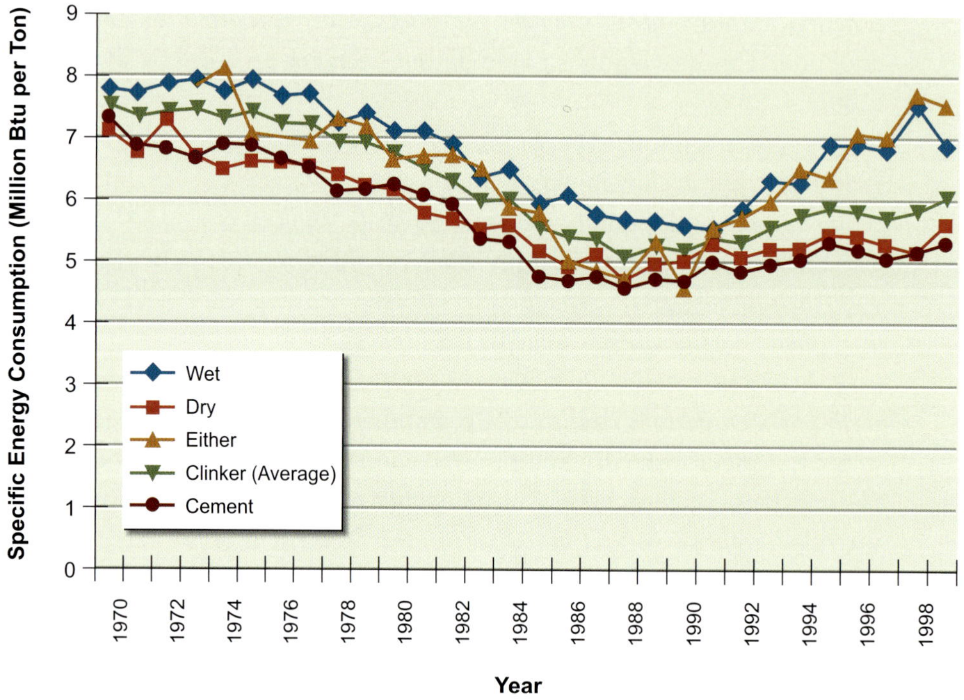

FIGURE 4.7 *Primary energy intensity of U.S. cement and clinker production, 1970–1999. Energy intensity is expressed in million Btu per ton, higher heating value. Source: Worrell and Galitsky, 2004.*

TABLE 4.12 Cement Industry Intensities, 1990 and 2000, and Mid-1990s Clinkers Factors, by World Region and Subregion (Primary Energy)

A. Cement Industry Energy Intensities by Region and Subregion

Region Energy Intensities			Subregion Energy Intensities		
	MJ per kg Clinker			MJ per kg Clinker	
Region Name	1990	2000	Subregion Name	1990	2000
I. North America	5.47	5.45	1. United States	5.50	5.50
			2. Canada	5.20	4.95
II. Western Europe	4.14	4.04	3. Western Europe	4.14	4.04
III. Asia	4.75	4.50	4. Japan	3.10	3.10
			5. Australia and New Zealand	4.28	4.08
			6. China	5.20	4.71
			7. Southeast Asia	5.14	4.65
			8. Republic of Korea	4.47	4.05
			9. India	5.20	4.71
IV. Eastern Europe	5.58	5.42	10. Former Soviet Union	5.52	5.52
			11. Other Eastern Europe	5.74	5.20
V. South and Latin America	4.95	4.48	12. South and Latin America	4.95	4.48
VI. Middle East and Africa	5.08	4.83	13. Africa	5.00	4.75
			14. Middle East	5.17	4.92

B. Cement Industry Mid-1990s Clinker Factors by Region and Subregion

Region Unit-based Emissions		Subregion Unit-based Emissions	
Region Name	Factor, kg CO_2/ kg Cement	Subregion Name	Factor, kg CO_2/ kg Cement
I. North America	0.88	1. United States	0.88
		2. Canada	0.88
II. Western Europe	0.82	3. Western Europe	0.81
III. Asia	0.85	4. Japan	0.80
		5. Australia and New Zealand	0.84
		6. China	0.83
		7. Southeast Asia	0.91
		8. Republic of Korea	0.96
		9. India	0.89
IV. Eastern Europe	0.83	10. Former Soviet Union	0.83
		11. Other Eastern Europe	0.83
V. South and Latin America	0.84	12. South and Latin America	0.84
VI. Middle East and Africa	0.89	13. Africa	0.87
		14. Middle East	0.89

Source: Battelle, 2002, substudy 8, p. 5.

plants, the combination of these activities appears to yield an improvement in energy use on the order of 10 percent. Recovering heat from the cooling stage yields substantial savings. If the heat is used for power generation, it can save up to half of the electricity used in the clinker process. Heat-recovery opportunities, however, are closely tied to the basic structure of the kiln: an advantage of the preheater, precalciner kilns is that waste heat is used in the first two stages. Thus, taking full advantage of the heat-recovery savings may require other major upgrades.

For the last (grinding) stage, technologies with short-payback periods yield modest improvements. Significant electricity savings are available from high-pressure roller presses, but these have payback periods of more than 10 years.

The most attractive and available technologies derive from changing the chemistry of the cement to reduce the need for calcination, thereby decreasing the high share of clinker characteristic of U.S. production. Blended cements include higher proportions of other cementitious materials, such as fly ash. Steel slag, which is already calcined, is an alternative to limestone for the production of clinker. The availability of these inputs and their probable price if they are used more widely in cement production raise concerns about how broadly they can penetrate the U.S. market. Worrell et al. (2004) identify potential energy savings of up to 20 percent from the deployment of blended cement technologies. Avoiding the production of clinker yields a double benefit in reduced CO_2 emissions because both chemical production and energy use are avoided.

Advanced technologies with a potential to further improve energy efficiency and reduce carbon emissions include a fluidized bed kiln, advanced comminution technologies, and the substitution of mineral polymers for clinker (Worrell et al., 2004). The Battelle (2002) study concludes that non-limestone-based binders may yield a reduction of 30 percent in CO_2 emissions. Additional advanced technologies aimed at reduction of CO_2 emissions include hybrid cement-energy plants, which are currently being investigated in the United States, and carbon capture and storage technology.[15]

---

[15]Cement plants present a much better carbon capture and storage opportunity than do advanced coal plants. Mitsubishi is currently designing carbon capture and storage technology suitable for advanced coal electric generating plants with an emission stream that is 12–15 percent CO_2. Cement plants, alternatively, have emission streams with 30 percent CO_2 content (Battelle, 2002). Current production of cement in the United States results in about 0.85 kg CO_2 per kilogram of cement. Thus, a carbon tax of $50 per tonne of CO_2 would increase the cost of a tonne of cement by $12.00, making the carbon tax one of the largest components in the cost of cement

Why have U.S. firms not adopted these technologies, and why do they significantly lag foreign firms? One component is simply the age of the installed plant. As discussed above, the major savings come with new plants and are routine in the facilities in countries whose capacities have grown rapidly in the past decade. U.S. consumption of cement increased during the 1990s, but by 2006 imports accounted for 20 percent of sales, so investment in new facilities has not mirrored the increase in demand.

Replacing an older, wet-kiln facility with a new, more energy-efficient dry-kiln facility raises challenges in the United States that may not be present in developing countries and may in part explain the vintage of domestic plants. Cement facilities are colocated by limestone quarries. A major upgrade, or new plant, has a very long depreciation period. Before undertaking such an investment, a company would typically want adequate limestone supplies for 50 years of operation. In many cases, this would mean that the desired location for a new dry-kiln plant would be at a different location from the older, wet-kiln plant. Permitting and meeting other requirements (e.g., environmental requirements and public hearings) at a new site can be extraordinarily difficult, time-consuming, and expensive, and such requirements indeed change the relative economics of maintaining the old, inefficient facility versus investing in a modern, efficient plant. McKinsey and Company (2008) identified regulatory restrictions as the major reason for the U.S. industry's apparent lack of new investment.

A second regulatory issue concerns the use of blended cements. The federal government has recently issued standards for blended cement, but widespread application requires actions by state standards boards. The existence of complex state and federal building codes and the expense of updating the codes are credited with slowing the introduction of blended cements that include a lower fraction of clinker. Regulatory restrictions and protracted government permitting and approval processes may be at the root of slow approval for the use of these cements. Alternatively, U.S. firms may not be eager to open the market to foreign manufacturers with greater experience and stronger track records than domestic sources possess.

In addition to regulatory restrictions, another component is plausibly the competitive environment in the industry. Until recently, import prices have only

---

(Battelle, 2002). It should be noted that carbon capture and storage technology requires substantial energy itself, so that its deployment would confer a significant energy efficiency penalty in the production of cement.

weakly pressured U.S. firms. Notwithstanding the North American Free Trade Agreement (NAFTA), antidumping tariffs protected domestic producers from low-priced Mexican imports until 2006, while transportation costs moderated competition from Asia. Over the past 20 years, foreign firms—most notably the Mexican firm CEMEX—have become major shareholders in more than 80 percent of U.S. domestic cement plants. With the change in ownership and the need for major reconstruction following the devastating hurricane season of 2005, political support in the United States for antidumping tariffs has waned.

If the tariffs are relaxed, foreign firms will face a choice: upgrade the plants in the United States or increase imports. Which option they choose, and the health of the domestic industry, may depend on opportunities to expand operations in the United States, including the ability to site new facilities in new locations.

Three studies have estimated the potential for energy savings in the cement industry. The highest estimated savings, by Worrell et al. (2004), is 0.29 quad, or 67 percent of projected energy consumption in 2020.[16] The CEF study (IWG, 2000) estimated the potential savings at 0.08 quad (19 percent), and McKinsey and Company (2008) estimated it at 0.1 quad (23 percent) in 2020.

4.4 CROSSCUTTING TECHNOLOGIES FOR IMPROVED ENERGY EFFICIENCY

One way to consider the potential impact of technological improvement in the industrial sector is to examine key crosscutting technologies that play a dominant role. These include combined heat and power systems, catalysis, pumps, motor and drive systems, design tools, and computational and other approaches to optimizing operations and maintenance (O&M). The obvious targets are the most energy-consuming processes. Separation processes, especially heat-driven separations, are a class of such energy-intensive processes, and these are discussed in some detail. Other high-temperature processes are also potentially capable of improvement. In addition, there are numerous examples of technology

---

[16]Based on an energy intensity of 4.61 million Btu/ton for wet-kiln processes and 2.89 million Btu/ton for dry-kiln processes, and a total production of 23.2 million tons of wet-process cement and 62.8 million tons of dry-process cement.

developments in which the impact on energy use is an indirect or a secondary consequence—for example, in sensor development and process controls.

To illustrate how industries are and could be improving the efficiency of their operations, the panel describes several specific examples of crosscutting technologies, including some that are already being introduced but that may have much greater application, some that are still in the development stage, and a brief mention of the fundamental manner in which energy efficiency may be approached to achieve changes not yet associated with specific devices.

The following seven subsections summarize these crosscutting technologies.

4.4.1 Combined Heat and Power

Combined heat and power units transform a fuel (generally natural gas) into electricity and then use the hot waste gas stream for processes such as space and hot-water heating or industrial and commercial processes. Large, central-station generators use only about one-third of the energy in the fuel to produce electricity, and the rest must be dispersed with cooling towers or transferred to rivers, lakes, or the ocean. By capturing and converting waste heat, CHP systems achieve effective electrical efficiencies of 50–80 percent (Casten and Ayres, 2007; Lovins, 2007). For applications in which there is a large demand for low-temperature steam, such as spray drying, CHP units can be nearly 100 percent efficient.

CHP facilities have been commonly established in energy-intensive industries such as food processing, pulp and paper, chemicals, metals, and oil refining. The levels of penetration of this technology depend on the availability and prices of natural gas as well as government policies. European countries, such as Finland and Denmark, are among the leaders in terms of installed CHP capacity, with 30–50 percent of their total electricity generated through CHP technologies. The Danish CHP success story is based on a package of strategies that evolved after the First Heat Supply Law was introduced in 1979. This law included planning regulations and financial incentives that worked together to create desirable market conditions for CHP (IEA, 2009).

For commercial and industrial installations that buy electricity and also use large quantities of natural gas for process heating, CHP could double energy efficiency and cut costs by half. Installations such as steel-rolling mills and paper mills could reap large benefits from CHP. Commercial installations such as hospitals and hotels could also benefit from CHP.

CHP offers two additional benefits. Since the electricity is generated on-site, there may be no need for transmission and distribution lines, thus saving the expense of building and operating the lines and also eliminating the 6–10 percent loss of electricity during transmission and distribution (King, 2006; Casten and Ayres, 2007). A much larger benefit for some customers is that generating electricity on-site eliminates the possibility of power disruptions from transmission or distribution line problems. Having a CHP unit backed up by central-station power gives much more reliable power. In this case, however, lines must be sized to accommodate peak power, and hence little capital would be saved on the network connection.

Despite the efficiency gains that it offers, CHP has been limited in its development owing to a variety of regulatory, structural, and economic factors, including local restrictions on air emissions; backup energy fees (i.e., for standby power) charged by utilities; costs associated with site-specific engineering and design; difficulty in obtaining a suitable natural gas supply contract; restrictions on selling electrical power to the grid; and the challenges of obtaining permits and meeting safety regulations. For small CHP installations, these barriers can be prohibitive (Sovacool and Hirsch, 2007).

A CHP system is normally sized to provide the heat needed at a facility. In general, the CHP unit generates less power than the facility needs, at least at peak times. When the facility seeks to buy power from the utility, it is charged a significant backup fee, since it is assumed that the CHP customer will be adding to peak load. The utility might also charge a standby fee, arguing that it must have reserve generation available in case the CHP unit stops operating. Insofar as the utility does incur extra costs due to the CHP unit, the CHP unit should pay. However, the costs should be assessed on the basis of how the CHP unit is operated and when it wants backup power. Similarly, the CHP unit ought to be paid the savings to the utility when it sells electricity to the system.

Within the commercial, institutional and industrial markets, engineering and plant staff may be resistant to CHP projects owing to limited familiarity, lack of internal coordination, concern over greater regulatory oversight, and competition for capital funding. Despite CHP's overall air emission benefits, modifications to a plant's industrial exhaust system may trigger federal or state review of the plant's air permits, which can inhibit some plant managers from pursuing CHP.

The three most critical technical challenges facing the market for CHP are that (1) original equipment manufacturers need to develop better prime movers (more efficient, cleaner, more reliable, more durable, and lower-cost systems); (2) original equipment manufacturers or aftermarket assemblers need to develop packaged CHP systems that facilitate plug-and-play installation; and (3) project developers need to demonstrate practical applications that can be replicated into "cookie-cutter" installations that yield attractive pricing for customers while maintaining acceptable profit margins for the technology vendors.

The potential for CHP is greatest in areas that have high electricity prices (tied to natural gas prices), availability natural gas, and large industrial and commercial thermal loads—most characteristic of the northeastern and southwestern United States. In the northeastern United States, natural gas is often used to fuel the marginal central power plant (i.e., the last unit turned on to supply demand) and sites' process and space-heating needs. Since the deployment of CHP often displaces natural gas used for process heat, water and space heating, and electricity generation, CHP can actually lower natural gas demand while lowering cost.

The best candidate sites for CHP have coincident need for electric and thermal energy. Large-scale CHP, such as at district energy campuses, can be efficient and cost-effective in the right setting and has already been implemented at many attractive sites.

Estimates of the cost-effective energy-savings potential with use of CHP nationwide range from 0.7 quad (based on McKinsey and Company, 2007) to 7.4 quads (based on Bailey and Worrell, 2005). The latter estimate includes nontraditional CHP technology opportunities such as the use of energy that is typically discarded from pressure-release vents or from the burning and flaring of waste streams, the use of gas turbines that are more complex than traditional ones, and the use of flue gases from CHP plants to power a furnace. The estimated potential savings of 7.4 quads includes an estimated 0.6 quad of opportunities outside the industrial sector, reducing the industrial-sector estimate by Bailey and Worrell (2005) to about 6.8 quads. An additional 2.4 quads of savings opportunities in the 6.8 quad total is associated with nontraditional equipment for recycling energy, including black liquor gasification and landfill gas recovery. Removing these 2.4 quads results in the range of 4.4–6.8 quads shown in Table 4.4. A third, intermediate estimate based on the CEF study (IWG, 2000) is that CHP could cost-effectively expand the energy savings by 2.0 quads (Lemar, 2001).

4.4.2 High-Temperature and Separation Processes

Many industrial processes involve high temperatures, and some separation processes have thermal efficiencies as low as 6 percent, making separation processes an attractive target for improving U.S. industry's energy efficiency. The energy used for separations totals about 4.5 quads per year, or 47 percent of all the energy used in manufacturing. Both distillation and membrane separation are described below to illustrate current and potential innovations that might reduce the energy intensity and total energy use of U.S. industry.

The petroleum industry consumes about 60 percent of the 2.4 quads of energy used each year in all industrial distillation processes. The distillation of chemicals and petroleum uses about 53 percent of the total energy used for industrial separations and is the largest energy-consuming process in industry. (Drying uses 20 percent less energy, and evaporation uses 60 percent less.) But distillation can be improved. Technologies that can significantly reduce the energy used in distillation processes include, for example, latent heat integration, multiple-effect distillation, and solution-thermodynamics-altering azeotropic or extractive distillation.

Other improvements include introducing heat exchangers along the distillation column, which can improve energy efficiency by about 25 percent. Analyzed and developed over about two decades, this process could be put into practice now (Jimenez et al., 2004; Mullins and Berry, 1984; Orlov and Berry, 1991; Schaller et al., 2001). Although in its fully optimized form this approach would probably be practical only with the construction of altogether new columns, side reboilers are one currently used approximation of this process.

Considerable attention has recently been paid to the development of separation processes based on membranes of many different kinds and on other porous materials. Membranes that separate different gases are particularly promising as means to improve energy efficiency. All membranes do need to be replaced from time to time, an aspect that must be included in assessments of their net utility (Martin et al., 2000b).

The ORNL and BCA, Inc. (2005) report *Materials for Separation Technologies: Energy and Emission Reduction Opportunities* identifies ways to reduce by about 240 trillion Btu/year (5 percent) the 4500 trillion Btu/year used for separations by U.S. industry, and it does not include improvements in distillation.

Material methods, notably membrane and micro- and nanoparticle separation methods, offer tantalizing possibilities. The challenges are in developing materials and methods with high throughput, high selectivity, low energy requirements, resistance to fouling, durability, and affordable costs. The ORNL and BCA, Inc. (2005) report emphasizes the importance of developing metrics for evaluation of the potential methods. According to that report, membrane separation is the most widely applicable of all technologies for reducing the energy of separation processes in the petroleum, chemical, and forest products industries (Nenoff et al., 2006; Banerjee et al., 2008). Zeolites are one of the kinds of materials that can achieve separations without requiring direct heat. However, the zeolite approach leaves the capturing material with the target material attached, so some removal process must be available if that material is valuable, or, if the material is considered a contaminant, either it must be removed or the used zeolite must be replaced and hence must have low cost.

Membranes can function to remove unwanted substances or to recover valuable material. For example, Amalgamated Research, Inc., and the Idaho National Laboratory are trying to develop a three-stage separation system for extracting valuable biomass products, and a project at the University of New South Wales, Australia, is working to develop a membrane for desalination (Chapman et al., 2004). Reported earlier at the 1995 Abu Dhabi Conference on Desalination (IDA, 1995) was a membrane desalination process in which water, heated to about 80°C, is reduced under pressure and passes as vapor through a membrane of polytetrafluoroethylene (PTFE: Teflon). Those proposing this membrane distillation process emphasize that the heat driving it may be waste heat or solar heat. Besides desalination, one of the most active areas of membrane development addresses the separation of hydrogen by capturing other substances in the membrane.

Still another membrane-based process for desalination is reverse osmosis, a process in which pressure is applied to the saline water, driving water molecules through the membrane, moving them opposite to the direction in which they would diffuse in the absence of the pressure. A variety of organic materials such as cellulose acetate, nylon, and polyamide have been used for reverse osmosis membranes. The reverse osmosis process developed for seawater by Energy Recovery, Inc. (ERI), is said to use 1.7 kWh per cubic meter of purified water produced. By including a pressure exchanging device with the reverse osmosis apparatus, ERI uses up to 60 percent less energy than is used in conventional reverse osmosis systems. Approximately 30 percent of the total cost of desalination is the cost of the required energy; the total cost of newly desalinated seawater falls between $3 and

$8 per 1000 gallons, so the cost of the energy would be between about $1 and less than $3 for that amount of product.[17] Electric fields are also used to drive membrane separation processes to remove ionic substances from water.

Membranes may be made of organic materials for relatively low temperature processes or of inorganic materials such as ceramics for high-temperature use. Some membranes are composed of both organic and inorganic materials. Some are made of metals—for example, palladium or palladium alloy—especially for hydrogen separation; RTI Corporation is active in this area. Polymeric membranes, the largest class, may also become important energy-efficient means for purifying hydrogen (Lin et al., 2006). Membranes are currently used successfully to separate light hydrocarbons as well as hydrogen from gas streams being used as fuel; the separated light hydrocarbons in many cases have industrial uses with a value considerably higher than that of the original fuel.

The market for membranes is growing rapidly. In 1999, it was more than $1 billion and was predicted to reach $3 billion by 2008. The energy savings in the chemical industries alone has been estimated at about 95 trillion Btu/year by 2025, a reduction of about 2.5 percent in the total energy consumption by those industries (Worrell et al., 2004).

The principal inhibitors to the adoption of membrane separation methods are a lack of selectivity, the limited range of conditions under which they can function (organic materials typically must be used between 45°C and 60°C and at a pH of 4–10), fragility and lack of durability, and cost. The use of ceramic or mixed-composition membranes is one of the most active approaches to address these limitations.

Combining membrane separation with distillation in a hybrid system is another approach under development. Typically, the distillation step provides a first-stage, partial separation, and the refinement is done with membranes.

4.4.3 Fabrication and Materials

Advanced materials are an important enabling technology for all sectors of the economy: industry, buildings, and transportation. The industrial sector needs advanced materials that resist corrosion, degradation, and deformation at high temperatures and pressures; inferential sensors, controls, and automation, with

---

[17] See http://www.hwsdesalination.com/index.htm, http://www.energyrecovery.com/ and http://www.membranes-amta.org/amta_media/pdfs/6_MembraneDesalinationCosts.pdf.

real-time nondestructive sensing and monitoring; and new computational techniques for modeling and simulating chemical pathways and advanced processes. The development and implementation of new materials can lead to the increased energy efficiency and decreased environmental impact of U.S. industrial systems. New materials may be developed to make processes more efficient: for example, membranes for separation, or to substitute less energy-intensive materials to provide specific properties and services such as composites and nanomaterials for structural applications in place of steel.

The U.S. Department of Energy (DOE) has conducted an industrial materials development program for more than 20 years in the Office of Industrial Technology Program. This has been in collaboration with industrial firms, universities, and national laboratories. Materials-related opportunities are also discussed in Section 4.4.2 and in Section 4.4.4. The materials portfolio has three focus areas: materials for degradation resistance, with an emphasis on materials that are ultrahard and low-friction; materials for energy systems, with an emphasis on advanced refractories and thermal insulators for waste-energy recovery and reuse; and materials for separations, with an emphasis on membrane materials development. In its fiscal year (FY) 2009 budget submission, the DOE estimates potential energy savings from these programs in the year 2020 of 103 trillion Btu and carbon savings of 1.5 million metric tons of carbon equivalent. One needs to be cautious in applying these savings numbers so as not to double count when they are implemented in a specific industrial sector.

Examples of current efforts include the following:

- *Nanocoatings* for high-efficiency industrial hydraulic and tooling systems: Widespread use of new superhard coatings will increase energy efficiency through diminished friction loss and increased seal reliability in hydraulic pumps. Additional savings are possible through extending the lifetime of optimum cutting performance in machine tooling. Increased system reliability with decreased downtime and replacement costs might provide economic benefits. Environmental benefits include reduced pollutant leakage through pump seals and reduced emissions due to energy efficiency. Degradation-resistant nanocoatings of several borides are being developed.
- *New material systems,* including super stainless steels, low-cost titanium, and magnesium are being developed. New super stainless steels have comparable cost and creep resistance to state-of-the-art advanced

austenitic stainless steels, but they have the potential for higher operating temperatures (up to 800–900°C) and durability under aggressive oxidizing conditions at a fraction of the cost. This represents a new class of heat-resistant alloys, because no alumina-forming, iron-based alloys with properties suitable for structural use above 600°C exist.

- *Novel refractory materials* for high-temperature, high-alkaline environments would be applied to boilers, furnaces, and gasifiers for use in the aluminum, chemical, forest products, and glass industries.
- *Fiber-reinforced aerogel-based pipe insulation systems* for industrial steam pipes would improve the high-temperature performance, durability, and life of aerogel materials. This insulation would be manufactured in blanket forms and have long-term water resistance. In addition to industrial steam pipes, this material could be applied to other components such as process heating lines, removable lids and covers, prejacketed insulation, and coker panels.
- *Hierarchical nanoceramics for industrial process sensors* are being developed in order to recover energy and water from industrial waste and process streams. An array of nanoceramic gas sensors for real-time burner balancing could increase combustion efficiency by at least 0.5 percent and be applicable to many processes.

Examples of the successfully demonstrated use of advanced materials include the following:

- An advanced heating system for high-performance aluminum forgings (Figure 4.8) that
  —Uses an optimized combination of radiant and convection heating for processing materials,
  —Decreases energy consumption by a factor of three,
  —Reduces heating times by an order of magnitude, and
  —Produces high-performance forgings with improved tensile and fatigue properties.

Field testing of the system in a full-scale production setup demonstrated a cost savings of 40–50 percent owing to reduced energy consumption, increased throughput, and improved consistency in the process and product:

FIGURE 4.8 *Continuous-belt infrared furnace for high-performance aluminum forgings. Testing of this system at Queen City Forging Company confirmed that it is more than three times more energy-efficient than current convection furnaces in preheating aluminum billets. It also provides grain refinement that enhances fatigue properties. Infrared preheated and forged components have been shown to last two times longer than conventionally preheated forgings.*
Source: DOE/EERE, 2005.

- Nickel aluminides for rolls in reheat furnaces made possible
  —The development of manufacturing procedures that enabled production of 115 nickel aluminide rolls for installation and testing,
  —The processing of more than 215,000 tons of steel during a 26-month period, and the elimination of more than 70 furnace shutdowns,

—An increase in both yield and product quality owing to the elimination of rolling-related downgrading of steel, and
—An increase of 35 percent in furnace energy efficiency.

Most of the materials that are currently in the R&D stage, such as those five listed above as examples of current efforts, will not be introduced in significant quantities until after 2020. Those that have been technically demonstrated successfully may be implemented within the next 10 years. To introduce a new material, particularly when substituting it for an existing material, the new material needs to be better than what it is replacing with regard to performance and cost. It often takes 10–20 years from "discovery" to have widespread impact in the marketplace. Part of the timeframe is the requirement to develop a database including the mechanical properties, performance results, life, and other characteristics so that designers will be comfortable using the new material. The nickel aluminides mentioned above, which increased furnace efficiency by an estimated 35 percent, have been under development since the mid-1980s and are just beginning to enter the marketplace in any significant amount. The development of new materials takes sustained funding, often from both the public and the private sector.

New fabrication processes are those crosscutting processes that support several energy-intensive industrial processes. The approach is to develop technologies that will improve yields per unit of energy cost for multiple elements of the manufacturing supply chain and will reduce waste and/or improve energy efficiency while demonstrating air- and water-neutral production methods. This is a relatively new focus for DOE, with areas of emphasis including the following: net and near-net design and manufacturing; advanced casting, forming, joining, and assembly; integrated predictive manufacturing and energy-efficient materials handling and plant operations; the engineering of functional materials and coatings; and nanomanufacturing, which would enable the mass production and application of nanoscale materials, structures, devices, and systems.

The U.S. materials and fabrication industries face the same competitive pressures as those facing the specific energy-intensive industries. There is strong cost and technology competition from foreign producers. The U.S. companies need to be at the cutting edge of developing and deploying new material to remain competitive.

4.4.4 Sensors and Process Controls

Much of the sensor development applicable to improved energy efficiency in the United States is conducted by DOE's Industrial Technologies Program (ITP), often in collaboration with industrial firms. In 2004 the sensors program had as a goal a reduction of energy use by 12 trillion Btu/year. Five components make up the program: Advanced Sensor Technologies, Next Generation Control and Automation, Improved Information Processing, Robotics, and Affordable Wireless Technologies. The adoption of these technologies would lead to highly automated processes with efficient, intelligent feedback control through continuous monitoring and diagnosis. DOE's approach involves automated monitoring to gather data, automated data analysis, automated feedback and control, and effective communication among the components. Sensors are used for inferential controls, real-time and nondestructive sensing and monitoring, wireless technology, and distributed intelligence. A goal is to have controls for plant production available by 2017.

There are many novel sensors for a wide range of applications. Of these, many are specialized for very specific tasks. In the papermaking industry, a fiber-optic sensor measures paper basis weight to improve wet-end control in papermaking and to make paper of a uniform basis weight and higher quality. It minimizes energy requirements. Another noncontacting laser sensor measures shear strength and bending stiffness by tracking the rate of propagation of ultrasonic shock waves in the paper. It is claimed that this device could save the U.S. paper industry approximately $200 million annually in energy costs.[18]

New kinds of sensors employ a variety of technologies that in the past were usually associated with individual measurements rather than with monitoring. For example, an X-ray diffraction sensor developed under the ITP allows online monitoring of the composition, specifically the phase, of steel as it is being manufactured. The sensor detects grain structure, orientation, and size, and it can measure stress as well. As the approach is developed, this monitoring technology may well be useful in the manufacture of aluminum, paper products, cement, semiconductors, pharmaceuticals, and ceramics as well. The power requirements are low, there are no moving parts, and the devices are insensitive to changes in position or temperature and to vibration.

Laser sensors are finding many applications. A laser sensor system developed by an energy research company to measure the constituents in a melt has

[18]See http://www.physorg.com/news4221.html.

been tested successfully by an aluminum manufacturer and could be used in other industries such as glass and steel manufacturing. Developed with support from DOE's Office of Energy Efficiency and Renewable Energy, the system has already been licensed and marketed, and sales have been made. Another laser sensor, using a tunable diode, detects temperature, nitrogen oxides, and carbon monoxide in furnace environments. Meant to optimize combustion processes, this system has been applied to a steel reheat furnace and an aluminum reverberatory furnace and might be applied in electric arc steelmaking furnaces as well. Another fiber-optic sensor measures the temperature profile of molten glass, up to depths of about 6 inches. Such profiles are critical for the precise temperature control necessary to shape manufactured objects such as bowls and jars, flat window glass, blown glass for lightbulbs, and fiberglass for insulation.

An important area in which efficient sensors are also being developed and used is in lighting. Reductions of 20–25 percent in lighting costs are claimed for systems that reduce voltage and current for fluorescent lighting (Globalight). Wal-Mart is using sensors to control lighting levels as one of its energy efficiency efforts.

Monitoring of gases in flow processes is another area in which new sensors are finding application. Pennsylvania State University, with technology from the Sandia National Laboratories, developed a solid-state sensor for monitoring hydrogen in gas streams that can operate over a range from 0.5 to 99 percent hydrogen. Hundreds of such units are now in use. In 2006, three Argonne National Laboratory scientists developed the fastest commercially producible hydrogen sensor, with its most probable near-term use in hydrogen-powered vehicles. The detector is made of nanoparticles of siloxane and palladium. The signal is the change, with hydrogen, of the resistivity of the nanobeads of the sensing material.

The monitoring of liquids offers many opportunities. Emerson Process Management has developed an energy-efficient pH meter to monitor boiler water, steam condensate, boiler feedwater, and other such fluids. This is based on replacing a diffusion junction with a very precise, very small capillary, to allow small amounts of fluid to flow without serious restriction. A detector has been developed to predict the flooding of distillation columns. This feature will allow increases of 2–5 percent in throughput.

Wireless networking for sensor systems is under very active development. The Wireless Industrial Networking Alliance is one vehicle stimulating such work. The DOE sponsored the publication in 2002 of *Industrial Wireless Tech-*

nology for the 21st Century (DOE, 2002) to review the state of the art and project its future.

The ITP sponsored its sixth annual conference in Rosemont, Illinois, on June 9–11, 2008, to review its research portfolio on sensors and automation. Topics covered included a microgas analyzer, imaging of surfaces for hot rolled steel bars, infrared temperature sensors, ultrasonic distance sensors, noncontacting speed measurement, digital sensor signals, the use of vibration power to supply the energy needed for sensor networks, and a variety of examinations of wireless sensor networks. The past project portfolio reviews in this program give an overview, through numerous samples, of advances in sensor methods whose development is supported by the DOE.

4.4.5 Steam and Process Heating

Process heating improvements and process and design enhancements can improve quality, reduce waste, reduce the intensity of material use, and increase in-process material recycling. Industrial facilities can eliminate some energy-intensive steps by implementing direct manufacturing processes, thereby reducing energy use, avoiding emissions, and enhancing productivity. This potential is vividly illustrated by the Save Energy Now assessments run by the ITP.

Save Energy Now was created to help reduce energy consumption at industrial process sites in the wake of Hurricane Katrina. It initially focused on steam and process heat in a select group of 200 of the nation's largest manufacturing facilities (there are about 6800 manufacturing facilities that use more than 1 trillion Btu annually) because these two areas make up about 74 percent of natural gas consumption in manufacturing. Facilities (companies) applied through an online process from November 2005 to January 2006; several requirements were set for applicant companies. Those facilities that did not make it into the 200 to receive assessments were still offered software or technical support from an industrial savings expert.

The Save Energy Now assessments at the participating industrial plants included the identification of ways to reduce natural gas use in steam and process heat as well as the on-site training of appropriate personnel to use the Save Energy Now software. Focused, rapid (lasting 3 days), and designed to closely involve plant personnel to achieve buy-in and capacity building for future in-house assessments, these assessments yielded an average of 8.8 percent energy savings annually with a payback in less than 2 years for most changes. In summary:

> By the end of 2006, DOE had completed all 200 of the promised assessments, identifying potential natural gas savings of more than 50 trillion Btu and energy cost savings of about $500 million. These savings, if fully implemented, could reduce CO_2 emissions by 4.04 million metric tons annually. These results, along with the fact that a large percentage of U.S. energy is used by a relatively small number of very large plants, clearly suggest that assessments are an expedient and cost-effective way to affect large amounts of energy use. (Wright et al., 2007, pp. i-ii)

The program was extended and expanded by ITP to include 250 facilities in 2007 and to focus on pumping, compressed air, and fan systems in addition to the main thrust of steam and process heat.

Save Energy Now built on what ITP was already doing:

- *Plant-wide assessments.* From 1999 through 2005, ITP conducted 49 of these (at a 50 percent cost-share basis). Each of these systematic assessments of plant-wide operations addressed a variety of generic and industry-specific technology areas and identified methods for optimizing plant processes.
- *Industrial assessment centers (IACs).* Currently, IACs are located at 27 universities, which conduct free assessments for small- to midsize facilities. Assessors include trained engineers and students. Over the past 20 years, the program has resulted in a database of information on nearly 100,000 cost-saving opportunities identified for industry as the result of 13,500 assessments. These assessments have saved IAC clients an average of more than $55,000 per year for each assessment, with paybacks typically averaging a year or less.
- *Best-practices decision software tools.* ITP has developed a suite of software-based decision tools to help industrial plant personnel identify energy efficiency improvements for plant process and utility systems. These are available free of charge and can be downloaded from the DOE's Web site.
- *Best-practices end-user training.* ITP provides daylong, year-round training on key areas to focus efforts for reducing energy waste in industrial processes. More than 10,000 people have attended these training sessions.
- *Best-practices specialist qualification training.* ITP has also instituted a training program that certifies individuals as experts on a particular

area of decision software (i.e. steam, process heat, pump); 460 people have been certified in this way.

- *Collaborative targeted assessments.* From 2001 through 2005, ITP conducted on-site training and assessments at 85 facilities. These assessments followed end-user training for plant personnel.

Figures 4.9 and 4.10, for steam and for process heat, respectively, show the average cost and payback for a select set of energy-saving actions.

On the supply side, industry can self-generate clean, high-efficiency power and steam and can create products and by-products that can serve as clean-burning fuels. The sector can also make greater use of coordinated systems that

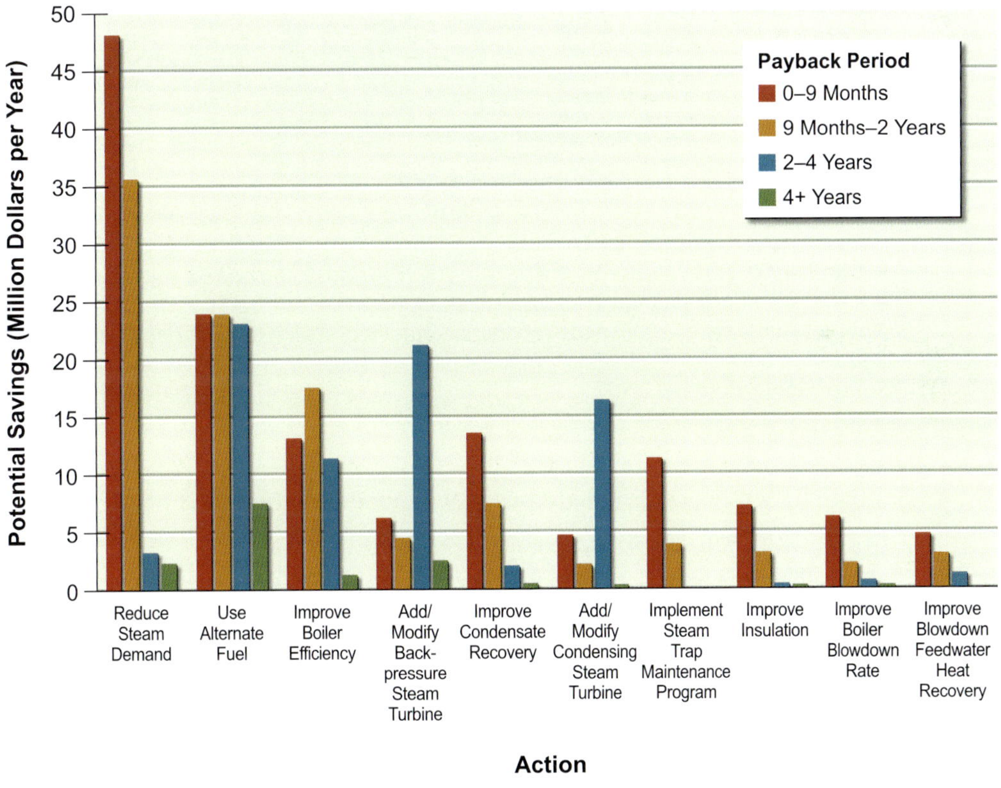

FIGURE 4.9 *Top-10 energy-saving actions for industrial steam systems. Shown are the estimated savings in industrial energy costs identified by the DOE's Save Energy Now assessments in 2006. Most of the savings had estimated payback periods of less than 2 years.*
Source: Wright et al., 2007.

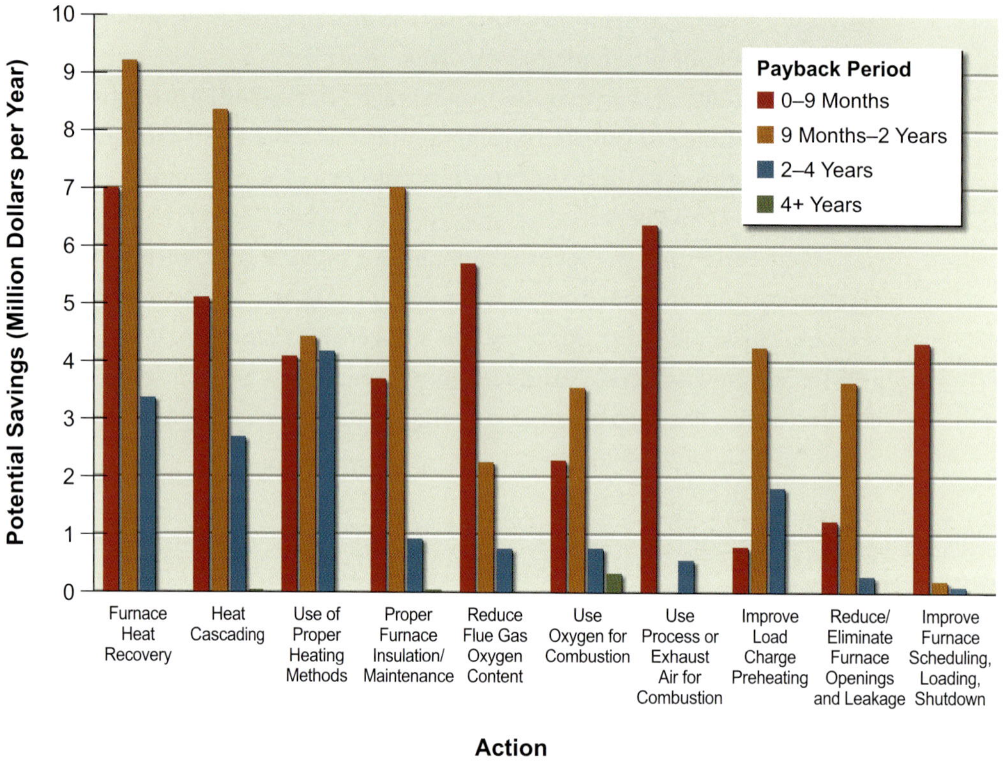

FIGURE 4.10 *Top-10 energy-saving actions for industrial process heating systems. Shown are the estimated savings in industrial energy costs identified by the DOE's Save Energy Now assessments in 2006. Most of the savings had estimated payback periods of less than 2 years.*
Source: Wright et al., 2007.

more efficiently use distributed-energy generation, combined heat and power, and heat integration.

Of these energy efficiency concepts, a number have been identified as suitable for near-term commercialization and deployment. Improvements are possible in steam boilers, direct-fired process heaters, and motor-driven systems, such as pumping and compressed air systems. For example, high-efficiency, low-NO_x-emission burners such as radiation stabilized burners, forced internal recirculation burners, ultralow-NO_x burners, and UltraBlue burners have improved efficiency over conventional burners. Real-time, continuous emissions monitors are available to measure common compounds as well as ones that are not typically measured, such as formaldehyde or ammonia, to better control overall operations. Many of these newer technologies have shown success in a limited number of commercial applications. Other near-term opportunities for increasing energy efficiency exist through the adoption of best energy-management practices; the adoption of more modern and efficient power- and steam-generating systems; integrated approaches that combine cooling, heating, and power needs; and the capture and use of waste heat. One example of a near-term opportunity is isothermal melting—a revolutionary aluminum melting technology with a continuous-flow system using an immersion heater that converts electricity to melting energy with 98 percent efficiency (Figure 4.11).

R&D opportunities suggest the possibility of further energy efficiency improvements to industrial steam and process heating. The development of ultra-high-efficiency boilers, in particular, could offer considerable efficiency gains over today's state-of-the-art boilers. These boilers employ a combination of advanced technologies such as high-efficiency burners (for example, forced internal recirculation burners, or ultralow-NO_x burners), heat-recovery components, and advanced sensors and controls, to achieve high efficiency and low levels of NO_x and CO_2 emissions. The first demonstration of a prototype industrial "super

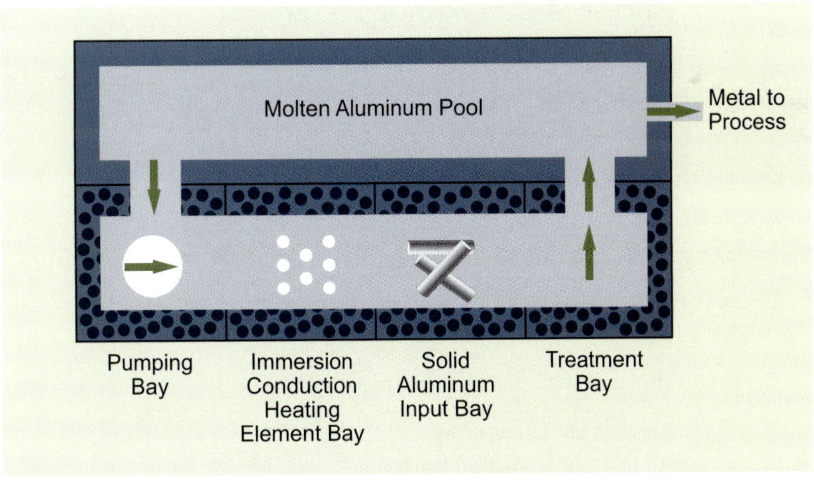

FIGURE 4.11 *Process improvement example: isothermal melting of aluminum. Most aluminum is melted in furnaces, which use radiant heating as the dominant heat transfer mechanism and have poor thermal efficiency. Isothermal melting of aluminum involves the use of immersion heaters in multiple heating bays, allowing electricity to be converted to heat that is conducted directly to the molten metal.*
Source: DOE/EERE, 2007.

boiler" has a projected payback of less than 2 years and is saving thousands of dollars in energy costs annually (Wright et al., 2007).

4.4.6 Basic Approaches from Thermodynamics

One approach that has already been very useful for some processes—such as the optimized distillation column described previously—is a general optimization procedure that can be carried out for virtually any process for which a thermodynamic description is available. Specifically, the use of optimal control analyses can sometimes reveal ways to introduce as control variables some of the quantities that may have been treated only as passive, secondary variables in the evolution of a process. The temperature profile of a distillation column is one example.

The modern application of basic thermodynamics to the optimization of real processes is described by Sieniutycz and Salamon (1990) and Berry et al. (1999). Such approaches describe, for example, means to optimize drying processes and solar-driven heat engines, active heat insulation, and adsorption-desorption processes. However, these are descriptions at a rather abstract level. In terms of real devices, there needs to be a bridge between this level of analysis and the design of testable operating systems.

4.4.7 Electric Motors and Drive Systems

Electric motors make up the largest single category of electricity end-use in the U.S. economy. They also offer considerable opportunity for electricity savings, especially in the industrial sector.

Based on an inventory of motor systems conducted in 1998 (Xenergy, Inc., 1998), it is estimated that industrial motor energy use could be reduced by 11–18 percent if facility managers undertook all cost-effective applications of mature, proven efficiency technologies and practices. Specifically, the implementation of all well-established motor system energy efficiency measures and practices that meet reasonable investment criteria could yield annual energy savings of 75–122 billion kWh. Many motor system efficiency improvements yield benefits in addition to energy cost reductions. The benefits include improved control over production processes, reduction in waste materials, and improved environmental compliance. This full energy efficiency potential cannot be captured all at once, since it would require roughly 10 percent of total new capital expenditures by all manufacturers (Xenergy, Inc., 1998).

Motor efficiency upgrades, improved methods of rewinding failed motors, and system efficiency improvements are all important sources of possible energy savings, but they vary in terms of the effort required. Most motor efficiency upgrades can be achieved fairly easily by selecting the most efficient available motor for the application at hand. System efficiency measures, however, often require a significant amount of expertise and effort on the part of industrial end users and their vendors to identify, design, implement, and maintain the upgraded systems. It is estimated that replacement of 80 percent of the current population of 1- to 200-horsepower motors will take 15–20 years (Xenergy, Inc., 1998). The challenge for government and utility efficiency programs is to assist in accelerating the pace of replacement.

A next generation of motor and drive improvements is also on the horizon, including motors with high-temperature superconducting materials. These advances are expected to be cost competitive in the midterm (i.e., in 2020 or later). Superconductors can be used to increase the magnetic field in a motor, thereby dramatically reducing motor size, weight, and energy losses. Cost and performance improvements are still needed, including reductions in the cost of cool superconducting materials.

4.5 BARRIERS TO DEPLOYMENT AND USE

The broader application of industrial technologies that are available for deployment is impeded by barriers such as the relative high risk and costs associated with new industrial technology, a lack of specialized knowledge relating to energy-efficient improvements, and an inadequate flow of information. These barriers and several others are discussed below. (See also Box 4.3.)

Companies must consider the *technical risks of adopting a new industrial technology*. Uncertainties about the benefits and impacts of new technology on existing product lines can be significant. Small technology changes, particularly in large, integrated process plants, can lead to major changes in process and product performance. In today's manufacturing environment with "24/7" operations, reliability and operational risks represent major concerns for industry when adopting new technologies. These perceived technical risks result in the more lengthy and larger-scale field testing of new technologies, more stringent investment criteria, and a slower pace of technology diffusion.

The conservatism about adopting new technologies is not unique to energy-saving technologies. For example, American steel companies continued to build open-hearth furnaces after World War II, despite the demonstration of superior basic oxygen furnaces. The old technology was familiar and the new technology was a risk. Most energy savings come in through new processes and improved products rather than as a result of investments focused on saving energy. The energy savings is one benefit from the new technology, and usually the most important benefit. Since energy costs are typically small relative to the costs of materials, labor, and plant and equipment, they usually are not the factor driving new investments.

Relatively high initial costs for industrial energy efficiency improvements can be an impediment to investments. New energy-efficient technologies often have longer payback periods than traditional equipment has, and they represent a more serious financial risk, since there is greater uncertainty in future energy prices. This aspect of risk slows technological change and can result in suboptimal choices. Moreover, the interest rates available for efficiency purchases are often much higher than the utility's cost of capital for new electricity-generating plants. Faced with uncertainty about future fuel prices, decision makers may simply avoid investments in new energy systems that require higher initial costs. Because changes in processes affect future cash flows—and often for long periods of time—it would be appropriate for plant managers to look at the net present value of discounted future cash flows rather than to focus on payback periods. A payback period does not properly account for the time value of money or risk, whereas net present value (if properly used) can account for both, as well as provide an accurate rate of return.

External benefits and costs are difficult to value quantitatively, and this inhibits industrial plant managers from investing in greenhouse gas mitigation and other pollution-abatement measures. Companies generally do so only when the investments are offset by lower energy or raw material costs or other cost benefits. External environmental benefits (e.g., benefits to society) are not usually considered in evaluating energy efficiency investments. Although they typically introduce innovations to the industrial sector, suppliers may be reluctant to expend resources in developing technologies for reducing greenhouse gas emissions, unless they have an assured market. On top of the typical risks posed by competing companies and products, uncertain demand can tip the scale toward unacceptable risk for potential financiers.

Distorted price signals also skew the demand for electricity in today's retail markets. While time-of-use pricing is available for many major industrial customers, electricity rates generally do not reflect the real-time costs of electricity production, which can vary by a factor of ten over a single day. Most customers in traditionally regulated markets buy electricity under time-invariant prices that are set months or years ahead of actual use. As a result, current market structures actually block price signals from reaching customers (Cowart, 2001, p. vii), such that consumers are unable to respond to the price volatility of wholesale electricity. Time-of-use pricing would encourage industrial customers to use energy more efficiently during periods when prices are high. According to Goldman (2006), 2700 commercial and industrial customers were enrolled in time-of-use pricing programs in 2003, representing 11,000 MW. Three programs in the Southeast (TVA, Duke Power, and Georgia Power) accounted for 80 percent of these participants, most of which used large amounts of energy. Thus, there would appear to be considerable room for expanding time-of-use pricing programs to other regions and to smaller enterprises.

Lack of specialized knowledge related to energy-efficient technologies and their relative benefits is an impediment to adoption. Industrial managers can be overwhelmed by the numerous products and programs that tout energy efficiency, especially in the absence of in-house energy experts, and may find it risky to rely on third-party information to guide investments. Energy consulting firms often lack the industry-specific knowledge to provide accurate energy and operational cost assessments, and many industrial operations do not have in-house engineering resources to sort through or analyze the information.

According to Neal Elliott, associate director for research at the American Council for an Energy Efficient Economy, "The number one issue with increasing end-use efficiency is the shortage of qualified energy managers and analysts" (Brown et al., 2008, p. 22). Business managers in commercial and industrial sectors are facing knowledge barriers, but commercial managers are more likely to adopt new technologies because the main efficiency improvements are related to common technologies, such as lighting and air-conditioning. Industrial plants often use very specific energy-consuming technologies that do not include off-the-shelf improvements. In addition, industrial sectors may distrust such companies as energy services companies (ESCOs), which specialize in energy efficiency technologies, because these companies do not have industry-specific knowledge as a basis for providing accurate estimates to the manager (Brown et al., 2008).

Incomplete and imperfect information is an impediment to the diffusion of energy-efficient industrial technologies and practices, such as CHP systems, materials substitution, recycling, and changes in manufacture and design. This barrier is exacerbated by the *high transaction costs for obtaining reliable information*

(Worrell and Biermans, 2005). Researching new technology and collecting other relevant information consume precious time and resources, especially for small firms, and many industries prefer to expend human and financial capital on other investment priorities. In some cases, industrial managers and decision makers are simply not aware of energy efficiency opportunities and low-cost ways to implement them.

This barrier is made more onerous by the limited government collection and analysis of energy use in the industrial sector. Consider, for example, the *Manufacturing Energy Consumption Survey* (MECS), a widely used publication that is published every 4 years by the DOE's Energy Information Administration (EIA). In it, one can find the fuel breakdown of the petroleum industry, but there is no estimation of how much energy is used in distillation columns or other separations. By contrast, for buildings, the EIA's *Residential Energy Consumption Survey, Buildings Energy Data Book,* and other such publications—many of which appear annually—contain substantially more detailed statistics than those available for manufacturing. More frequent and comprehensive collection and publication of such data and analyses are needed.

Investments in industrial energy efficiency technologies are hindered by *market risks caused by uncertainty* about future electricity and natural gas prices and unpredictable long-term product demand.

The *high cost of capital and constrained credit markets* are also significant barriers to energy efficiency improvements in industry. New technologies have to compete for financial and technical resources against projects that achieve other company goals and against familiar technologies. Financial constraints can hinder the diffusion of technologies within industries; a technology may not spread across its potential market owing to the constraints of expected adopters, which do not all have the same ability to raise capital (Canepa and Stoneman, 2004). In addition, if the technology involved is new to the market in question, even if it is well demonstrated elsewhere, the problem of raising capital may be further exacerbated.

Capital market barriers can inhibit efficiency purchases. Although, in theory, firms might be expected to borrow capital any time that a profitable investment opportunity presents itself, in practice firms often ration capital—that is, they impose internal limits on capital investment. The result is that mandatory investments (e.g., required by environmental or health regulations) and those that are most central to the firms' product line often are made first. Projects to increase capacity or bring new products to the market typically have priority over energy cost-cutting investments.[19]

In the United States, firms can face fiscal policies unfavorable to investments in end-use efficiency. The current federal tax code discourages capital investments in general, as opposed to the direct expensing of energy costs. More specifically, tax credits designed to encourage technology adoption are limited by alternative minimum tax rules, tax credit ceilings, and limited tax credit carryover to following years; these limitations prevent the credits from being used to their full potential by qualified companies. Furthermore, outdated tax depreciation rules require firms to depreciate energy efficiency investments over a longer period of time than many other investments (e.g., only 5 years for a new data center), making these investments less cost-effective than other investment options (Brown and Chandler, 2008). This is partly because energy-efficient products have long depreciable lives, such as 15 years for a new motor or a new industrial boiler. An illustration of the consequence of these depreciation rules is that a new backup generator would be depreciated over 3 years, whereas a CHP system would be depreciated over 20 years. The CHP system would provide both reliability and energy efficiency; the backup generator, however, provides reliability at the expense of energy efficiency and clean air. This is another case of legislation lagging behind (and inhibiting) technological progress. Federal depreciation schedules were put into place more than two decades ago as part of the IRS Reform Act of 1986 (P.L. 99-514), and they have not kept up with technological innovations. A modification of depreciation schedules would remove a significant barrier to industrial efficiency investments, but it would require legislative action (Brooks et al., 2006).

Regulatory barriers can also inhibit energy-saving improvements at industrial facilities. For example, as part of the 1977 Clean Air Act Amendments (P.L. 95-95; 91 Stat. 685), Congress established the New Source Review (NSR) program and modified it in the 1990 amendments, but it exempted old coal plants and industrial facilities from the New Source Performance Standards (NSPS) intended to promote the use of the best air pollution control technologies, taking into account the cost of such technology and any other non–air quality, health, and environmental impact and energy requirements. However, investment in an upgrade could trigger an NSR, and the threat of such a review has prevented many upgrades. NSR thus imposes pollution controls where they are least needed

---

[19]Sergio Dias, Northwest Energy Efficiency Alliance, personal communication, November 8, 2006.

and artificially inflates the value of the dirtiest plants. Altogether, these effects have led some critics to question whether the NSR program and the NSPS have resulted in higher levels of pollution than would have occurred in the absence of regulation (Brown and Chandler, 2008).

4.6 THE BUSINESS CASE FOR ENERGY EFFICIENCY

Other than subsidies, regulation, and supporting public policies, what might motivate industry to improve its energy efficiency? What are the most important leverage points for motivating efficiency improvements? Some of the most important of these drivers are described below.

- *Rising energy prices.* The sustained pain of rising oil, coal, natural gas, and electricity prices is motivating a renewed interest in energy efficiency. To remain competitive, industry must find ways to reduce its energy consumption, and higher energy costs can make efficiency investments more cost-competitive. Of course, the cost of efficiency investments can also rise with energy costs (perhaps lagged by a few years). Thus, the excitement over finally being able to justify an alternative-fuel or energy-reduction project because of recent energy cost increases is often dampened by the discovery of an accompanying rise in the cost of equipment and materials.
- *Environmental concerns and regulations.* Many states are allowing industry to use energy efficiency to qualify for NO_x and SO_2 offsets in non-attainment zones. With a lowering of acceptable ozone concentrations, many additional counties in the United States are going to be in non-attainment. Title IV SO_2 allowances are now trading at less than \$100/ton, and NO_x is trading at less than \$1000/ton. At higher prices, these allowances could provide a lucrative stream of payments for many industrial efficiency investments.[20] Most energy policy analysts forecast

[20]The average weighted price for a ton of SO_2 in 2009 was \$69.74 (http://www.epa.gov/airmarkt/trading/2009/ 09summary.html). The average price for a ton of NO_x in March 2009 was roughly \$625/ton (seasonal) (http://www.ferc.gov/ market-oversight/othr-mkts/emiss-allow/othr-emns-no-so-pr.pdf).

that there will be tradable allowances for CO_2 sometime in the next several years.

- *Demand charges and demand-response incentives.* The ability of industry to cut peak electric loads is a motivator for utilities to incentivize demand response (shifting loads to off-peak periods) in industry. Industrial energy efficiency measures that reduce energy demand (or slow its growth) can also help utilities meet energy needs, so promoting such savings can be in the utilities' interest. In combination with peak-load pricing for electricity, energy efficiency and demand response can be a lucrative enterprise for industrial customers.
- *Collateral benefits.* Secondary or collateral benefits such as increased productivity, improved product quality, reduced labor costs, and enhanced reliability are often strong drivers for energy efficiency improvements (Worrell et al., 2003). This was illustrated effectively in *Cool Companies: How the Best Companies Boost Profit and Productivity by Cutting Greenhouse Gas Emissions* (Romm, 1999), which describes the many ways that corporations have benefited from increasing the energy efficiency of their operations.
- *International competition.* If a company cannot sell its products because of the cost of the energy needed to produce them relative to the costs of domestic or international competitors, attention may turn to energy efficiency improvements in the manufacturing process.
- *Corporate sustainability.* Voluntarily reducing greenhouse gas emissions and implementing climate change mitigation strategies offer ways to boost shareholder and investor confidence, profit from future legislation, access new markets, lower insurance costs, avoid liability, offer competitive benefits, and prevent and prepare for physical and market damage caused by further climate impact. Almost all of the Fortune 500 companies are publishing corporate responsibility reports. Many companies are setting energy efficiency goals (e.g., Johnson and Johnson, BP, Exxon, Dupont). Similarly, ISO 14000 certification informs the public about the nature of the production processes and is being required by DOE, Dow, and others.
- *Shareholder activism, good corporate governance, and reputation management* are other potential drivers of energy efficiency in industry. ENERGY STAR® designations and other government programs that recognize outstanding environmental performance by corporate

America have proven to be strong motivators of resource and energy conservation.

- *Insurance access and costs, legal compliance, and concerns regarding fiduciary duty* are additional business case drivers for managing greenhouse gas emission reductions through energy efficiency (Natural Edge Project, 2005).

4.7 FINDINGS

The following findings derive from the panel's analysis of industrial efficiency summarized in this chapter.

I.1 Independent studies using different approaches agree that the economic potential for improved energy efficiency in industry is large. Of the 34.3 quads of energy forecasted to be consumed by U.S. industry in 2020 (EIA, 2008), 14–22 percent could be saved through cost-effective energy efficiency improvements (those with an internal rate of return of at least 10 percent or that exceed a company's cost of capital by a risk premium). These innovations would save 4.9–7.7 quads annually by 2020.

I.2 Additional efficiency investments could become cost-competitive through energy RD&D. Enabling and crosscutting technologies, such as advanced sensors and controls, microwave processing of materials, nanoceramic coatings, and high-temperature membrane separation, can provide efficiency gains in many industries as well as throughout the energy system—for example, in vehicles, feedstock conversion, and electricity transmission and distribution.

I.3 Industry has experienced a significant shift to offshore manufacturing of components and products. If the net energy embodied in imports and exports is taken into account, the energy consumption attributable to industry would be increased by 5 quads.

I.4 Energy-intensive industries such as aluminum, steel, and chemicals have devoted considerable resources to increasing their energy efficiency. For many other industries, energy represents 10 percent or less of costs and is not a priority. Energy efficiency investments compete for human and financial resources with other goals such as increased production, improved productivity, introduction of new products, and compliance

with environment, safety, and health requirements. Outdated capital depreciation schedules, backup fees for combined heat and power systems, and other policies also hamper energy efficiency investment.

I.5 More detailed data, collected more frequently, are needed to better assess the status of and prospects for energy efficiency in industry. Proprietary concerns will have to be addressed to achieve this.

I.6 Drivers for energy efficiency in industry include rising and volatile energy prices, intense competitive pressure to lower costs, and an increased focus on corporate sustainability.

4.8 REFERENCES

AFPA (American Forest and Paper Association). 2007. Talking Points for Key Provisions Opposed by AF&PA in H.R. 3221. 2007. Available at http://www.bipac.net/page.asp?g=afpa&content=one-pager-Talking_Points_for_House_Energy_Bill&parent=AFPA.

AISI (American Iron and Steel Institute). 2005. Saving One Barrel of Oil per Ton. Washington, D.C.: AISI. October.

AISI. 2006. Annual Statistical Report. Washington, D.C.: AISI.

American Chemistry Council. 2008. Guide to the Business of Chemistry. Available at http://www.americanchemistry.com/store/detail.aspx?ID=243.

ANL (Argonne National Laboratory). 1990. Environmental Consequences of, and Control Processes for, Energy Technologies. Pollution Technology Review Series, No. 181. Norwich, N.Y.: William Andrew.

Bailey, O., and E. Worrell. 2005. Clean Energy Technologies: A Preliminary Inventory of the Potential for Electricity Generation. Report LBNL-57451. Berkeley, Calif.: Lawrence Berkeley National Laboratory. September.

Banerjee, R., A. Phan, B. Wang, C. Knobler, H. Furukawa, M. O'Keeffe, and O.M. Yaghi. 2008. High-throughput synthesis of zeolitic imidazolate frameworks and application to CO_2 capture. Science 319:939-943.

Battelle (Battelle Memorial Institute). 2002. Toward a Sustainable Cement Industry: Climate Change, Substudy 8, An Independent Study Commissioned by the World Business Council for Sustainable Development. Columbus, Ohio: Battelle Memorial Institute. March.

Berry, R.S, V.A. Kazakov, S. Sieniutycz, Z. Szwast, and A.M. Tsirlin. 1999. Thermodynamic Optimization of Finite-Time Processes. Chichester, U.K.: John Wiley & Sons.

Brooks, S., B. Elswick, and R.N. Elliott. 2006. Combined Heat and Power: Connecting the Gap Between Markets and Utility Interconnection and Tariff Practices (Part 1). Washington, D.C.: American Council for an Energy-Efficient Economy. March.

Brown, M., and S. Chandler. 2008. Governing confusion: How statutes, fiscal policy, and regulations impede clean energy technologies. Stanford Law and Policy Review 19(3):472-509.

Brown, M., J. Chandler, M. Lapsa, and B. Sovacool. 2008. Carbon Lock-In: Barriers to Deploying Climate Change Mitigation Technologies. Report TM-2007/124. Oak Ridge, Tenn.: Oak Ridge National Laboratory. Available at http://www.ornl.gov/sci/btc/pdfs/brown_doc7435_tm124_ 08.pdf.

Brown, M., M. Levine, W. Short, and J. Koomey. 2001. Scenarios for a clean energy future. Energy Policy 29(14):1179-1196.

Canepa, A., and P. Stoneman. 2004. Comparative international diffusion: Patterns, determinants and policies. Economics of Innovation and New Technology 13(3):279-298.

Casten, T., and R. Ayres. 2007. Energy myth eight—Worldwide power systems are economically and environmentally optimal. Pp. 201-238 in Energy and American Society—Thirteen Myths (B.K. Sovacool and M.A. Brown, eds.). New York: Springer.

Chapman, S., G. Leslie, and I.B. Law. 2004. Membrane bioreactors (MBR) for municipal wastewater treatment—An Australian perspective. Journal of the Australian Water Association 4:65.

Cowart, R. 2001. Efficient Reliability: The Critical Role of Demand-Side Resources in Power Systems and Markets. Report to the National Association of Regulatory Utility Commissioners by the Regulatory Assistance Project. Available at http://www.raponline.org/Pubs/General/EffReli.pdf.

DOE (Department of Energy). 2002. Industrial Wireless Technology for the 21st Century. Industrial Wireless Workshop, San Francisco, July 30, 2002. Washington, D.C.: DOE, Office of Energy Efficiency and Renewable Energy.

DOE. 2004. Energy Use, Loss, and Opportunities Analysis: U.S. Manufacturing and Mining. Prepared by Energetics Incorporated and E3M Incorporated. Washington, D.C.: DOE, Industrial Technologies Program. December. Available at http://www1.eere.energy.gov/industry/energy_systems/analysis.html.

DOE. 2005a. Energy and Environmental Profile of the U.S. Pulp and Paper Industry. Washington, D.C.: DOE, Industrial Technologies Program.

DOE. 2005b. U.S. Climate Change Technology Program—Technology Options for the Near and Long Term. Washington, D.C.: DOE, Climate Change Technology Program. August. Available at http://www.climatetechnology.gov.

DOE. 2006a. Chemical Bandwidth Study: Energy Analysis: A Powerful Tool for Identifying Process Inefficiencies in the U.S. Chemical Industry. Draft Summary Report. Study conducted for the U.S. Department of Energy by JVP International, Incorporated, and Psage Research, LLC. Summary report prepared by Energetics Incorporated. Washington, D.C.: DOE. December. Available at http://www1.eere.energy.gov/industry/chemicals/bandwidth.html.

DOE. 2006b. Energy Bandwidth for Petroleum Refining Processes. Prepared by Energetics Incorporated. October. Washington, D.C.: DOE. Available at http://www1.eere.energy.gov/industry/petroleum_refining/bandwidth.html.

DOE. 2006c. Pulp and Paper Industry Energy Bandwidth Study. Prepared by Jacobs Engineering Group and Institute of Paper Science and Technology for the American Institute of Chemical Engineers and DOE. Washington, D.C.: DOE. August. Available at http://www1.eere.energy.gov/industry/forest/bandwidth.html.

DOE. 2006d. Strategic Plan. Climate Change Technology Program. Washington, D.C.: DOE. September.

DOE. 2007a. Energy and Environmental Profile of the U.S. Petroleum Refining Industry. Prepared by Energetics Incorporated. Washington, D.C.: DOE, Industrial Technologies Program. November. Available at http://www1.eere.energy.gov/industry/petroleum_refining/analysis.html.

DOE. 2007b. Energy Technology Solutions: Public-Private Partnerships Transforming Production. Washington, D.C.: DOE, Energy Efficiency and Renewable Energy. October.

DOE. 2008. U.S. Steel Industry Energy Efficiency Fact Sheet. Washington, D.C.: Climate Vision Partnership, Sector Initiatives: Iron and Steel. Available at http://www.climatevision.gov/sectors/steel/index.html. August 7.

DOE/EERE (Department of Energy/Office of Energy Efficiency and Renewable Energy). 2005. Infrared Heating Technology: Crosscutting Applications, Widespread Benefits. March. Available at http://www1.eere.energy.gov/industry/imf/pdfs/ irheatingmarch05.pdf.

DOE/EERE. 2007. IMPACTS—Industrial Technologies Program: Summary of Program Results for CY 2005. Washington, D.C.: Department of Energy, Office of Energy Efficiency and Renewable Energy. February. Available at http://www1.eere.energy.gov/industry/ about/pdfs/impact2005.pdf.

DOE/EERE. 2008. Energy Intensity Indicators in the U.S., Highlights of Trends, Industrial Total Energy Consumption. April 14. Available at http://www1.eere.energy.gov/ba/pba/intensityindicators/total_industrial.html.

EIA (Energy Information Administration). 1998. Annual Energy Outlook 1998. Washington, D.C.: Department of Energy, Energy Information Administration.

EIA. 1999. Annual Energy Outlook 1999. Washington, D.C.: Department of Energy, Energy Information Administration.

EIA. 2002. 2002 Manufacturing Energy Consumption Survey. Washington, D.C.: Department of Energy. Available at http://www.eia.doe.gov/emeu/efficiency/mecs_trend_9802/ mecs_trend9802.html.

EIA. 2004. Monthly Energy Review. Washington, D.C.: Department of Energy. Available at http://www.eia.doe.gov/emeu/mer/contents.html.

EIA. 2007. Annual Energy Outlook 2007. DOE/EIA-0383(2007). Washington, D.C.: Department of Energy, Energy Information Administration.

EIA. 2008. Annual Energy Outlook 2008. DOE/EIA-0383(2008). Washington, D.C.: Department of Energy, Energy Information Administration.

EIA. 2009a. International Energy Outlook. Washington, D.C.: Department of Energy, Energy Information Administration.

EIA. 2009b. Monthly Energy Review (July). Washington, D.C.: Department of Energy, Energy Information Administration.

Expert Group on Energy Efficiency. 2007. Realizing the Potential of Energy Efficiency: Targets, Policies, and Measures for G8 Countries. United Nations Foundation Expert Report. Washington, D.C.

Fruehan, R.J. 2008. Future Steelmaking Processes. Materials Science and Engineering Department, Carnegie Mellon University. Available at http://www.osti.gov/bridge/servlets/purl/840930-yRKP7T/webviewable/840930.PDF.

Fruehan, R., O. Fortini, H. Paxton, and R. Brindle. 2000. Theoretical Minimum Energies to Produce Steel for Selected Conditions. Report prepared by Carnegie Mellon University for the Department of Energy. March. Available at http://www1.eere.energy.gov/industry/steel/pdfs/theoretical_minimum_energies.pdf.

Goldman, C., Lawrence Berkeley National Laboratory. 2006. Utility Experience with Real-Time Pricing. Presentation at the Workshop on Smart Meters and Time-Based Rates, Montepelier, Vt., March 15. Available at http://www.state.vt.us/psb/document/ElectricInitiatives/GoldmanVermontPSBpricingworkshop031506.pdf/.

IDA (International Desalination Association). 1995. Proceedings of International Desalination Association World Congress, International Desalination and Water Reuse Conference, Abu-Dhabi, United Arab Emirates, 1995. Topsfield, Mass.: IDA.

IEA (International Energy Agency). 2004. Oil Crises and Climate Challenges—30 Years of Energy Use in IEA Countries. Paris: Organisation for Economic Cooperation and Development (OECD)/IEA.

IEA. 2007. Tracking Industrial Energy Efficiency and CO_2 Emissions. Paris, France: IEA.

IEA. 2009. Cogeneration and District Energy: Sustainable Energy Technologies for Today and Tomorrow. Paris, France: IEA. Available at http://www.iea.org/files/CHPbrochure09.pdf.

IPCC (Intergovernmental Panel on Climate Change) 2007. Climate Change 2007: Mitigation of Climate Change. Contribution of Working Group III to the Fourth Assessment Report of the Intergovernmental Panel on Climate Change. B. Metz, O.R. Davidson, P.R. Bosch, R. Dave, L.A. Meyer, eds. Cambridge, United Kingdom, and New York: Cambridge University Press.

IWG (Interlaboratory Working Group on Energy-Efficient and Clean Energy Technologies). 2000. Scenarios for a Clean Energy Future. ORNL/CON-476 and LBNL-44029. Oak Ridge National Laboratory and Lawrence Berkeley National Laboratory. November.

Jimenez, E.S, P. Salamon, R. Rivero, C. Rendon, K.H. Hoffmann, M. Schaller, and B. Andresen. 2004. Optimization of a diabatic distillation column with sequential heat exchangers. Industrial and Engineering Chemistry Research 43:7566-7571.

KEMA, Inc. 2006. California Industrial Existing Construction Energy Efficiency Potential Study, Volumes 1 and 2. Final Report to Pacific Gas and Electric Company. Prepared with assistance from Lawrence Berkeley National Laboratory and Quantum Consulting. Arnhem, The Netherlands: KEMA, Inc. May.

King, D. 2006. Electric Power Micro-grids: Opportunities and Challenges for an Emerging Distributed Energy Architecture. Doctoral Thesis. Pittsburgh, Pa.: Carnegie Mellon University. May.

Laitner, J.A., and M.A. Brown. 2005. Emerging industrial innovations to create new energy efficient technologies. In Proceedings of the Summer Study on Energy Efficiency in Industry. Washington, D.C.: American Council for an Energy-Efficient Economy.

Lave, L.B. 2009. The potential of energy efficiency: An overview. The Bridge 39(2):5-14.

LBNL (Lawrence Berkeley National Laboratory). 2005. Energy Efficiency Improvement and Cost Saving Opportunities for Petroleum Refineries, An ENERGY STAR® Guide for Energy and Plant Managers. LBNL-57260-Revision. Prepared by C. Galitsky, S. Chang, E. Worrell, and E. Masanet. Berkeley, Calif.: LBNL. February.

Lemar, P.L. 2001. The potential impact of policies to promote combined heat and power in U.S. industry. Energy Policy 29(14):1243-1254.

Lin, H., E. Van Wagner, B.D. Freeman, L.G. Toy, and R.P. Gupta. 2006. Plasticization-enhanced hydrogen purification using polymeric membranes. Science 311:639-642.

Lovins, A.B. 2007. Energy myth nine—Energy efficiency improvements have already reached their potential. Pp. 239-264 in Energy and American Society—Thirteen Myths. B.K. Sovacool and M.A. Brown, eds. New York: Springer.

Martin, N., N. Anglani, D. Einstein, M. Khrushch, E. Worrell, and L.K. Price. 2000a. Opportunities to Improve Energy Efficiency and Reduce Greenhouse Gas Emissions in the U.S. Pulp and Paper Industry. LBNL-46141. Berkeley, Calif.: Lawrence Berkeley National Laboratory.

Martin, N., W. Worrell, M. Ruth, L. Price, R.N. Elliott, A.M. Shipley, and J. Thorne. 2000b. Emerging Energy-Efficient Industrial Technologies. Berkeley, Calif.: Lawrence Berkeley National Laboratory. October.

McKinsey and Company. 2007. The Untapped Energy Efficiency Opportunity in the U.S. Industrial Sector. Report prepared for DOE, Office of Energy Efficiency and Renewable Energy. New York: McKinsey and Company.

McKinsey and Company. 2008. The Untapped Energy Efficiency Opportunity of the U.S. Industrial Sector: Details of Research, 2008. New York: McKinsey and Company.

Mullins, O.C., and R.S. Berry. 1984. Minimization of entropy production in distillation. Journal of Physical Chemistry 88:723-728.

NRC (National Research Council). 2007. Prospective Evaluation of Applied Energy Research and Development at DOE (Phase Two). Washington, D.C.: The National Academies Press.

NREL (National Renewable Energy Laboratory). 2002. Chemical Industry of the Future: Resources and Tools for Energy Efficiency and Cost Reduction Now. Report DOE/GO-102002-1529 and NREL/CD-840-30969. October. Available at http://www.nrel.gov/docs/fy03osti/30969.pdf.

Natural Edge Project. 2005. Prospering in a Carbon Constrained World. Queensland, Australia: Natural Edge Project. May.

Nenoff, T.M., M. Ulutagay-Kartin, R. Bennett, K. Johnson, G. Gray, T. Anderson, M. Arruebo, R. Noble, J. Falconer, X. Gu, and J. Dong. 2006. Novel Modified Zeolites for Energy-Efficient Hydrocarbon Separations. SAND2006-6891. Albuquerque, N.Mex.: Sandia National Laboratories.

Optimal Energy, Inc. 2003. Energy Efficiency and Renewable Energy Resource Development Potential in New York State. Final Report. Prepared for the New York State Energy Research and Development Authority. August. Available at http://www.nyserda.org/energy_Information/otherdocs.asp#EERER.

Orlov, V.N., and R.S. Berry. 1991. Estimation of minimal heat consumption for heat driven separation processes via methods of finite-time thermodynamics. Journal of Physical Chemistry 95:5624-5628.

ORNL (Oak Ridge National Laboratory) and BCA, Inc. 2005. Materials for Separation Technologies: Energy and Emission Reduction Opportunities. Washington, D.C.: Department of Energy, Office of Energy Efficiency and Renewable Energy. Available at http://www.eere.energy.gov/industry/imf/analysis.html.

Pace Energy Project. 2002. Combined Heat and Power Market Potential in New York State. Prepared for the New York State Energy Research and Development Authority, Albany, N.Y. May. Available at http://www.icfi.com/markets/energy/doc_files/eea-chp-ny.pdf.

Romm, J.J. 1999. Cool Companies: How the Best Companies Boost Profit and Productivity by Cutting Greenhouse Gas Emissions. New York, N.Y.: Island Press.

Schaller, M., K.H. Hoffmann, G. Siragusa, P. Salamon, and B. Andresen. 2001. Optimized performance in diabatic distillation columns. Computers and Chemical Engineering 25:1537-1548.

Schipper, L. 2004. International comparisons of energy end use: Benefits and risks. Pp. 1-27 in Encyclopedia of Energy. Volume 3. C.J. Cleveland, ed. Amsterdam: Elsevier.

Shipley, A., A. Hampson, B. Hedman, P. Garland, and P. Bautista. 2008. Combined Heat and Power: Effective Energy Solutions for a Sustainable Future. ORNL/TM-2008/224. Oak Ridge, Tenn.: Oak Ridge National Laboratory. December. Available at http://www1.eere.energy.gov/industry/distributedenergy/.

Sieniutycz, S., and P. Salamon, eds. 1990. Finite-Time Thermodynamics and Thermoeconomics, Advances in Thermodynamics. Volume 4. New York: Taylor and Francis.

Sovacool, B.K., and R.F. Hirsh. 2007. Energy myth six: The barriers to new and innovative energy technologies are primarily technical: The case of distributed generation (DG). Pp. 145-169 in Energy and American Society—Thirteen Myths. B.K. Sovacool and M.A. Brown, eds. New York: Springer.

Thorp IV, B.A., and L.D. Murdock-Thorp. 2008. Compelling case for integrated biorefineries. Paper presented at the 2008 PAPERCON Conference, May 4-7, Dallas, Texas. Norcross, Ga.: Technical Association of the Pulp and Paper Industry.

U.S. Census Bureau. 2007. 2006 Annual Survey of Manufacturers (ASM). Data available at http://www.census.gov/mcd/asmhome.html.

van Oss, H.G. 2005. Background Facts and Issues Concerning Cement and Cement Data. Open-File Report 2005-1152. Washington, D.C.: U.S. Department of the Interior, USGS.

Weber, C. 2008. Energy Use and Flows in the U.S. Economy, 1997-2002. Prepared for the National Academy of Sciences by Carnegie Mellon University, Pittsburgh, Pa. August.

Worrell, E., and G. Biermans. 2005. Move over! Stock turnover, retrofit, and industrial energy efficiency. Energy Policy 33:949-962.

Worrell, E., and C. Galitsky. 2004. Energy Efficiency Improvement and Cost Saving Opportunities for Cement Making—An ENERGY STAR® Guide for Energy and Plant Managers. LBNL-54036. Berkeley, Calif.: Lawrence Berkeley National Laboratory.

Worrell, E., and M. Neelis. 2006. World's Best Practice Energy Intensity Values for Selected Industrial Sectors. LBNL-62806. Berkeley, Calif.: Lawrence Berkeley National Laboratory.

Worrell, E., and L. Price. 2001. Policy scenarios for energy efficiency improvement in industry. Energy Policy 29(14):1223-1241.

Worrell, E., C. Galitsky, and L. Price. 2008. Energy Efficiency Improvement Opportunities for the Cement Industry. Berkeley, Calif.: Lawrence Berkeley National Laboratory. January.

Worrell, E., J. Laitner, M. Ruth, and H. Finman. 2003. Productivity benefits of industrial energy efficiency measures. Energy, the International Journal 28(11):1081-1098.

Worrell, E., L. Price, and C. Galitsky. 2004. Emerging Energy-Efficient Technologies in Industry: Case Studies of Selected Technologies. LBNL-54828. Berkeley, Calif.: Lawrence Berkeley National Laboratory. May.

Worrell, E., L. Price, N. Martin, C. Hendriks, and L.O. Meida. 2001. Carbon dioxide emissions from the global cement industry. Annual Review of Energy and the Environment 26:303-329.

Wright, A., M. Martin, B Gemmer, P. Scheihing, and J. Quinn. 2007. Results from the U.S. DOE 2006 Save Energy Now Assessment initiative: DOE's Partnership with U.S. Industry to Reduce Energy Consumption, Energy Costs, and Carbon Dioxide Emissions. ORNL/TM-2007/138. Oak Ridge, Tenn.: Oak Ridge National Laboratory.

WSA (World Steel Association) Statistical Archive. 2008. Available at http://www.worldsteel.org/?action=stats_search. Accessed December 2, 2008.

Xenergy, Inc. 1998. United States Industrial Electric Motor Systems Market Opportunities Assessment. Washington, D.C.: Department of Energy, Office of Energy Efficiency and Renewable Energy. Available at http://www1.eere.energy.gov/industry/bestpractices/pdfs/mtrmkt.pdf.

5 Overarching Findings and Lessons Learned from Federal and State Energy Efficiency Policies and Programs

The opportunities described in Chapters 2 through 4 to improve energy efficiency in, respectively, U.S. residential and commercial buildings, the U.S. transportation sector, and U.S. industrial manufacturing are summarized here in Table 5.1, which presents the panel's conservative and optimistic estimates for cost-effective annual energy savings available in these three sectors in 2020 and 2030.[1] The panel's estimates are not projections; they reflect its assessments of technology potential assuming a rapid rate of deployment, but a rate nonetheless consistent with past deployment rates. If society were to give a higher priority to efficiency, perhaps because of higher energy prices, energy shortages, or concern about greenhouse gas emissions, deployment rates would be faster and the savings would be greater.

To achieve the energy-savings potential outlined in Table 5.1, the manner in which Americans use energy will have to be transformed, and policy actions will doubtless be an integral part of this transformation. Although policy recommendations are outside the scope of this study, in order to inform the policy debate and contribute to a better understanding of how impediments can be overcome, the panel reviewed some of the experience with—and importantly, lessons learned from—policies and programs aimed at influencing energy use in the United States.

---

[1]As discussed in Chapter 3, "Energy Efficiency in Transportation," the focus of that assessment relates to technologies that could power the nation's cars and light trucks. If other categories, such as heavy-duty vehicles and aviation, had been included in the analysis, the panel's estimate of the total savings would be greater. Forthcoming National Research Council reports will provide estimates for these two categories.

TABLE 5.1 Panel Estimate of the Potential for Cost-Effective Annual U.S. Energy Savings (in quads) Achievable with Energy Efficiency Technologies in 2020 and 2030

	Conservative Estimate		Optimistic Estimate	
	2020	2030	2020	2030
Buildings, primary (source) electricity	9.4	14.4	9.4	14.4
Residential	4.4	6.4	4.4	6.4
Commercial	5.0	8.0	5.0	8.0
Buildings, natural gas	2.4	3.0	2.4	3.0
Residential	1.5	1.5	1.5	1.5
Commercial	0.9	1.5	0.9	1.5
Transportation, light-duty vehicles	2.0	8.2	2.6	10.7
Industry, manufacturing	4.9	4.9	7.7	7.7
Total	18.6	30.5	22.1	35.8

Note: Savings are relative to the reference scenario of the EIA's *Annual Energy Outlook 2008* (EIA, 2008) or, for transportation, a similar scenario developed by the panel. See Table 1.2 for more information on the baselines used in the panel's analysis of the buildings, transportation, and industry sectors.

5.1 OVERARCHING FINDINGS

On the basis of its estimates of the energy savings potential outlined in Table 5.1, the panel presents the following overarching finding:

Overarching Finding 1

Energy-efficient technologies for residences and commercial buildings, transportation, and industry exist today, or are expected to be developed in the normal course of business, that could potentially save 30 percent of the energy used in the U.S. economy while also saving money. If energy prices are high enough to motivate investment in energy efficiency, or if public policies are put in place that have the same effect, U.S. energy use could be lower than business-as-usual projections by 19–22 quadrillion Btu (17–20 percent) in 2020 and by 30–36 quadrillion Btu (25–31 percent) in 2030.[2,3]

[2]The basis for comparison for the buildings and industry sectors is the reference scenario of the U.S. Department of Energy's Annual Energy Outlook 2008, produced by the Energy Information Administration (EIA, 2008), and the panel's similar but slightly modified baseline for the transportation sector.

[3]The Committee on America's Energy Future report (NAS-NAE-NRC, 2009) estimated the amount of possible savings as 15–17 quads (about 15 percent) by 2020 and 32–35 quads (about

A savings of the amount of energy estimated in Overarching Finding 1 would reverse the growth in energy use forecasted by the Department of Energy's Energy Information Administration (EIA, 2008). Instead of increasing from 99 quadrillion Btu (99 quads) in 2008 to 111 quads in 2020 and 118 quads in 2030, as forecast by the EIA (2008), full deployment of cost-effective, energy-efficient technologies would cause U.S. energy use to fall to 89–92 quads in 2020 and 82–88 quads in 2030.

Table 5.1 shows that reducing electricity use in buildings provides the greatest opportunity for energy savings. In fact, these potential savings are so large that, as indicated in Overarching Finding 2, the effects on electricity generation could be dramatic.

Overarching Finding 2

The full deployment of cost-effective, energy-efficient technologies in buildings alone could eliminate the need to add to U.S. electricity generation capacity. Since the estimated electricity savings in buildings from Table 5.1 exceeds the EIA forecast for new net electricity generation in 2030, implementing these efficiency measures would mean that no new generation would be required except to address regional supply imbalances, replace obsolete generation assets, or substitute more environmentally benign generation sources.

The potential savings summarized above are very attractive. As discussed in Chapters 2 through 4, however, many barriers to the deployment of energy-efficient technologies exist, even though the adoption of such technologies is projected to save money over time. These barriers include potentially high up-front costs, alternative uses for investment capital deemed more attractive, the volatility of energy prices leading to uncertainty with respect to the payback time, and the lack of information available to consumers about the relative performance and costs of technology alternatives.

---

30 percent) by 2030. Since the release of that report, further analysis by the panel refined the amount of possible savings in 2020 to 17–20 percent.

Overarching Finding 3

The barriers to improving energy efficiency are formidable. Overcoming these barriers will require significant public and private support, as well as sustained initiative. The experience of leading states provides valuable lessons for national, state, and local policy makers in the leadership skills required and the policies that are most effective.

One valuable lesson learned is that the long lifetimes of buildings and some capital equipment present a particularly important barrier to implementing energy-efficient technologies. These investments—particularly buildings—can last for decades or even centuries, blocking the implementation of more efficient substitutes.

Overarching Finding 4

Long-lived capital stock and infrastructure can lock in patterns of energy use for decades. Thus, it is important to take advantage of opportunities (during the design and construction of new buildings or major subsystems, for example) to insert energy-efficient technologies into these long-lived capital goods.

In the rest of this chapter the panel discusses this and other examples of valuable experience gained from the implementation of federal and state policies aimed at overcoming barriers to energy savings. The review below concentrates on federal actions, but it also covers some actions taken in two large states, California and New York, as well as some policies adopted by electric utilities.

5.2 ENERGY EFFICIENCY POLICIES AND PROGRAMS

Between 1975 and 1980 the federal government adopted a number of laws that established educational efforts and financial incentives for energy efficiency, and it authorized the setting of efficiency standards. More recent legislation has established minimum efficiency standards for a wide range of household appliances and equipment used in the commercial and industrial sectors, as well as tax incentives to stimulate the commercialization and adoption of highly efficient products and buildings. Over the past 30 years the federal government has also devoted billions of dollars to energy efficiency research and development. In addition, many states have implemented building energy codes, utility-based energy efficiency programs, and other policies to complement these federal initiatives.[4]

5.2.1 Vehicle Efficiency Standards

In 1975 the United States adopted energy efficiency standards—known as corporate average fuel economy (CAFE) standards—for cars and light trucks. These standards played the leading role in the near doubling of the average fuel economy of new cars and the 55 percent increase in light-truck fuel economy from 1975 to 1988 (Greene, 1998). In addition, a tax on inefficient "gas guzzlers" contributed to the rise in vehicle fuel economy during the late 1970s and 1980s (Geller and Nadel, 1994). Had these efficiency improvements not been implemented, the U.S. car and light truck fleet would have consumed an additional 2.8 million barrels per day (bbl/d) of gasoline in 2000 (NRC, 2002). The gasoline savings meant lower levels of oil imports and consequently lower trade deficits in the United States compared with what they would have been otherwise. The CAFE standards were met mainly through technological improvements in engines and drivetrains, as well as through vehicle weight reduction (NRC, 2002).

The original CAFE standards for cars reached their maximum level in 1985; small increases in the standards for light trucks have been adopted since then.[5] With no further increase in standards, the average fuel economy of each type of vehicle (cars and light trucks) remained nearly constant during the 1987–2007 period. In fact, the combined average fuel economy of new cars and light trucks actually declined from a high of 22.0 miles per gallon (mpg) in 1987 to 20.2 miles per gallon in 2006–2007 (estimated on-road performance, not rated fuel economy), due mainly to the shift from cars toward less-efficient sport utility vehicles (SUVs), pickup trucks, and minivans (EPA, 2007a). As a result of declining new-vehicle fuel economy and increasing vehicle-miles traveled, U.S. gasoline consumption increased 31 percent from 1986 through 2006 (EIA, 2007).

---

[4]This review does not consider energy tax increases that have been enacted over the past 30 years, because such increases have been very modest. The federal tax on gasoline, for example, was increased incrementally from 4¢/gal in 1973 to a total of 18.4¢/gal by 1993, but it has not been increased since then. Corrected for inflation, the gasoline tax in 2006 was only 26 percent greater than it was in 1973.

[5]Small increases in the light-truck standards were adopted through 2004. More significant but still modest increases were administered starting in 2005. The Energy Independence and Security Act of 2007 mandated more substantial increases, slated to amount to at least a 40 percent increase over the 2005 level by 2020.

One of the flaws in the original CAFE standards was the lower standards for SUVs and other trucks relative to standards for cars, thereby encouraging manufacturers to redesign trucks to serve as passenger vehicles (Gerard and Lave, 2003). However, other factors also contributed to the shift from cars to light trucks, making it difficult to determine the role of CAFE in this regard (Greene, 1998; NRC, 2002).

Auto manufacturers blocked efforts to increase the standards for many years despite numerous studies showing that raising the standards was technically and economically feasible (NRC, 2002; Difiglio et al.,1990; Greene and DeCicco, 2000). Pressure to raise the standards grew, however, as energy security concerns increased. The U.S. Congress enacted the first significant increase in the CAFE standards in more than 30 years as part of the Energy Independence and Security Act (EISA; Public Law 110-140), which was signed into law by President George W. Bush in December 2007. EISA requires the Department of Transportation to set tougher fuel-economy standards starting in 2011 until the standards reach at least 35 mpg for cars and light trucks combined in 2020—a 40 percent increase over the current standards.[6] EISA also gradually phases out the fuel-economy credits for dual-fuel vehicles, a policy that reduced the effectiveness of the CAFE standards without significantly increasing the use of alternative fuels.

It is estimated that the new CAFE standards will save 1.0 million bbl/d of gasoline by 2020 and 2.4 million bbl/d by 2030, while providing more than $50 billion in net economic benefits for consumers (ACEEE, 2007). These estimates include a "rebound effect"—that is, the increase in travel demand due to the reduction in the cost per mile driven as vehicle fuel economy improves. This effect is generally thought to be real but small (Greene, 1998; NRC, 2002; Small and Van Dender, 2007).

5.2.2 Appliance Efficiency Standards

Appliance efficiency standards were first enacted by states—including California, New York, Massachusetts, and Florida—during the late 1970s and early 1980s (Nadel, 2002). Appliance manufacturers, disturbed by the patchwork of state standards, then supported the adoption of uniform national standards. National

---

[6]The Obama administration recently proposed that these requirements, specified by Subtitle A of EISA 2007, be accelerated.

standards, developed through negotiations between manufacturers and energy efficiency advocates, first became law in 1987. These standards led to dramatic improvements in the energy efficiency of new refrigerators, air conditioners, clothes washers, and other appliances sold in the United States.

In 1992, minimum efficiency standards were extended to motors, heating and cooling equipment used in commercial buildings, and some types of lighting products. In 2005, standards were adopted for a variety of "second-tier" products, including torchiere light fixtures, commercial clothes washers, exit signs, distribution transformers, ice makers, and traffic signals. With the addition of these new products, national minimum efficiency standards were in place for more than 40 types of products.

Appliance efficiency standards eliminate the least efficient products from the marketplace. At times, such standards have been technology forcing—meaning that few if any products could meet the standard at the time that it was established. This was the case for the standards for refrigerators and clothes washers set by the U.S. Department of Energy (DOE) (Nadel, 2002; Goldstein, 2007). The DOE is authorized to strengthen the minimum efficiency standards on a particular product if it determines that doing so is technologically feasible and economically justified.

It is estimated that national appliance efficiency standards saved 88 terawatt-hours (TWh) of electricity in 2000, or 2.5 percent of national electricity use that year (Nadel, 2002). The retirement of less efficient, older appliances, combined with the adoption since 2000 of new and updated standards, is expected to result in energy savings of 268 TWh in 2010, or 6.9 percent of projected national electricity use in that year, and 394 TWh by 2020, or 9.1 percent of projected national electricity use in that year (Nadel et al., 2006).

The appliance standards laws include initial energy performance requirements, but they also direct the DOE to review them periodically and to adopt more stringent standards if technically feasible and economically justified. For example, the standards on refrigerators and freezers first adopted in 1987 have been significantly strengthened twice since then. As shown in Figure 5.1, the combination of federal and state standards resulted in a 70 percent reduction in the average electricity use of new refrigerators sold in the United States from 1972 through 2001; during this period the price (in constant dollars) also fell by 62 percent, while the refrigerated volume actually increased. New standards on fluorescent lighting ballasts were adopted in 2000, followed by new standards on water heaters, clothes washers, and central air conditioners and heat pumps. Despite

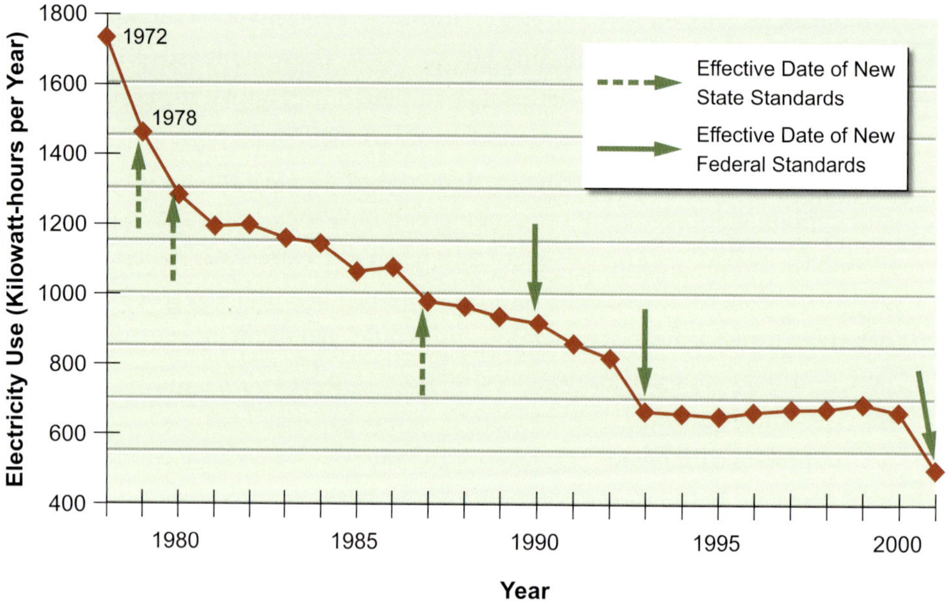

FIGURE 5.1 *Average annual electricity consumption of new refrigerators sold in the United States, 1972–2001.*
Source: Geller, 2003.

completing these revisions, the DOE has missed legal deadlines for updating standards for about 20 other products. These delays have reduced the energy savings and economic benefits of appliance efficiency standards.

Additional appliance efficiency standards were included in EISA. Most noteworthy are those on general-service lamps, standards that will make it illegal to sell ordinary incandescent lamps after the standards take effect. In Phase One, which takes effect in three stages from 2012 to 2014, manufacturers will be able to produce and sell improved incandescent lamps as well as compact fluorescent lamps (CFLs) and light-emitting diode (LED) lamps that meet the efficiency requirements—that is, the minimum lumens of light output per watt of power consumption. In Phase Two, which takes effect in 2020, only CFLs and LED lamps will qualify unless manufacturers are able to roughly triple the efficiency of incandescent lamps. It is estimated that these new standards will save 59 TWh per year by 2020, in addition to the savings from standards on other products (ACEEE, 2007).

5.2.3 Building Energy Codes

Federal legislation passed in 1976 called for the adoption of national standards for building energy efficiency (also known as building energy codes). The building industry strongly opposed this policy, however, and it was eventually converted to voluntary guidelines and design tools (Clinton et al., 1986). Meanwhile, many states and localities adopted mandatory energy codes for new homes and commercial buildings. Model codes, such as the International Energy Conservation Code (IECC) and American Society of Heating, Refrigerating, and Air-Conditioning Engineers (ASHRAE) Standard 90.1, are widely followed by states and localities, thereby bringing some uniformity to building energy codes. The model codes are updated periodically through a consensus-seeking process. As of 2008, 19 states had adopted the 2006 version of the IECC or a more stringent code for new homes, and 27 states had adopted the ASHRAE 90.1-2004 or 90.1-2006 code or a more stringent code for new commercial buildings (DOE, 2008).

Building energy codes are enforced at the local level throughout the country. There is some evidence that code enforcement and compliance have been weak in various regions (Halverson et al., 2002; Kinney et al., 2003; Khawaja et al., 2007), and a number of jurisdictions have taken steps to simplify their energy codes in order to facilitate compliance. Training architects, builders, contractors, and local code officials can significantly improve code compliance and can also be very cost-effective in terms of energy savings per program dollar (Stone et al., 2002). The DOE provides software tools, technical assistance, and grants to support code adoption and implementation.

It is estimated that the influence of building energy codes on new homes and commercial buildings constructed during the 1990s reduced U.S. energy use by 0.54 quad in 2000 (Nadel, 2004). This is a conservative estimate of the impact of energy codes in that it does not consider buildings constructed before 1990 or after 1999. The DOE estimates that if all states adopted the update to the model commercial building energy code approved by ASHRAE in 1999, building owners and occupants would save about 0.8 quad over 10 years (DOE, 2007a). Even more energy savings would result if all states adopted a more recent version of the ASHRAE model standard, such as the 2007 version. Energy codes in general are very cost-effective, with any extra first cost for complying with the code usually paid back through energy savings in 7 or fewer years (WGA, 2006).

5.2.4 Government-Funded Research, Development, and Demonstration

From 1978 to 2000, the DOE spent more than $7 billion (1999 dollars) on energy efficiency research, development, and demonstration (RD&D) programs, and as estimated by a report from the National Research Council, some of the most successful RD&D programs are yielding net economic benefits to the nation of around $30 billion (NRC, 2001).

DOE-funded research has contributed to the development and commercialization of a number of energy-efficient building technologies, including high-efficiency appliances, electronic lighting ballasts, and low-emissivity windows. RD&D programs tend to be most effective (Geller and McGaraghan, 1998; Alic et al., 2003) when they:

- Involve collaboration between public research institutions (such as universities and DOE national laboratories) and the private sector,
- Focus on multiple technologies and designs,
- Contribute to all stages of the innovation and product development process, and
- Are complemented by other policies, such as financial incentives or regulations that stimulate market demand.

In contrast to the building technology program, DOE's transportation technology RD&D program has had very little effect on the vehicle marketplace. This result is attributed to the fact that the DOE initially chose to focus on a limited number of advanced engines and power systems, such as Stirling engines, gas turbines, and battery-powered electric vehicles—none of which proved viable because of technological problems, lack of industry interest, and/or lack of market acceptance. The more recent focus on hybrid-electric power trains and fuel cells also has not influenced commercial vehicles so far, although considerable technical progress has been made and these technologies show great promise (NRC, 2008). This experience demonstrates that RD&D projects should be carefully selected and designed, taking into account technological, institutional, and market barriers.

The Department of Energy operates a number of programs to promote greater energy efficiency in industry. Until 2007, the DOE funded RD&D mainly in partnership with nine energy-intensive sectors—agriculture, aluminum, chemicals, forest products, glass, metal casting, mining, petroleum, and steel. More recently the DOE has shifted RD&D toward crosscutting "technology platforms" such as industrial reactions and separations, waste-heat minimization and recovery, and high-temperature processing. The DOE recently identified nearly 100 technologies that it supported in the past decade that are now commercially available and saving energy to some degree. These technologies are estimated to have saved about 1.1 quads of energy cumulatively and about 0.1 quad in 2005 alone (DOE, 2007b).

5.2.5 Federal Incentives and Grants

Federal tax credits were provided for energy efficiency measures purchased by households and businesses in the late 1970s and early 1980s. The credit amounted to 15 percent of the measure cost[7] for households and 10 percent of the measure cost for businesses. However, studies were not able to document that the tax credits expanded the adoption of energy efficiency measures (Clinton et al., 1986; OTA, 1992). This result was attributed to the small size of the credits and the fact that the credits applied to commonplace efficiency measures such as home insulation and weather stripping, which had already been widely adopted before the credits took effect. These tax incentives cost the U.S. Treasury around $10 billion and were discontinued in 1985.

Based in part on this experience, new tax credits were enacted in 2005 for innovative energy efficiency measures that included hybrid, fuel cell, and advanced diesel vehicles; highly efficient new homes and commercial buildings; and efficient appliances. These tax credits were intended to support the commercialization and market development of these innovative technologies but not necessarily to save a significant amount of energy. In addition, a 10 percent tax credit of up to $500 was adopted for energy retrofits to the building envelope of existing homes. Except for the tax credits for advanced vehicles, these new tax credits were slated to expire at the end of 2007, but most were extended as part of the American Recovery and Reinvestment Act of 2009 (ARRA; Public Law 111-5). It is still too early to evaluate the impact of the 2005 tax credits.

Low-income households typically spend 16 percent of their total annual income on home energy costs, compared to 5 percent or less for middle- and upper-income households (DOE, 2006). The DOE provides grants to improve the energy efficiency of low-income housing through the Weatherization Assistance

---

[7] *Measure cost* is the full cost for an add-on measure such as insulation or a variable-speed motor drive, but the incremental cost for a higher-efficiency pump or motor.

Program (WAP). The WAP has helped more than 6.2 million families reduce their energy consumption and energy cost burden since 1976. The DOE estimates that these households experienced a $1.6 billion reduction in energy costs during the winter of 2005 as a result of the weatherization efforts (DOE, 2006).

Energy efficiency researchers developed improved home audit and retrofit techniques during the course of the WAP. These techniques include instrumented audits using blower doors, advanced air-sealing techniques, and greater emphasis on heating system improvements. As a result, the WAP lowered space-heating energy consumption in participating households by an average of 31 percent for homes weatherized from 1993 through 2002 (Berry and Schweitzer, 2003). This is significantly greater than the energy savings realized in homes weatherized during the 1970s and 1980s.

In recent years, federal funding for the WAP has enabled the weatherization of about 100,000 homes per year. The ARRA provides $5 billion in additional funds for home weatherization, as well as more flexible qualification and spending criteria. This one-time funding will enable the retrofitting of approximately 1 million homes.

5.2.6 State and Utility Programs

States and electric utilities have played a significant role in advancing energy efficiency in the United States. Many state utility regulatory commissions or state legislatures require electric utilities to operate energy efficiency programs, also known as demand-side management (DSM) programs. Most of these programs are funded through a small surcharge on electricity sales. In some states, utilities are allowed to earn more profit from the kilowatt-hours of electricity use saved through their energy efficiency programs than from the kilowatt-hours of new electricity delivered as a result of building new power plants or other energy supply facilities (not accounting for the relative size of the different resource options) (Kushler et al., 2006). This approach removes the financial disincentive that discourages utilities from promoting energy efficiency measures to their customers. In a few states, these programs are implemented by independent entities or state agencies rather than by utilities.

Utility and state efficiency programs provide information, technical assistance, and financial incentives to end users in order to encourage their greater adoption of cost-effective energy efficiency measures. Rebates are provided for a wide range of measures ranging from energy-efficient lightbulbs to heat pumps to commercial and industrial energy efficiency projects. Many programs promote ENERGY STAR® products and buildings and support the implementation of building energy codes as well as encourage new construction that exceeds minimum code requirements. Program managers have made considerable progress in developing and refining strategies for maximizing energy savings and program cost-effectiveness over the past 25 years (Nadel and Geller, 1996; NAPEE, 2006).

Although the funding for utility DSM programs was cut during the mid-1990s as utility deregulation and restructuring began, it has rebounded in recent years. The funding for these programs nationwide increased from about $0.9 billion in 1997 to $2.0 billion in 2006 (CEE, 2007). There is considerable variation in the commitment to—and funding for—utility energy efficiency programs among states. As of 2004, the leading states were spending more than $15 per capita on these programs, whereas the median state was spending only $1.64 per capita. The top 20 states (in terms of spending per capita) accounted for 88 percent of nationwide spending on utility energy efficiency programs in 2004 (Eldridge et al., 2007).

Some states have adopted energy-savings requirements, sometimes known as energy efficiency resource standards, for utility and state energy efficiency programs. Connecticut, for example, is requiring electricity providers to achieve 1 percent electricity savings per year from end-use efficiency efforts, combined heat and power (CHP) plants, and waste-heat recovery. Minnesota has enacted overall energy-savings goals of 1.5 percent per year, with at least 1 percent coming from utility efficiency programs. Texas now requires electricity providers to offset 20 percent of projected load growth through end-use efficiency programs. Other states including Hawaii, Nevada, and North Carolina allow utilities to count the energy savings from efficiency programs as well as renewable energy generation toward meeting overall clean-energy standards (Nadel, 2007).

Utilities have collaborated to stimulate the development and commercialization of advanced technologies and superior efficiency levels. One such example is the "golden carrot" incentive program that led to the introduction of new, highly efficient refrigerators during the 1990s and also paved the way for next-generation efficiency standards (Geller and Nadel, 1994). In this case, participating utilities paid the winning manufacturer incentives—up to $30 million—as it sold qualifying highly efficient models in their service areas. Utilities have also collaborated to stimulate the development and commercialization of highly efficient air conditioners and heat pumps, heat-pump water heaters, air duct testing and sealing techniques, and other new energy efficiency technologies (Nadel et al., 2003).

In addition to the funding of energy efficiency programs by utilities, the DOE provides funding to states through the State Energy Program (SEP). This funding, a total of $49.5 million in fiscal year 2007, is used by state energy offices for loans and grants, energy audits, codes and standards efforts, training, and other educational activities related to advancing energy efficiency and renewable energy use. Researchers have determined that the state-level programs are very cost-effective, with more than $7 in energy-bill savings on average for each dollar provided by the DOE (Tonn and Peretz, 2007). The programs also yield substantial nonenergy benefits, including local economic development, reduction in pollutant emissions, and improved public health and safety. The ARRA of 2009 provides $3.1 billion in one-time funding for the SEP.

Utility and state energy efficiency programs as a whole reduced electricity use by about 74 TWh in 2004, or 2.0 percent of electricity sales nationwide (York and Kushler, 2006). In leading states such as California, Connecticut, Minnesota, Vermont, and Washington, however, efficiency programs reduced electricity use in 2004 by 7 to 9 percent, considering the cumulative impact of the programs in these states. Furthermore, energy savings have risen since 2004 because overall DSM funding has increased. National energy savings reached approximately 90 TWh as of 2006.

5.2.7 Promotion of Combined Heat and Power Systems

In addition to their other accomplishments, policy initiatives have also improved the efficiency of energy conversion and supply, specifically by expanding the use of combined heat and power systems, also known as cogeneration. The Public Utility Regulatory Policies Act of 1978 (PURPA; Public Law 95-617) includes a section mandating that utilities buy power from cogenerators (and other qualifying, smaller power producers) at avoided costs. PURPA also exempts qualifying facilities from regulatory oversight under the Public Utilities Holding Company Act (PUHCA; Public Law 74-333) and from constraints on natural gas use imposed by the Power Plant and Industrial Fuel Use Act (Public Law 95-620). Some states, such as California and New York, set attractive avoided costs and contract terms, and CHP capacity nationwide expanded from less than 10 gigawatts (GW) in 1980 to almost 44 GW by 1993 (Elliott and Spurr, 1999).

The expansion of CHP capacity slowed during the 1990s because of declining avoided costs and the onset of utility deregulation and restructuring, but the Clinton administration launched a new CHP initiative in late 1998, setting a goal of 92 GW of installed CHP capacity by 2010. The DOE and the Environmental Protection Agency (EPA) support efforts to remove barriers and promote greater awareness and adoption of CHP systems. Partly in response to this national effort, some states, including California, New York, New Jersey, Connecticut, and Texas, have removed regulatory barriers, adopted favorable emissions regulations, and provided financial incentives for CHP systems (Eldridge et al., 2007).

Survey data indicate that installed CHP capacity in the United States reached 85 GW at more than 3,300 sites as of 2006 (Shipley et al., 2008). CHP plants generated 506 TWh of electricity in 2006, almost 12 percent of all electricity produced nationwide. It is estimated that the use of CHP systems resulted in about 1.9 quads of primary-energy savings that year (Shipley et al., 2008). Given that over 85 percent of the CHP capacity in the country was installed after 1980, the panel attributes 1.62 quads of energy savings in 2006 to PURPA and other policy initiatives aimed at stimulating the adoption of CHP systems. (The effects of PURPA and other energy initiatives are summarized in Table 5.2.)

5.2.8 Consumer Education, Training, and Technical Assistance

Complementing the minimum efficiency standards discussed above, the ENERGY STAR® product-labeling program informs U.S. consumers of the most efficient products in the marketplace at any particular time. The ENERGY STAR® label exists for a wide range of products, including personal computers and other types of office equipment, kitchen and laundry appliances, air conditioners and furnaces, windows, commercial appliances, and lighting devices. Energy-efficient commercial buildings and new homes also can qualify for the ENERGY STAR® label. The ENERGY STAR® label helps consumers by reducing uncertainties about energy performance and lowering transaction costs for obtaining such information.

Figure 5.2 shows the growth in the market share (percentage of new sales) for various ENERGY STAR® appliances over the past decade. Market shares for clothes washers, dishwashers, and room air conditioners greatly increased during this time period. Revisions in the ENERGY STAR® qualification level temporarily reduced the ENERGY STAR® market share for refrigerators and room air conditioners from 2000 to 2001.

It is estimated that the ENERGY STAR® program in aggregate has resulted in about 175 TWh of electricity savings as of 2006 (EPA, 2007b). However, some of this savings is also counted by utility and state-based energy efficiency pro-

TABLE 5.2 Estimates of Annual Energy Savings from Major Energy Efficiency Policies and Programs

Policy or Program	Electricity Savings (TWh/yr)	Primary Energy Savings (Quads/yr)	Year	Source
CAFE vehicle efficiency standards	—	4.80	2006	NRC, 2002[a]
Appliance efficiency standards	196	2.58	2006	Nadel et al., 2006[b]
PURPA and other CHP initiatives	—	1.62	2006	Shipley et al., 2008[c]
ENERGY STAR® labeling and promotion	132	1.52	2006	EPA, 2007b[d]
Building energy codes	—	1.08	2006	Nadel, 2004[e]
Utility and state end-use efficiency programs	90	1.06	2006	York and Kushler, 2006[f]
DOE industrial efficiency programs	—	0.40	2005	DOE, 2007[b]
Weatherization assistance program	—	0.14	2006	DOE, 2006[g]
Federal energy management program	—	0.11	2005	FEMP, 2006[h]
Total	—	13.32	—	—

[a]Extrapolation of fuel savings estimated by the NRC to 2006, and assuming 75 percent of the energy savings from vehicle efficiency improvements are due to the CAFE standards.

[b]Interpolates between savings estimates by ACEEE for 2000 and 2010.

[c]Assumes that 85 percent of the energy savings from all CHP systems installed in 2006 was due to PURPA and other policy initiatives.

[d]Assumes only 75 percent of the energy savings estimated by U.S. EPA in order to avoid double counting savings with utility and state programs.

[e]Increases the energy savings estimate for new buildings constructed during 1990–1999 from Nadel (2004) by 100 percent to account for the impact of codes prior to 1990 and post-1999.

[f]Extrapolates the 2004 national electricity savings estimate to 2006 based on national DSM budget estimates for 2005 and 2006.

[g]Assumes 5.6 million weatherized households and average energy savings of 25 million Btu/yr per household, from Berry and Schweitzer (2003).

[h]Based on the reported reduction in energy use per square foot of floor area during 1985–2005 and actual primary energy use in federal buildings as of 2005 (i.e., excluding energy use by transport vehicles and equipment).

grams. The ENERGY STAR® program has achieved the most energy savings in the areas of commercial building improvements and personal computers, monitors, and other types of office equipment. The ENERGY STAR® program continues to develop criteria and adopt labeling for additional products—for example, for television sets and water heaters. In some cases, the ENERGY STAR® program paves the way for minimum efficiency standards by bringing new energy efficiency measures into wide production and use.

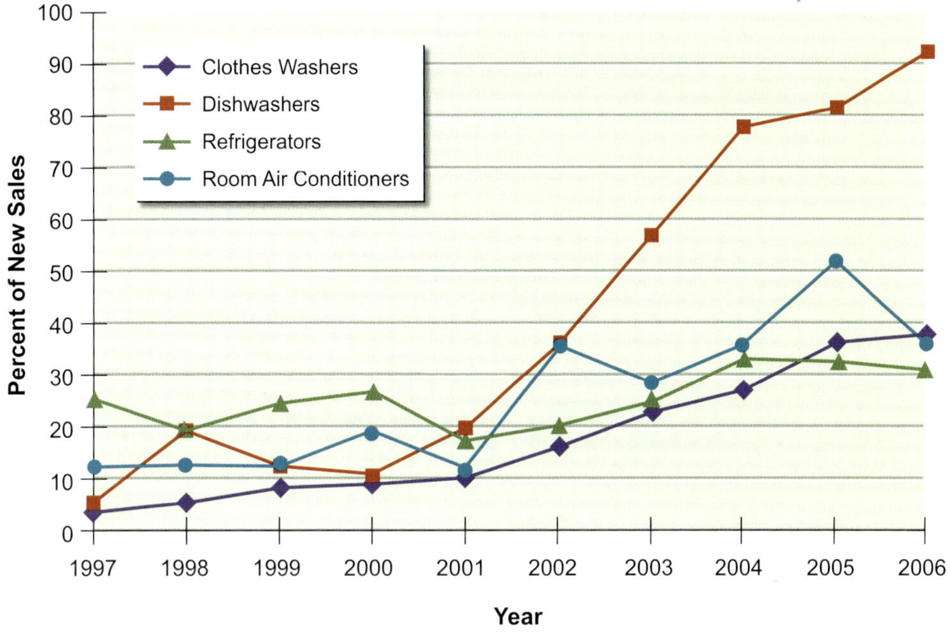

FIGURE 5.2 *ENERGY STAR® appliance market shares (percent of new sales), 1997–2006.*
Note: AC = air conditioner.
Source: Karney, 2006.

The DOE operates a number of education, training, and technical assistance programs for industries. These programs promote energy efficiency improvements in motors and in pumping and compressor systems. They also provide energy assessments for larger industrial plants and conduct energy audits for small- and medium-size manufacturers through university-based Industrial Assessment Centers. In combination, it is estimated that these programs have reduced industrial energy use by 1.75 quads cumulatively and 0.3 quad in 2005 alone (DOE, 2007b).

The DOE operates a program known as the Federal Energy Management Program (FEMP) to improve energy efficiency in federal facilities. Starting in 1991, executive orders instructed federal agencies to reduce their energy use per square foot of floor space. The FEMP provides technical assistance, training, and help with innovative approaches to project financing and implementation. Many federal agencies are using energy service companies and performance contracts to implement efficiency projects with support from the FEMP. In response to the relevant executive orders and the FEMP, site energy use per square foot of floor area in federal buildings declined nearly 30 percent from 1985 to 2005. During this period, however, there was growing electrification in federal buildings, with the result that primary energy use per square foot of floor area declined by only 16 percent (FEMP, 2006).

The federal government also implements some significant information and education programs aimed at increasing energy efficiency in the transportation sector. These include the light-vehicle fuel economy information program (www.fueleconomy.gov) and the EPA's SmartWay program for both passenger and freight vehicles (www.epa.gov/smartway/). In 2006, the EPA updated the test methods for measuring vehicle fuel economy as well as the label that appears on new vehicles. This was done to bring the estimates of miles per gallon on the label closer to actual fuel economy. The new label includes more prominently displayed fuel-cost information and a graphic comparing the fuel economy of various models in a particular vehicle category.

5.2.9 Overall Energy Savings

Table 5.2 provides estimates of the annual energy savings resulting from most of the policies and programs discussed above. In some cases (i.e., for CAFE standards and PURPA), the savings reflect a judgment of the relative importance of the policies and market forces. The total energy savings, 13.3 quads per year, was equivalent to more than 13 percent of national energy use in 2006—more than the energy supplied by nuclear power and hydroelectric power combined. It was also more than five times the increase in renewable energy supply in the United States between 1973 and 2006.

It should be noted, however, that these policies and programs provided only a moderate amount of the total energy savings associated with the 50 percent decline in national energy intensity during the 1973–2007 period (see Chapter 1). Increasing energy prices, ongoing technological change, and structural change also contributed to the steep decline in energy intensity in the past 35 years.

A comparison of the energy savings across the various policies and programs in Table 5.2 shows that regulatory initiatives such as the CAFE standards, appliance efficiency standards, and PURPA provided the largest amount of energy savings. It should be recognized that some energy efficiency policy initiatives, such as RD&D efforts in the buildings sector, are not included in Table 5.2 in order to avoid double counting of savings.

5.3 THE CALIFORNIA EXPERIENCE

Figure 5.3 illustrates electricity use per capita from 1960 to 2006 in California, New York, and the United States as a whole. California maintained nearly flat per capita electricity consumption from 1975 to 2006, and in 2006 its per capita use was about 40 percent less than that in the United States as a whole, although the two were nearly the same in the 1960s. The shaded wedge in Figure 5.3 depicts the growth in U.S. per capita electricity consumption (since 1973). The U.S. and California trend lines started to diverge in 1974 when the United States experienced its first energy crisis.

Many factors contributed to the difference between the California and U.S. trend lines. California began setting its building and appliance efficiency standards earlier, and its electricity prices increased more rapidly, than was the case in most of the United States. And, of course, compared with the entire United States, other factors such as a different mix of industries and differences in climate also contributed (Sudarshan and Sweeney, 2008).

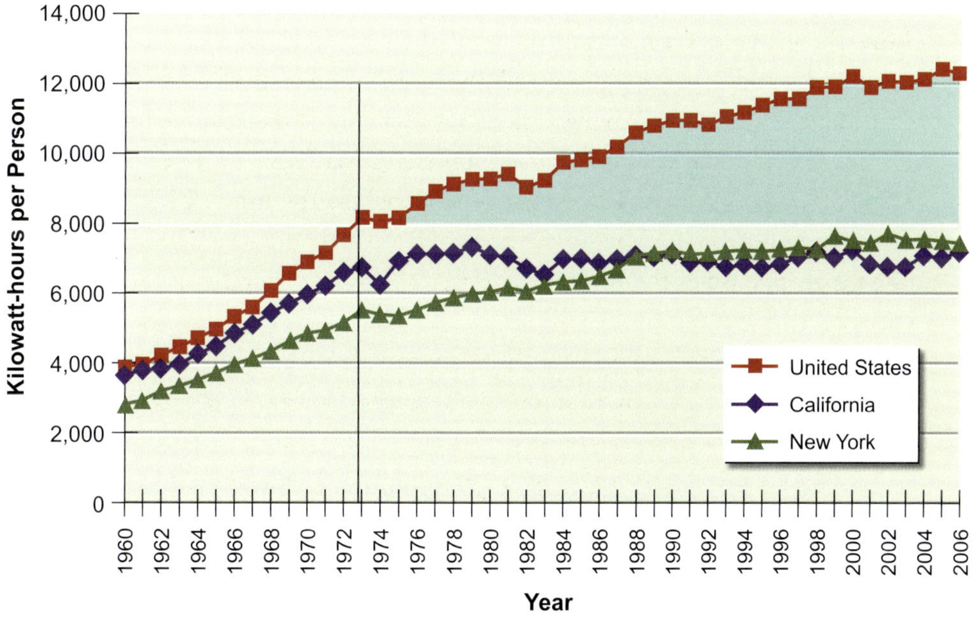

FIGURE 5.3 *Per capita electricity consumption (not including on-site generation) in California, New York, and the United States, 1960–2006.*
Source: U.S. Department of Energy, Energy Information Administration. State Energy Data System, State Energy Consumption, Price, and Expenditure Estimates, available at http://www.eia.doe.gov/emeu/states/_seds.html.

Table 5.3 compares 2005 per capita electricity consumption in California and the United States as a whole, broken out by major sectors (residential, commercial, and industrial) of the economy. The residential sector and the industrial sector each accounted for about 40 percent of the difference in consumption, and the commercial sector accounted for the remaining 20 percent. Sudarshan and Sweeney (2008) show that the types of industries in California are less electricity-intensive than those in the country as a whole, a factor that explains most of the difference in the industrial sector. Likewise, California has less commercial floor space per capita than does the United States as a whole, causing most of the difference in the commercial sector. These and other structural, price, and climatic effects can explain about 75 percent of the difference in electricity intensity between California and the country as a whole (Sudarshan and Sweeney, 2008).

In addition to the various policies (discussed below) that California has pursued to increase the efficiency of electricity consumption, the impact of price on consumption is certainly worth noting. Figure 5.4 illustrates on a state-by-state basis the connection between the average price of electricity and per capita consumption in 2006.

Although certainly not the whole story, consumption is clearly price elastic, and so the straight-line fit (with an R^2 of 41 percent) has a negative slope of –811 kWh per capita per 1¢/kWh. Both New York and California have high-priced electricity and below-average per capita consumption, as indicated in Figure 5.4, and both have a relatively low percentage of total consumption associated with the industrial sector—New York just 11 percent and California only 19 percent, compared to the national average of 29 percent.

Regarding policies that influenced the trend shown in Figure 5.3, California first enacted efficiency standards for major types of appliances and for new residential and commercial buildings in the mid-1970s. The state has updated its

TABLE 5.3 Comparison of Per Capita Electricity Consumption in the United States and in California in 2005

	United States (kWh/person)	California (kWh/person)	Difference (kWh/person)	Difference (%)
Residential	4,586	2,369	2,216	42
Commercial	4,302	3,253	1,048	20
Industrial	3,438	1,391	2,048	39
Total	12,326	7,013	5,312	100

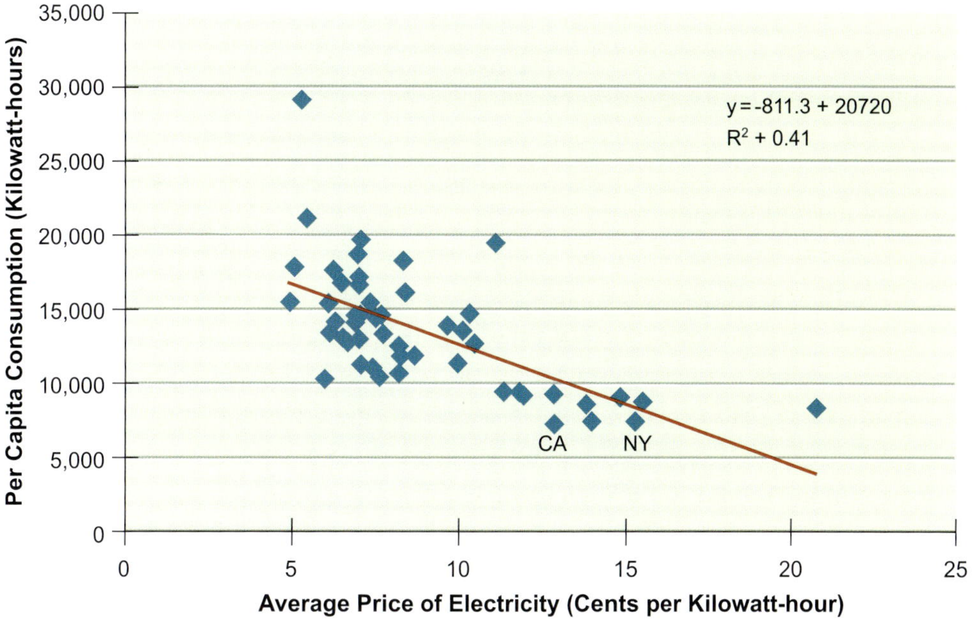

FIGURE 5.4 *Average price of electricity and per capita consumption, 2006; 50-state data. Source: U.S. Department of Energy, Energy Information Administration, Electric Power Annual. Data Tables, 2006 State Data Tables (EIA-816). Population estimates from the U.S. Census Bureau, Annual Estimates of the Population for the U.S., Regions, States and Puerto Rico: April 1, 2000, to July 1, 2007 (NST-EST2007-01).*

energy efficiency requirements for new buildings many times since then and has adopted minimum efficiency standards for additional appliances as well. After action by California and other states, the federal government also began to enact standards, at times even adopting those of California. Over the years, most of the appliance standards have migrated to national standards and therefore have resulted in efficiency gains in all states. The national energy savings are substantial, as noted previously.

In addition to appliance standards and energy codes for new buildings, California has implemented substantial state and utility energy efficiency programs. Figure 5.5 provides annual funding levels for investment in energy efficiency by California's investor-owned utilities for the years 1976–2004, with forecasts for 2005–2012.[8] Funding has varied considerably over time due to factors such as the

[8]These utilities provide service to about 75 percent of the state's population. The remainder is served by municipal utilities and other public agencies.

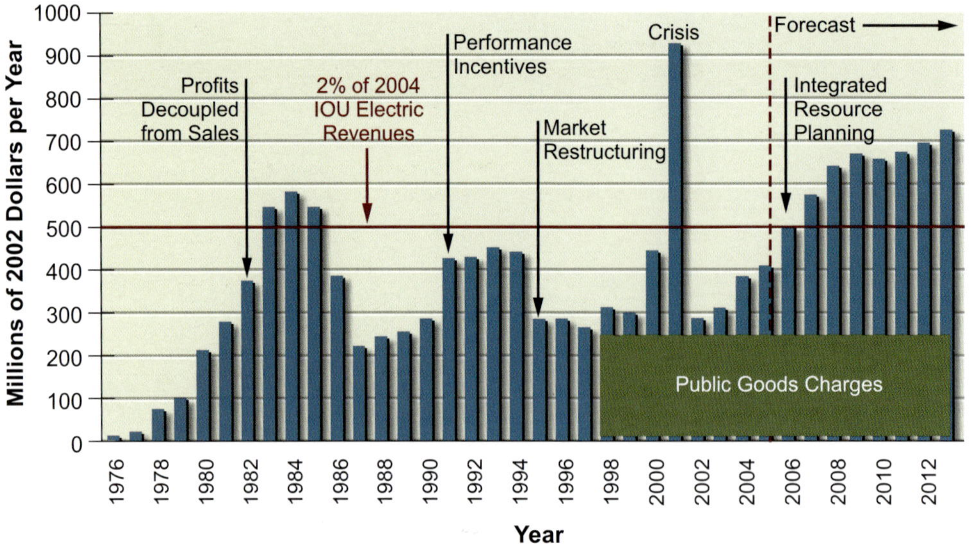

FIGURE 5.5 *Funding (in constant 2002 dollars) for investor-owned utility energy efficiency programs in California, 1976–2004, and forecasted for 2005–2012. Also shown are key policies enacted by the state since the early 1980s. Public-good charges of 0.3¢/kWh sold were imposed beginning in 1998 to fund energy efficiency and other public-benefits activities.*
Note: IRP = integrated resource planning; "crisis" refers to the temporary power shortages and severe electricity price spikes experienced in 2001.
Source: Rosenfeld and McAuliffe, 2008.

movement toward deregulation and restructuring during the 1990s, but energy efficiency funding has rebounded in recent years and is expected to continue to rise in the near future.

California enacted a number of policies to stimulate vigorous utility efficiency programs. The state has now placed energy efficiency as its most preferred resource and has committed to the aggressive funding of these efforts for the next few years, as Figure 5.5 illustrates. Figure 5.5 also highlights some of the following important policies that the state has adopted since the early 1980s:

- *1982:* Decoupling of utility profits from sales in order to eliminate the negative incentives associated with reduced sales;
- *1991:* Providing of performance incentives to utilities that met or exceeded efficiency savings, a policy that was then abandoned during California's initial efforts at restructuring its electricity sector;

- *1998:* Implementation of a mandatory charge of 0.3¢/kW sold in order to fund energy efficiency and other "public benefits" activities;
- *2001:* Inclusion of efficiency as a part of integrated resource planning, and the direct comparison of energy savings to other options for meeting future load and load growth requirements, including other policy considerations;
- *2004:* Establishment of energy efficiency goals of about 1 percent per year (about 2.3 TWh per year) through 2013; and
- *2007:* Reinstatement of performance incentives. California established specific efficiency targets with a risk/reward incentive mechanism, as illustrated in Figure 5.6. Utilities are allowed a bonus, in addition to cost recovery, if they achieve at least 85 percent of their savings goals.

California updated its building efficiency standards and approved more than $500 million for additional energy efficiency programs in 2001. These actions were taken in order to help the state respond to temporary power shortages and severe electricity price spikes caused by flaws in the state's utility restructuring policy. These emergency efforts were very successful. In total, compared to levels of the previous year, California reduced its electricity use by about 7 percent and peak demand by 10 percent in the summer of 2001 (CEC, 2002). These savings enabled the state to avoid further power shortages during the summer of 2001.

The combination of standards and programs has resulted in considerable electricity savings since the inception of these efforts. Figure 5.7 illustrates the cumulative effects of appliance standards, building energy codes, utility efficiency programs, and what is termed market transformation—that is, longer-term market impacts due to previous state and utility efficiency programs—in California from 1975 though 2003. In total, it is estimated that these initiatives saved about 40 TWh per year (1.13 MWh per capita) as of 2003, equivalent to about 15 percent of actual electricity use in the state that year.

California continues to promote and advance energy efficiency throughout all sectors of the economy. Energy efficiency, the top priority of the state's Energy Action Plan, will be heavily relied on if the state is to meet its goal of reducing greenhouse gas emissions in 2020 to levels experienced in 1990 (CEC, 2008).

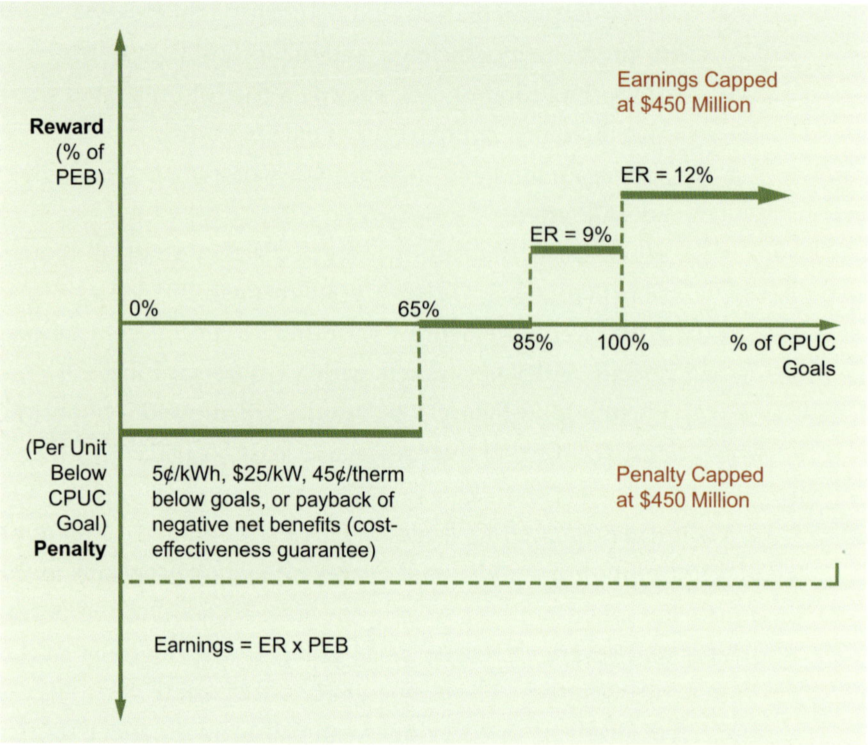

FIGURE 5.6 *Utility energy efficiency incentive mechanism adopted by the California Public Utilities Commission (CPUC) in 2007. The CPUC established efficiency targets along with a risk/reward mechanism under which utilities are allowed both cost recovery and a bonus (or penalty) for exceeding (or falling short of) the targets.*
Note: PEB = performance earnings basis; ER = earnings rate(s), the percentage of PEB that will accrue to shareholders.
Source: CPUC, 2007.

5.4 THE NEW YORK EXPERIENCE

New York State has a long history of implementing policy actions to encourage more efficient use of energy across all sectors. Policy makers have recognized that cost-effective strategies to reduce energy use are critical for stimulating and maintaining economic growth. New York's energy efficiency efforts include the system benefits charge (SBC) described below; the adoption and continual updating of energy building codes and appliance standards; executive orders directing state agencies and authorities to improve energy efficiency; and well-funded research and development programs.

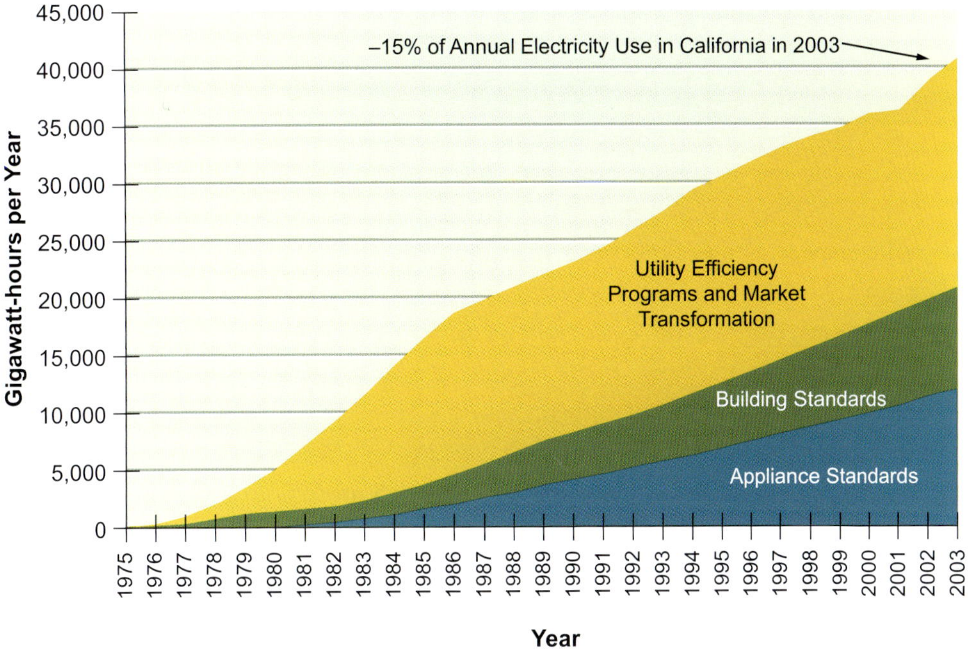

FIGURE 5.7 *Annual electricity savings from key energy efficiency policies and programs implemented in California, 1975–2003.*
Source: CEC, 2007.

Although the past three decades have been characterized by increases in population, greater demands for energy-using technologies, and increasing lifestyle expectations, New York State has maintained a relatively flat level of total energy use per capita (about 36 percent lower than the national average in 2005). New York is currently the second-least energy-intensive state in the continental United States on a per capita basis (after Rhode Island). Its relatively low energy use per capita is due in part to its highly energy-efficient urban transportation, which includes subways, commuter rail, buses, and ferries, as well as to structural changes such as the shift away from a heavy industrial base and toward a service and information economy.

New York's electricity use per capita relative to that of the United States (see Figure 5.3) is an indicator of the overall success of the state's continuing efforts to promote the efficient use of electricity. Between 1960 and 2006, the difference between New York and the United States in annual electricity use per capita widened from about 1100 kWh to about 4900 kWh. In 2006 the residential sector accounted for 41 percent of the difference, the industrial sector for 53 percent, and the commercial and transportation sectors for the remaining 6 percent (Table 5.4).

New York's energy efficiency efforts began in the late 1970s with federal funding for a State Energy Conservation Program (SECP). The funding was small relative to the need, but the efforts initiated through the New York State Energy Office (NYSEO) represented an important beginning in achieving greater energy efficiency and conservation savings and provided experience for government programs working in concert with the private sector. Over the years, the NYSEO was able to develop a diverse portfolio of programs serving the residential, business, and government sectors. These programs took another step forward in the 1980s as a result of receiving significant funding from a legal settlement against Exxon and other oil companies for charging excessive prices for their crude oil in the late 1970s. By 1989, New York State had received more than $335 million, including interest, from this funding source.

New York's energy efficiency efforts directed at the electric utility sector began in earnest in 1984, driven largely by concerns about the construction delays and escalating costs that were plaguing the Shoreham and Nine Mile Point 2 nuclear power plants and the Somerset coal plant. At the time, DSM programs were viewed by New York's Public Service Commission (PSC) as potential alternatives to continued investment in new, central-station power-generation projects. As a result, investor-owned utilities were required by the PSC to develop pilot-scale DSM programs that included energy efficiency and load management. The programs were initially funded at approximately $25 million annually, representing approximately one-quarter of 1 percent of gross annual utility revenue.

Following an assessment of the pilot programs in 1987, the PSC concluded

TABLE 5.4 Comparison of Per Capita Electricity Use in the United States and in New York in 2006

	United States (kWh/person)	New York (kWh/person)	Difference (kWh/person)	Difference (%)
Residential	4,514	2,508	2,006	41
Commercial	4,341	3,938	403	8
Industrial	3,378	776	2,602	53
Transportation	25	145	−121	−2
Total	12,258	7,367	4,890	100

that DSM programs were a viable and economic alternative to new energy supply resources and that DSM should be considered on an equal footing with supply resources in integrated resource planning. At a minimum, it was recognized that DSM could delay the need for peaking capacity, even if the need for new baseload power supplies could not be completely eliminated. The job creation and environmental benefits associated with reducing electricity use were also identified and quantified as further justification for investment in DSM. Utilities were directed by the PSC to assess DSM potential, identify cost-effective programs, establish DSM goals, and develop long-range DSM plans, including incentive, information, and education programs.

In the early 1990s, the PSC implemented a revenue decoupling mechanism to allow utilities to recover revenues lost from energy efficiency reductions (determined by the amount by which actual sales revenue fell below the forecast adopted in the most recent rate case). Along with the revenue decoupling mechanism, the PSC approved financial incentives for achieving energy efficiency goals, as well as financial penalties for falling short of goals. The incentive scheme proved to be effective and was successfully adapted to each investor-owned utility (DeCotis, 1989).

By 1993, DSM spending by investor-owned utilities in New York State reached \$280 million (equivalent to about \$400 million in 2007 dollars; Figure 5.8), a dramatic increase from the initial \$25 million spent in 1984. Additional DSM spending by the state's energy authorities raised the state's annual investment in energy efficiency resources in 1993 to about \$330 million (about \$470 million in 2007 dollars).

New York began the process of restructuring its electricity industry in 1996. A key element of this effort was that investor-owned utilities were required to sell generation assets to independent power producers. As a result, New York's investor-owned utilities were transformed into transmission and distribution companies. With the transition to wholesale market competition, the responsibilities for administering DSM programs were transferred to the New York State Energy Research and Development Authority (NYSERDA). The utilities' current role, following the divestiture of their generation assets, is to collect program funds from ratepayers through a system benefits charge. The funds are provided to NYSERDA, under the oversight of the PSC, to administer energy efficiency, load management, environmental protection, and research and development programs. NYSERDA has been administering statewide SBC programs in cooperation with the New York Power Authority and the Long Island Power Authority since 1998.

By 2007, annual investment in energy efficiency by New York's energy-related authorities increased to nearly $300 million (see Figure 5.8). Accounting for the cumulative annual impact of programs implemented since 1990, New York has lowered its annual electricity use by nearly 12 TWh, or about 8 percent of end-use sales (Figure 5.9). This 12 TWh of demand-side resources has reduced New York's CO_2 emissions by about 6.5 million tons per year, equivalent to removing about 1.3 million cars from the roads annually. All SBC energy efficiency programs (administered by NYSERDA) are required by the PSC to be cost-effective, which means that the present value of estimated lifetime monetary benefits exceeds the costs of implementing the programs. Through year's end in 2007, the benefit-cost ratio, counting only direct utility system benefits for New York's portfolio of SBC-funded energy efficiency programs, is 6.2 (on a present-value basis). Including nonenergy benefits, such as improved comfort, safety, and productivity, the benefit-cost ratio increases to 9.9, and adding macroeconomic

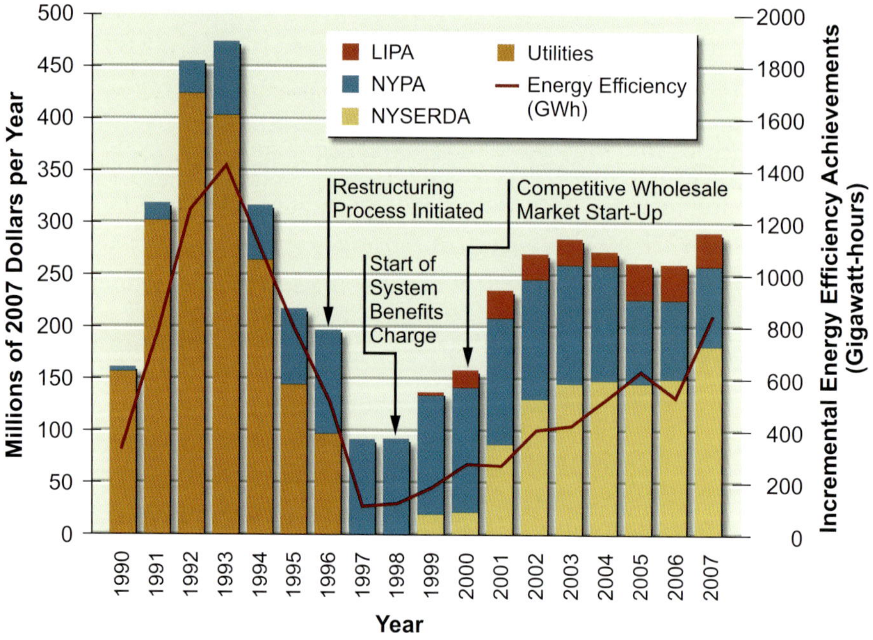

FIGURE 5.8 *New York State's annual energy efficiency expenditures (in constant 2007 dollars) and achievements, 1990–2007.*
Note: EE = energy efficiency; GWh = gigawatt-hours; LIPA = Long Island Power Authority; NYPA = New York Power Authority; NYSERDA = New York State Energy Research and Development Authority.
Source: Courtesy of NYSERDA.

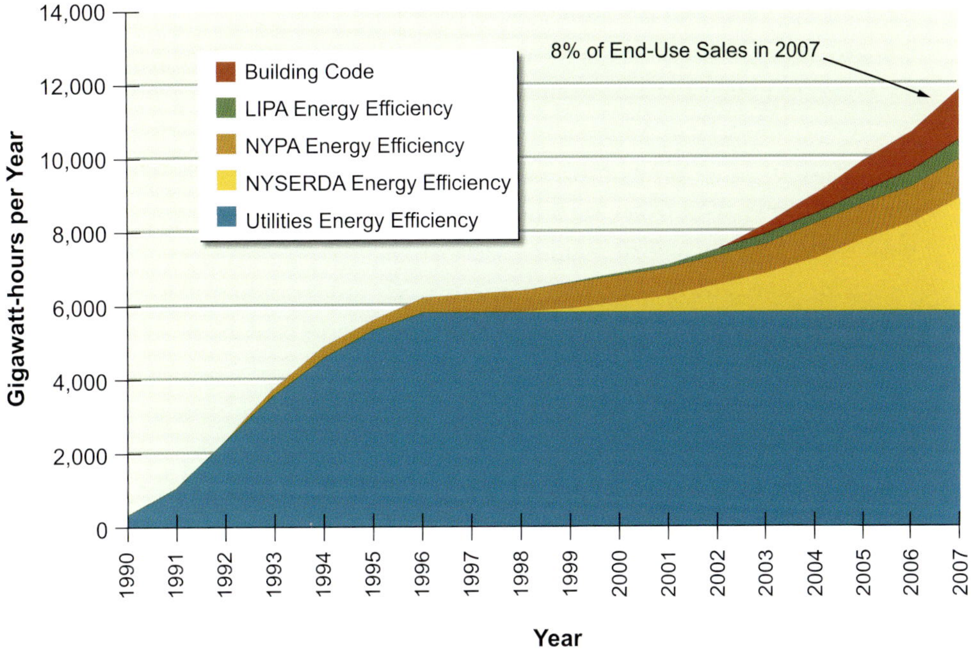

FIGURE 5.9 *New York State's energy efficiency achievements, 1990 through 2007: annual electricity use.*
Note: EE = energy efficiency; LIPA = Long Island Power Authority; NYPA = New York Power Authority; NYSERDA = New York State Energy Research and Development Authority.
Source: Courtesy of NYSERDA.

benefits (e.g., valuing added employment) increases the ratio to 13.2 (NYSERDA, 2008).

In April 2007, then-New York Governor Eliot Spitzer initiated an energy efficiency program of unparalleled proportions, known as the "15 by 15" program, by calling for a 15 percent reduction in electricity use in 2015 compared to the business-as-usual projected level of electricity use for that year (Spitzer, 2007).

5.5 LESSONS LEARNED

What lessons can be drawn from the wide-ranging experience encapsulated in this chapter regarding policies and programs aimed at increasing energy efficiency at both the national and the state level? Most importantly, the experience demonstrates that well-designed policies can result in substantial energy savings. This is clear from the fact that the policies taken together reduced national energy use in 2006 by more than 13 percent according to the estimates in Table 5.2. Also, leading states such as California and New York have been able to increase energy efficiency more than other states have, resulting in greater benefits for citizens, businesses, and the environment.

The experience shows that minimum efficiency standards can be a very effective strategy for stimulating energy efficiency improvements on a large scale, especially if standards are updated periodically. Minimum efficiency standards have been a key part of both federal and state energy efficiency efforts. Such standards should be technically and economically feasible and should provide manufacturers with enough lead time to phase out the production of nonqualifying products in an orderly manner.

Government-funded RD&D contributed to the development and commercialization of a number of important energy efficiency technologies. Experience has demonstrated that RD&D can take many years to pay off, and that attention should be devoted to commercialization and market development as well as to technological advancement. Also, a prudent RD&D portfolio includes high-risk, potentially high-payoff projects as well as those involving lower-risk, incremental improvements (NRC, 2001).

Although there is evidence that energy prices influence energy efficiency and levels of energy consumption, as illustrated in Figure 5.4, neither the federal government nor states have used energy taxes to any significant degree as a strategy for stimulating greater energy efficiency.

Financial incentives, including those provided by utilities, can increase the adoption of energy efficiency measures. Financial incentives should be carefully designed, however, avoiding costly efforts that have little or no incremental impact on the marketplace. One way to avoid this outcome is to provide incentives for newly commercialized technologies—in particular those with a high first cost but with good prospects for cost reduction as demand grows, production expands, and learning occurs.

Information dissemination, education, and training can increase the awareness of energy efficiency measures and improve know-how with respect to energy management. The ENERGY STAR® labeling program exemplifies the impact that a well-conceived, widely promoted labeling and education effort can have. Education and training are also important for the successful implementation of building energy codes.

In general, energy efficiency policies and programs work best if they are integrated into market transformation strategies, addressing the range of barriers that are present in a particular situation (Geller and Nadel, 1994). In the appliance market, for example, all of the following are being carried out simultaneously: government-funded RD&D helps to develop and commercialize new technologies; product labeling educates consumers; efficiency standards eliminate inefficient products from the marketplace; and incentives offered by some utilities and states encourage consumers to purchase products that are significantly more efficient than the minimum standards. This combination of actions has led to dramatic improvements in the efficiency of refrigerators and other types of appliances, and the efficiency gains and energy savings are continuing today.

The experience described above suggests that energy efficiency policies should be kept in place for a decade or more in order to ensure an orderly development of energy efficiency markets. At the same time, policies such as efficiency standards and targets, product labeling, and financial incentives should be revised periodically. This will increase their effectiveness and reduce program costs, for example, by phasing out incentives as particular technologies become well established in the marketplace. Dynamic policies steadily improved residential appliance efficiency, whereas stagnant policies failed to maintain car and light-truck efficiency improvements during the 1990s and the early part of this decade.

5.6 CHANGING CONSUMER BEHAVIOR

The energy efficiency policies and programs discussed in this chapter focus primarily on increasing the energy efficiency of buildings, appliances, vehicles, and industrial operations. Less attention has been devoted to changing consumer behavior—for example, encouraging people to drive less or buy fewer and/or smaller vehicles, appliances, or homes. Consumer behavior can be influenced in a number of ways (PIEE, 2007), including the following:

- Offering convenient alternatives such as practical and high-quality mass transit services;

- Using financial incentives such as taxing energy, taxing carbon dioxide and other pollutant emissions, or taxing inefficient devices more heavily;
- Increasing awareness, for example by educating people about the environmental consequences of their lifestyle choices; and
- Providing feedback on energy consumption—for example, by including easy-to-understand comparative information on energy use on monthly utility bills.

It remains to be seen if changing behavior can play a larger role in energy efficiency efforts in the coming decades.

5.7 REFERENCES

ACEEE (American Council for an Energy-Efficient Economy). 2007. Energy Bill Savings Estimates as Passed by the Senate. Washington, D.C.: ACEEE. Available at http://www.aceee.org/energy/national/EnergyBillSavings12-14.pdf.

Alic, J.A., D.C. Mowery, and E.S. Rubin. 2003. U.S. Technology and Innovation Policies: Lessons for Climate Change. Arlington, Va.: Pew Center on Global Climate Change.

Berry and Schweitzer, 2003. Metaevaluation of National Weatherization Assistance Program Based on State Studies, 1993-2002. ORNL/CON-488. Oak Ridge, Tenn.: Oak Ridge National Laboratory.

CEC (California Energy Commission). 2002. The Summer 2001 Conservation Report. Sacramento, Calif.: CEC.

CEC. 2007. Table 42, California Energy Demand 2008-2018: Staff Revised Forecast, Final Staff Forecast, 2nd ed., CEC-200-2007-015-SF2. Sacramento, Calif.: CEC. November 27.

CEC. 2008. 2008 Energy Action Plan Update. CEC-100-2008-001. Sacramento, Calif.: CEC; San Francisco, Calif.: California Public Utilities Commission. Available at http://www.fypower.org/pdf/cpuc_eap_update.pdf.

CEE (Consortium for Energy Efficiency). 2007. U.S. Energy-Efficiency Programs—A $2.6 Billion Industry. Boston, Mass.: CEE.

Clinton, J., H. Geller, and E. Hirst. 1986. Review of government and utility energy conservation programs. Annual Review of Energy 11:95-142.

CPUC (California Public Utilities Commission). 2007. CPUC Decision 07-09-043, September 20, 2007. Interim Opinion on Phase 1 Issues: Shareholder Risk/Reward Mechanism for Energy Efficiency Programs. Document COM/DGX, ALJ/MEG/rbg, issued September 25, 2007. Available at http://docs.cpuc.ca.gov/published/final_decision/73172.htm.

DeCotis, P.A. 1989. Balancing shareholder and customer interests in incentive ratemaking. Electricity Journal 2(10):16-23.

Difiglio, C., K.G. Duleep, and D. Greene. 1990. Cost effectiveness of future fuel economy improvements. Energy Journal 11(1):65-68.

DOE (U.S. Department of Energy). 2006. Weatherization Works! Fact Sheet. Washington, D.C.: DOE Office of Energy Efficiency and Renewable Energy, Weatherization Assistance Program.

DOE. 2007a. Energy Department Determines That Model Commercial Building Code Will Save Energy and Benefit Consumers. Washington, D.C.: DOE Office of Energy Efficiency and Renewable Energy, Industrial Technologies Program. Available at http://www.energycodes.gov/implement/determinations_com_news.stm.

DOE. 2007b. Impacts. Industrial Technologies Program: Summary of Program Results for CY 2005. Washington, D.C.: DOE Office of Energy Efficiency and Renewable Energy, Industrial Technologies Program.

DOE. 2008. Status of Energy Codes. Washington, D.C.: DOE. Available at http://www.energycodes.gov/implement/state_codes/index.stm.

EIA (Energy Information Administration). 2007. Annual Energy Review 2006. DOE/EIA-0384(2006). Washington, D.C.: Department of Energy, Energy Information Administration.

EIA. 2008. Annual Energy Outlook 2008. DOE/EIA-0383(2008). Washington, D.C.: Department of Energy, Energy Information Administration.

Eldridge, M., B. Prindle, D. York, and S. Nadel. 2007. The State Energy Efficiency Scorecard for 2006. Washington, D.C.: American Council for an Energy-Efficient Economy.

Elliott, R.N., and M. Spurr. 1999. Combined Heat and Power: Capturing Wasted Energy. Washington, D.C.: American Council for an Energy-Efficient Economy.

EPA (Environmental Protection Agency). 2007a. Light-Duty Automotive Technology and Fuel Economy Trends: 1975 Through 2007. EPA420-R-07-008. Washington, D.C.: EPA Office of Transportation and Air Quality.

EPA. 2007b. Energy Star® and Other Climate Protection Partnership Programs: 2006 Annual Report. Washington, D.C.: EPA Office of Air and Radiation.

FEMP (Federal Energy Management Program). 2006. Annual Report to Congress on Federal Government Energy Management and Conservation Programs: Fiscal Year 2005. Washington, D.C.: DOE Federal Energy Management Program.

Geller, H. 2003. Energy Revolution: Policies for a Sustainable Future. Washington, D.C.: Island Press.

Geller, H., and S. McGaraghan. 1998. Successful government-industry partnership: The U.S. Department of Energy's role in advancing energy-efficient technologies. Energy Policy 26:167-177.

Geller, H., and S. Nadel. 1994. Market transformation strategies to promote end-use efficiency. Annual Review of Energy and the Environment 19:301-346.

Gerard, D., and L.B. Lave. 2003. The Economics of CAFE Revisited: A Response to CAFE Critics and a Case for Fuel Economy Standards. Washington, D.C.: AEI-Brookings Joint Center for Regulatory Studies.

Goldstein, D.B. 2007. Saving Energy, Growing Jobs: How Environmental Protection Promotes Economic Growth, Profitability, Innovation, and Competition. Point Richmond, Calif.: Bay Tree Publishing.

Greene, D.L. 1998. Why CAFE worked. Energy Policy 26:595-614.

Greene, D., and J. DeCicco. 2000. Engineering-economic analyses of automotive fuel economy potential in the United States. Annual Review of Energy and Environment 25:477-535.

Halverson, M., J. Johnson, D. Weitz, R. Majette, and M. LaLiberte. 2002. Making residential energy codes more effective: Building science, beyond code programs, and effective implementation strategies. Pp. 2.111-2.122 in Proceedings of the 2002 ACEEE Summer Study on Energy Efficiency in Buildings. Washington, D.C.: American Council for an Energy-Efficient Economy.

Karney, R. 2006. ENERGY STAR® appliance market update. Presentation at 2006 ENERGY STAR® Appliance Partner Meeting, September 26-28, Newport, Rhode Island. Available at http://www.energystar.gov/index.cfm?c=partners.pt_meetings_archive.

Khawaja, M.S., A. Lee, and M. Levy. 2007. Statewide Codes and Standards Market Adoption and Noncompliance Rates. Report prepared for Southern California Edison. Portland, Oreg.: Quantec, LLC. May 10.

Kinney, L., H. Geller, and M. Ruzzin. 2003. Increasing Energy Efficiency in New Buildings in the Southwest: Energy Codes and Best Practices. Boulder, Colo.: Southwest Energy Efficiency Project. August.

Kushler, M., D. York, and P. Witte. 2006. Aligning Utility Interests with Energy Efficiency Objectives: A Review of Recent Efforts at Decoupling and Performance Incentives. Washington, D.C.: American Council for an Energy-Efficient Economy. October.

Nadel, S. 2002. Appliance and equipment efficiency standards. Annual Review of Energy and the Environment 27:159-192.

Nadel, S. 2004. Supplementary Information on Energy Efficiency for the National Commission on Energy Policy. Washington, D.C.: American Council for an Energy-Efficient Economy.

Nadel, S. 2007. Energy Efficiency Resource Standards Around the U.S. and World. Washington, D.C.: American Council for an Energy-Efficient Economy. September.

Nadel, S., and H. Geller. 1996. Utility DSM: What have we learned? Where are we going? Energy Policy 24:289-302.

Nadel, S., J. Thorne, H. Sachs, B. Prindle, and R.N. Elliott. 2003. Market Transformation: Substantial Progress from a Decade of Work. Washington, D.C.: American Council for an Energy-Efficient Economy. April.

Nadel, S., A. de Laski, M. Eldridge, and J. Kleisch. 2006. Leading the Way: Continued Opportunities for New State Appliance and Equipment Standards. Report ASAP-6/ACEEE-A062. (Updated from and supersedes report ACEEE-051.) Washington, D.C.: American Council for an Energy-Efficient Economy; Boston, Mass.: Appliance Standards Awareness Project. March.

NAPEE (National Action Plan for Energy Efficiency). 2006. National Action Plan for Energy Efficiency. Washington, D.C.: Department of Energy and Environmental Protection Agency. July.

NAS-NAE-NRC (National Academy of Sciences-National Academy of Engineering-National Research Council). 2009. America's Energy Future: Technology and Transformation. Washington, D.C.: The National Academies Press.

NRC (National Research Council). 2001. Energy Research at DOE: Was It Worth It? Energy Efficiency and Fossil Energy Research 1978 to 2000. Washington, D.C.: National Academy Press.

NRC. 2002. Effectiveness and Impact of Corporate Average Fuel Economy (CAFE) Standards. Washington, D.C.: National Academy Press.

NRC. 2008. Review of the Research Program of the FreedomCar and Fuel Partnership: Second Report. Washington, D.C.: The National Academies Press.

NYSERDA (New York State Energy Research and Development Authority). 2008. New York Energy $mart^{SM} Program, Evaluation and Status Report, Year Ending December 31, 2007. Final Report. Albany, N.Y.: NYSERDA. March.

OTA (Office of Technology Assessment). 1992. Building Energy Efficiency. Washington, D.C.: OTA.

PIEE (Precourt Institute on Energy Efficiency). 2007. Presentations from the 2007 Behavior, Energy, and Climate Change Conference, Sacramento, Calif., November 7-9. Available at http://piee.stanford.edu/cgi-bin/htm/Behavior/2007_becc_conference.php.

Rosenfeld, Arthur H., and Patrick McAuliffe. 2008. Opportunities in the Building Sector: Managing Climate Change. Staff paper CEC-999-2008-016. California Energy Commission. June. Available at http://www.energy.ca.gov/2008publications/cec-999-2008-016/cec-999-2008-016.pdf.

Shipley, A., A. Hampson, B. Hedman, P. Garland, and P. Bautista. 2008. Combined Heat and Power: Effective Energy Solutions for a Sustainable Future. ORNL/TM-2008/224. Oak Ridge, Tenn.: Oak Ridge National Laboratory.

Small, K., and K. Van Dender. 2007. Fuel efficiency and motor vehicle travel: The declining rebound effect. Energy Journal 28:25-51.

Spitzer, E. 2007. "15 by 15"—A Clean Energy Strategy for New York. Speech by Governor Eliott Spitzer, April 19. Available at http://www.ny.gov/governor/keydocs/CleanEnergySpeech-final.pdf.

Stone, N., D. Mahone, P. Eilert, and G. Fernstrom. 2002. What's a utility codes and standards program worth, anyway? Pp. 9.341-9.351 in Proceedings of the 2002 ACEEE Summer Study on Energy Efficiency in Buildings. Washington, D.C.: American Council for an Energy-Efficient Economy.

Sudarshan, A., and J. Sweeney. 2008. Deconstructing the "Rosenfeld Curve." Working Paper. Draft. Palo Alto, Calif.: Stanford University. June 1. Available at http:// peec.stanford.edu/modeling/research/Deconstructing_the_Rosenfeld_Curve.php.

Tonn, B., and J.H. Peretz. 2007. State-level benefits of energy efficiency. Energy Policy 35:3665-3674.

WGA (Western Governors' Association). 2006. Energy Efficiency Task Force Report. Denver, Colo.: WGA. Available at http://www.westgov.org/wga/initiatives/cdeac/Energy%20Efficiency-full.pdf.

York, D., and M. Kushler. 2006. A nationwide assessment of utility sector energy efficiency spending, savings, and integration with utility system resource acquisition. In Proceedings of the 2006 ACEEE Summer Study on Energy Efficiency in Buildings. Panel 8, Number 24. Washington, D.C.: American Council for an Energy-Efficient Economy.

Appendixes

A America's Energy Future Project

In 2007, the National Academies initiated the America's Energy Future (AEF) project (Figure A.1) to facilitate a productive national policy debate about the nation's energy future. The Phase I study, headed by the Committee on America's Energy Future and supported by the three separately constituted panels whose members are listed in this appendix, will serve as the foundation for a Phase II portfolio of subsequent studies at the Academies and elsewhere, to be focused on strategic, tactical, and policy issues, such as energy research and development priorities, strategic energy technology development, policy analysis, and many related subjects.

A key objective of the AEF project is to facilitate a productive national policy debate about the nation's energy future.

COMMITTEE ON AMERICA'S ENERGY FUTURE

HAROLD T. SHAPIRO, Princeton University, *Chair*
MARK S. WRIGHTON, Washington University in St. Louis, *Vice Chair*
JOHN F. AHEARNE, Sigma Xi and Duke University
ALLEN J. BARD, University of Texas at Austin
JAN BEYEA, Consulting in the Public Interest
WILLIAM F. BRINKMAN, Princeton University
DOUGLAS M. CHAPIN, MPR Associates
STEVEN CHU,[1] Lawrence Berkeley National Laboratory

[1]Resigned from the committee on January 21, 2009.

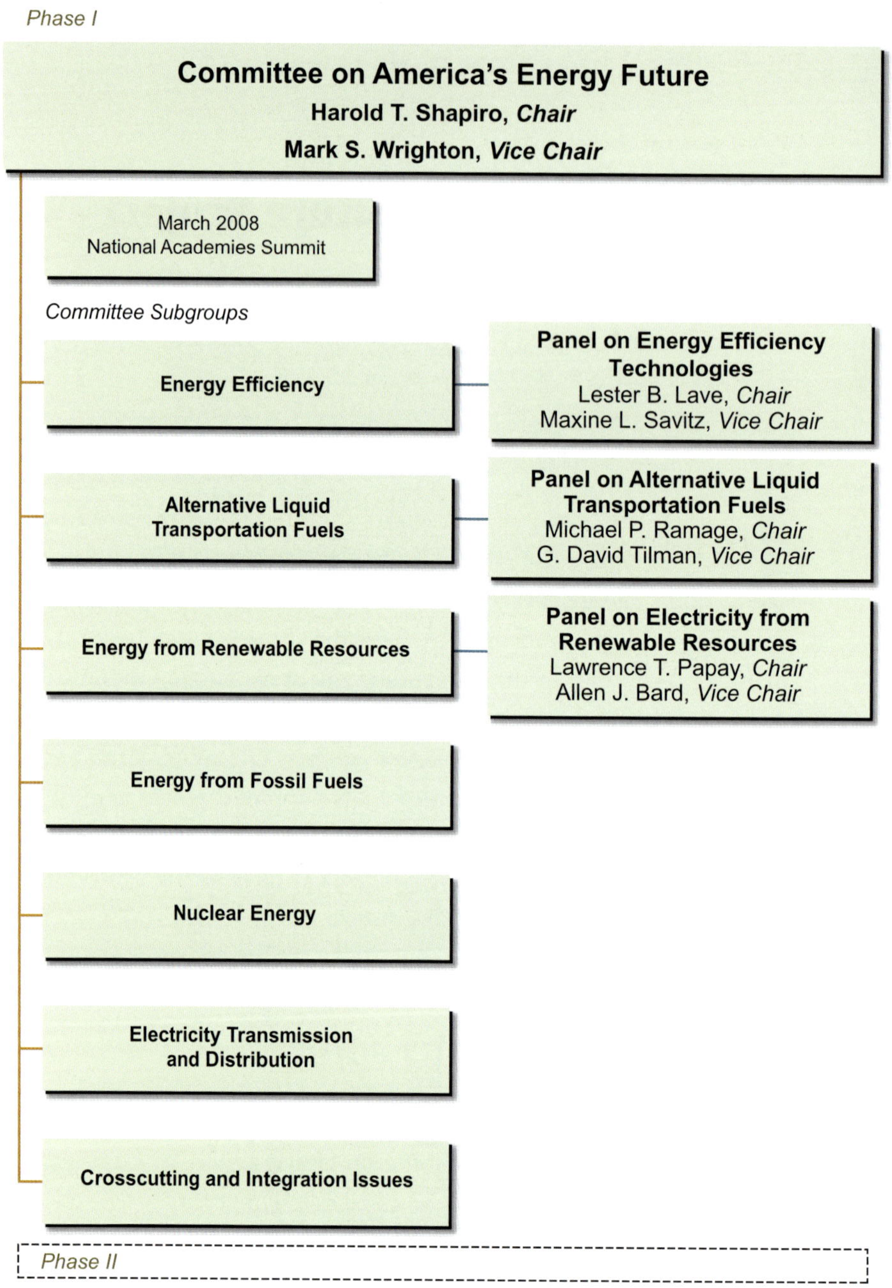

FIGURE A.1 *America's Energy Future Project.*

CHRISTINE A. EHLIG-ECONOMIDES, Texas A&M University
ROBERT W. FRI, Resources for the Future
CHARLES H. GOODMAN, Southern Company (retired)
JOHN B. HEYWOOD, Massachusetts Institute of Technology
LESTER B. LAVE, Carnegie Mellon University
JAMES J. MARKOWSKY, American Electric Power Service Corp. (retired)
RICHARD A. MESERVE, Carnegie Institution for Science
WARREN F. MILLER, JR., Texas A&M University
FRANKLIN M. ("Lynn") ORR, JR., Stanford University
LAWRENCE T. PAPAY, PQR LLC
ARISTIDES A.N. PATRINOS, Synthetic Genomics, Inc.
MICHAEL P. RAMAGE, ExxonMobil (retired)
MAXINE L. SAVITZ, Honeywell, Inc. (retired)
ROBERT H. SOCOLOW, Princeton University
JAMES L. SWEENEY, Stanford University
G. DAVID TILMAN, University of Minnesota, St. Paul
C. MICHAEL WALTON, University of Texas at Austin

PANEL ON ENERGY EFFICIENCY

LESTER B. LAVE, Carnegie Mellon University, *Chair*
MAXINE L. SAVITZ, Honeywell, Inc. (retired), *Vice Chair*
R. STEPHEN BERRY, University of Chicago
MARILYN A. BROWN, Georgia Institute of Technology
LINDA R. COHEN, University of California, Irvine
MAGNUS G. CRAFORD, Philips LumiLeds Lighting
PAUL A. DECOTIS, Long Island Power Authority
JAMES DEGRAFFENREIDT, JR., WGL Holdings, Inc.
HOWARD GELLER, Southwest Energy Efficiency Project
DAVID B. GOLDSTEIN, Natural Resources Defense Council
ALEXANDER MACLACHLAN, E.I. du Pont de Nemours and Company (retired)
WILLIAM F. POWERS, Ford Motor Company (retired)
ARTHUR H. ROSENFELD, E.O. Lawrence Berkeley National Laboratory
DANIEL SPERLING, University of California, Davis

PANEL ON ALTERNATIVE LIQUID TRANSPORTATION FUELS

MICHAEL P. RAMAGE, ExxonMobil Research and Engineering Company (retired), *Chair*
G. DAVID TILMAN, University of Minnesota, St. Paul, *Vice Chair*
DAVID GRAY, Noblis, Inc.
ROBERT D. HALL, Amoco Corporation (retired)
EDWARD A. HILER, Texas A&M University (retired)
W.S. WINSTON HO, Ohio State University
DOUGLAS R. KARLEN, U.S. Department of Agriculture, Agricultural Research Service
JAMES R. KATZER, ExxonMobil Research and Engineering Company (retired)
MICHAEL R. LADISCH, Purdue University and Mascoma Corporation
JOHN A. MIRANOWSKI, Iowa State University
MICHAEL OPPENHEIMER, Princeton University
RONALD F. PROBSTEIN, Massachusetts Institute of Technology
HAROLD H. SCHOBERT, Pennsylvania State University
CHRISTOPHER R. SOMERVILLE, Energy Biosciences Institute
GREGORY STEPHANOPOULOS, Massachusetts Institute of Technology
JAMES L. SWEENEY, Stanford University

PANEL ON ELECTRICITY FROM RENEWABLE RESOURCES

LAWRENCE T. PAPAY, Science Applications International Corporation (retired), *Chair*
ALLEN J. BARD, University of Texas, Austin, *Vice Chair*
RAKESH AGRAWAL, Purdue University
WILLIAM L. CHAMEIDES, Duke University
JANE H. DAVIDSON, University of Minnesota, Minneapolis
J. MICHAEL DAVIS, Pacific Northwest National Laboratory
KELLY R. FLETCHER, General Electric
CHARLES F. GAY, Applied Materials, Inc.
CHARLES H. GOODMAN, Southern Company (retired)
SOSSINA M. HAILE, California Institute of Technology
NATHAN S. LEWIS, California Institute of Technology
KAREN L. PALMER, Resources for the Future

JEFFREY M. PETERSON, New York State Energy Research and Development Authority
KARL R. RABAGO, Austin Energy
CARL J. WEINBERG, Pacific Gas and Electric Company (retired)
KURT E. YEAGER, Galvin Electricity Initiative

AMERICA'S ENERGY FUTURE PROJECT DIRECTOR

PETER D. BLAIR, Executive Director, Division on Engineering and Physical Sciences

AMERICA'S ENERGY FUTURE PROJECT MANAGER

JAMES ZUCCHETTO, Director, Board on Energy and Environmental Systems (BEES)

AMERICA'S ENERGY FUTURE PROJECT STAFF

KEVIN D. CROWLEY, Director, Nuclear and Radiation Studies Board (NRSB), *Study Director*
DANA G. CAINES, Financial Manager, BEES
SARAH C. CASE, Program Officer, NRSB
ALAN T. CRANE, Senior Program Officer, BEES
GREG EYRING, Senior Program Officer, Air Force Studies Board
K. JOHN HOLMES, Senior Program Officer, BEES
LaNITA JONES, Administrative Coordinator, BEES
STEVEN MARCUS, Editorial Consultant
THOMAS R. MENZIES, Senior Program Officer, Transportation Research Board
EVONNE P.Y. TANG, Senior Program Officer, Board on Agriculture and Natural Resources
MADELINE G. WOODRUFF, Senior Program Officer, BEES
E. JONATHAN YANGER, Senior Program Assistant, BEES

B Panel Biographical Information

Lester B. Lave, *Chair,* is the Harry B. and James H. Higgins Professor of Economics and University Professor at Carnegie Mellon University. He is also director, Carnegie Mellon Green Design Initiative, and codirector, Carnegie Mellon Electricity Industry Center. His teaching and research interests include applied economics, political economy, quantitative risk assessment, safety standards, modeling the effects of global climate change, public policy concerning greenhouse gas emissions, and understanding the issues surrounding the electricity transmission and distribution system. He is a recipient of the Distinguished Achievement Award of the Society for Risk Analysis. Dr. Lave is a member of the Institute of Medicine.

Maxine L. Savitz, *Vice Chair,* is a director of the Washington Advisory Group. Dr. Savitz is a former deputy assistant secretary for Conservation, U.S. Department of Energy (DOE). She received the Outstanding Service Medal from DOE in 1981. Prior to her DOE service, she was program manager for Research Applied to National Needs at the National Science Foundation. Following her government service, Dr. Savitz served in executive positions in the private sector, including: president of Lighting Research Institute, assistant to the vice president for engineering at The Garrett Corporation, and general manager of AlliedSignal Ceramic Components. She recently retired from the position of general manager for Technology Partnerships at Honeywell. Dr. Savitz is a member of the American Association for the Advancement of Science. She was appointed to the National Science Board in 1998. She is a member of the Secretary of Energy Advisory Board, the DOE's Laboratory Operations Board, and advisory bodies for Oak Ridge National Laboratory (ORNL) and Pacific Northwest National Laboratory.

Dr. Savitz also serves on the board of directors of the Electric Power Research Institute and the American Council for an Energy Efficient Economy. Dr. Savitz is a member of the National Academy of Engineering (NAE). She received a B.A. in chemistry from Bryn Mawr College and a Ph.D. in organic chemistry from the Massachusetts Institute of Technology.

R. Stephen Berry is the James Franck Distinguished Service Professor Emeritus of Chemistry at the University of Chicago and holds appointments in the College, the James Franck Institute, and the Department of Chemistry. He has also held an appointment in the School of Public Policy Studies at the University of Chicago and has worked on a variety of subjects ranging from strictly scientific matters to a variety of topics in policy. He spent 1994 at the Freie Universität Berlin as an awardee of the Humboldt Prize. In 1983 he was awarded a MacArthur Fellowship. His experimental research includes studies of negative ions, chemical reactions, detection of transient molecular species, photoionization, and other laser-matter interactions. Other research has involved interweaving thermodynamics with economics and resource policy, including efficient use of energy. Since the mid-1970s, Dr. Berry has worked on issues of science and the law, and with management of scientific data, activities that have brought him into the arena of electronic media for scientific information and issues of intellectual property in that context. Dr. Berry is a member of the National Academy of Sciences (NAS). He attended Harvard University, where he received an A.B. and an A.M. in chemistry and a Ph.D. in physical chemistry.

Marilyn A. Brown is a professor of public policy at the Georgia Institute of Technology. Previously, she was the interim director of the Engineering Science and Technology Division at ORNL. During her 22 years at ORNL, Dr. Brown researched the impacts of policies and programs aimed at advancing the market entry of sustainable energy technologies and led several energy technology and policy scenario studies. Prior to serving at ORNL, she was a tenured associate professor in the Department of Geography at the University of Illinois, Urbana-Champaign, where she conducted research on the diffusion of energy innovations. She has authored more than 150 publications and has been an expert witness in hearings before committees of both the U.S. Senate and the House of Representatives. A recent study that she co-led, *Scenarios for a Clean Energy Future,* was the subject of two Senate hearings, has been cited in proposed federal legislation, and has had a significant role in international climate change debates. She serves on the board of directors of several energy, engineering, and environmental organizations, including the Alliance to Save Energy and the American Council for an Energy Efficient Economy, and she serves on the editorial board of the *Journal of Technology Transfer*. Dr. Brown is a member of the National Commission on Energy Policy. She has a Ph.D. in geography from Ohio State University and a master's degree in resource planning from the University of Massachusetts.

Linda R. Cohen is a professor of economics and associate dean for research and graduate studies for the School of Social Sciences at the University of California, Irvine. She is a fellow and former council member of the California Council for Science and Technology, and was a member of the Advisory Panel for the Public Interest Energy Research Program for the California Energy Commission. She recently served on National Research Council (NRC) committees on the benefits of DOE programs in energy efficiency and fossil energy and on the American Physical Society Panel on Public Affairs' Committee on Energy and Environment. In 2004 Dr. Cohen held the Gilbert White Fellowship at Resources for the Future of Washington, D.C. Her energy-related publications include *The Technology Pork Barrel,* "When Can Government Subsidize Research Joint Ventures? Politics, Economics and Limits to Technology Policy," "Is U.S. Science Policy at Risk? Trends in Federal Support for R&D," and *Prospective Evaluation of Applied Energy Research and Development at DOE (Phase One): A First Look Forward,* with coauthors. She received a Ph.D. in social sciences from the California Institute of Technology.

Magnus G. Craford is the chief technology officer of Philips LumiLeds Lighting. Dr. Craford began his professional career as a research physicist at Monsanto Chemical Company. His initial research dealt with the development of optoelectronics materials and devices using a variety of compound semiconductor materials. In 1979, he joined Hewlett Packard Company as a manager in the Optoelectronics Division, responsible for the development of technology and processes for manufacturing visible light emitting diodes. In 1999, Dr. Craford assumed his current position as chief technical officer of LumiLeds Lighting, then a joint venture of Agilent Technologies and Philips Lighting, now owned by Philips. He is a fellow of the Institute of Electrical and Electronics Engineering (IEEE). He has received the MRS Medal, the IEEE Morris N. Liebmann Award, the Holonyak Award of the Optical of America, the Welker Award of the International Symposium on Compound Semiconductors, the Electronics Division Award of the

Electrochemical Society, and the Distinguished Alumni Award of the University of Illinois College of Engineering. He has published more than 50 papers and book chapters. Dr. Craford is a member of the NAE. He received a B.A. in physics from the University of Iowa and a Ph.D. in physics from the University of Illinois.

Paul A. DeCotis is vice president of power markets at Long Island Power Authority, where he oversees strategic resource planning; fuel, energy, and capacity purchases and sales; power project development and management; and participation in the region's wholesale power markets. Prior to this he was deputy secretary for energy in New York, serving as senior energy advisor to Governor Spitzer and Governor Paterson. He was also chair of the State Energy Planning Board. Mr. DeCotis previously served as director of energy analysis for the New York State Energy Research and Development Authority and before that was chief of policy at the State Energy Office in New York. Until his appointment as deputy secretary, he was president of a management consulting business, specializing in executive and board development, strategy, and mediation. Since 1985, he has served as an adjunct faculty member at several colleges and universities, including Cornell University, Rochester Institute of Technology, and Sage Graduate School. Mr. DeCotis is a member of the Board on Energy and Environmental Systems of the National Research Council; a member of the Energy Working Group of the Coalition of Northeastern Governors; a member of the Energy Resources Board of the American University at Kosovo; an editorial board member of the *Energy Efficiency Journal*; an executive committee member of the New York Reliability Council; and a member of the New York Smart Grid Consortium. He has served on and chaired many professional organizations and associations and has extensive community service experience. Mr. DeCotis received his B.A. in international business management from the State University College at Brockport, his M.A. in economics from the University at Albany, and his M.B.A. in finance from the Sage Graduate School at Russell Sage College.

James H. DeGraffenreidt, Jr., is chairman of the board and chief executive office of the WGL Holdings, Inc., the parent company of Washington Gas. He also serves as chairman and CEO of Washington Gas, the natural gas utility serving more than 980,000 customers in the Washington, D.C., metropolitan area and surrounding region. After practicing law as a partner or associate at different law firms and as assistant people's counsel in Maryland, he joined WGL Holdings, Inc., as senior managing attorney. Mr. DeGraffenreidt also serves on numerous boards, including the American Gas Association Alliance to Save Energy, MedStar Health, Harbor Bankshares Corporation, Maryland Science Center, the Walters Art Museum, and the Massachusetts Mutual Life Insurance Company. He received a B.A. from Yale College and a J.D. and an M.B.A. from Columbia University.

Howard Geller is the executive director of the Southwest Energy Efficiency Project (SWEEP), a public interest venture he founded in 2001. Based in Boulder, Colorado, SWEEP promotes policies and programs to advance energy efficiency in a six-state region that includes Arizona, Colorado, Nevada, New Mexico, Utah, and Wyoming. He is the former executive director of the American Council for an Energy-Efficient Economy (ACEEE). He established ACEEE's Washington, D.C., office in 1981, stepping down as executive director in February 2001. Dr. Geller has advised and conducted energy efficiency studies for utilities, governmental organizations, and international agencies. He has testified before the U.S. Congress on energy issues many times and has influenced key energy legislation, including the National Appliance Energy Conservation Act of 1987 and the Energy Policy Act of 1992. He is author or coauthor of four books. His most recent book, *Energy Revolution: Policies for a Sustainable Future,* was published in December 2002 by Island Press. Dr. Geller has spent significant time working on energy efficiency issues in Brazil, where he helped to start and frequently advises Brazil's National Electricity Conservation Program. He was awarded the 1998 Leo Szilard Award for Physics in the Public Interest by the American Physical Society in recognition of his contributions to national appliance efficiency standards and more efficient energy use in general. Dr. Geller is a member of the editorial advisory board for the journal *Energy Policy* and was the associate editor for energy efficiency for the Macmillan *Encyclopedia of Energy*. He received a Ph.D. in energy policy from the University of Sao Paulo in Brazil and a master's degree in mechanical engineering from Princeton University.

David B. Goldstein codirects the energy program of the National Resources Defense Council. He has worked on energy efficiency and policy since the early 1970s. In 2002, the MacArthur Foundation recognized his achievement in the field by awarding him one of its prestigious 5-year fellowships. Dr. Goldstein has worked toward the development of energy efficiency standards for new buildings and appliances at the regional and national levels, both in the United States and in Russia. He negotiated the agreement that led to the National Appliance Energy Conservation Act of 1987 and has helped design and direct energy efficiency programs with utilities and state regulatory agencies. Dr. Goldstein also created the Location Efficient Mortgage, a program designed to reduce urban sprawl and car use. He was a founding director of the Consortium for Energy Efficiency and the New Buildings Institute. Dr. Goldstein is a fellow of the American Physical Society and a recipient of its Leo Szilard Award for Physics in the Public Interest. He received a Ph.D. in physics from the University of California at Berkeley.

Alexander MacLachlan retired from E.I. du Pont de Nemours & Company in 1993 after more than 36 years of service. He had been senior vice president for research and development (R&D) and chief technical officer since 1986. In 1994, he joined the DOE as deputy undersecretary for technology partnerships and in 1995 was made deputy undersecretary for R&D management. He left DOE in 1996 but remained on its Secretary of Energy Advisory Board, Laboratory Operations Board, Sandia President's Advisory Council, and the National Renewable Energy Laboratory's Advisory Council until 2003. He has participated in several studies for the NRC, including *Containing the Threat from Illegal Bombings* (1998), *Technology Commercialization: Russian Challenges, American Lessons* (1998), *Building an Effective Environmental Management Science Program* (1997), and most recently, *Countering the Threat of Improvised Explosive Devices* (2007). He was also chair for the Committee to Review the Department of Transportation's Intelligent Vehicle Initiative. He served recently on the NRC's Board on Radioactive Waste Management. Dr. MacLachlan is a member of Phi Beta Kappa and a member of the NAE. He received a B.S. in chemistry from Tufts University and a Ph.D. in physical organic chemistry from the Massachusetts Institute of Technology.

William F. Powers retired as vice president of research, Ford Motor Company. He has extensive expertise in advanced R&D of automotive technology. His approximately 20 years at Ford included positions as director, Vehicle, Powertrain and Systems Research; director, Product and Manufacturing Systems; program manager, Specialty Car Programs; and executive director, Ford Research Laboratory and Information Technology. Prior positions also include professor, Department of Aerospace Engineering, University of Michigan, during which time he consulted with NASA, Northrop, Caterpillar, and Ford; research engineer, University of Texas; and mathematician and aerospace engineer, NASA Marshall Space Flight Center. He is a fellow, IEEE; fellow, Society of Automotive Engineers; fellow, American Society of Mechanical Engineers; member, NAE; and foreign member, Royal Swedish Academy of Engineering Sciences. He received a B.S. in aerospace engineering from the University of Florida and a Ph.D. in engineering mechanics from the University of Texas at Austin.

Arthur H. Rosenfeld is a professor of physics at Lawrence Berkeley National Laboratory (LBNL). He is also a member of the California Energy Commission. After completing his graduate studies, Dr. Rosenfeld went to the University of California at Berkeley, where he joined, and eventually led, the Nobel Prize–winning particle physics group of Luis Alvarez at LBNL until 1974. At that time, he changed to the new field of efficient use of energy, formed the Center for Building Science at LBNL, and led it until 1994. The center developed electronic ballasts for fluorescent lamps (which led to compact fluorescent lamps), low-emissivity windows, and the DOE-2 computer program for the energy analysis and design of buildings. He received the Szilard Award for Physics in the Public Interest in 1986 and the Carnot Award for Energy Efficiency from the DOE in 1993. In 2006, Dr. Rosenfeld received the Enrico Fermi Award, the oldest and one of the most prestigious science and technology awards given by the U.S. government. Dr. Rosenfeld is a cofounder of the ACEEE, the University of California's Institute for Energy Efficiency, and the Washington-based Center for Energy and Climate Solutions. From 1994 to 1999 Dr. Rosenfeld served as senior advisor to the DOE's Assistant Secretary for Energy Efficiency and Renewable Energy. He received a Ph.D. in physics from the University of Chicago.

Daniel Sperling is a professor of civil engineering and environmental science and policy and director of the Institute of Transportation Studies at the University of California, Davis. Dr. Sperling has done extensive studies on alternative transportation fuels, fuel cell vehicles, and sustainable transportation, and has authored 200-plus technical papers and eight books. He has been a member of several NRC committees related to transportation, including the Committee to Review the R&D Strategy for Biomass-Derived Ethanol and Biodiesel Transportation Fuels, the Committee on Alternative and Strategies for Future Hydrogen Production and Use, and the Committee for the Study of the Long-Term Viability of Fuel Taxes for Transportation Finance. He is the chair of the Transportation Research Board's (TRB's) Sustainability and Transportation Committee and a former chair of the TRB's Alternative Transportation Fuels Committee. Dr. Sperling was elected a National Associate of the National Academies in 2004. He received a B.S. in civil engineering from Cornell University and a Ph.D. in transportation engineering from the University of California at Berkeley.

C Presentations and Panel Meetings

The panel met five times in the course of its work. It heard from outside experts in open sessions at the first two meetings.

FIRST PANEL MEETING: OCTOBER 22–23, 2007, WASHINGTON, D.C.

Presentations in Open Session

Lee Schipper, Director of Research, World Resources Institute, WRI Center for Sustainable Transport

Jaana Remes, Senior Fellow, McKinsey Global Institute

Mark Levine, Senior Staff Scientist, Lawrence Berkeley Laboratory

David Rodgers, Deputy Assistant Secretary for Energy Efficiency and Renewable Energy, U.S. Department of Energy

Steve Nadel, American Council for an Energy Efficient Economy

Alan Crane, Senior Program Officer, NRC, on the NRC's 2001 CAFE report

Bob Simon, Staff Director, Senate Energy and Natural Resources Committee

SECOND PANEL MEETING: DECEMBER 12–13, 2007, WASHINGTON, D.C.

Douglas Kaempf, Director, Office of Industrial Technologies, U.S. Department of Energy

Kathleen Hogan, Director, Climate Protection Partnership Division, U.S. Environmental Protection Agency

Steven Smith, Joint Global Change Research Institute, Pacific Northwest National Laboratory

Revis W. James, Director, Energy Technology Assessment Center, Electric Power Research Institute

Jonathan Creyts, Principal, McKinsey & Company

Fred Moore, Director, Manufacturing & Technology, Energy, Dow Chemical Company

K. John Holmes, Senior Program Officer, NRC, on the latest NRC vehicle fuel efficiency technology study

John Heywood, panel member, on transportation work at MIT

THIRD PANEL MEETING: MARCH 11–12, 2008, WASHINGTON, D.C.

Closed meeting

FOURTH PANEL MEETING: MAY 3–4, 2008, WASHINGTON, D.C.

Closed meeting

FIFTH MEETING: JUNE 26, 2008, WASHINGTON, D.C.

Closed meeting

D Definitions of Energy Efficiency

The term *energy efficiency* is used in several ways. The definition perhaps most often used is based simply on how much of a given task or product (be it the heating of a building for a specified time, the miles driven by a car, or the tons of iron smelted) is achieved per unit of energy expended for that task or product. For example, the number of tons of iron, t, that can be recovered from ore per Btu of energy, E, used in the smelting process, t/E, is one possible measure of energy efficiency.

Another definition is based on the *total* energy, E_{tot}, required to provide a product. According to this definition, the energy efficiency for making a ton of iron would be the tons of iron, t, per Btu of total energy required, including mining, transportation, smelting, and any other input, t/E_{tot}.

Both of the measures of energy efficiency defined above would be termed *first-law efficiency* (derived from the first law of thermodynamics), being based simply on actual energy use and not taking into account such things as the excess entropy due to the irreversibility of real processes. Hence, in many situations, one may use a *second-law efficiency* (derived from the second law of thermodynamics), which, instead of energy, uses the *free energy,* usually the Gibbs free energy, G, where $G = H - TS$. H is the enthalpy, and $H = E + pV$, where p is pressure and V is the volume of the system—in this case the volume of the iron produced. T is the temperature and S is the entropy. Because most processes are carried out at constant pressure, enthalpy H is the most appropriate measure, and one uses H rather than energy E. If one wishes to use the second-law efficiency, one simply replaces E, the energy used, with G, the free energy, in the expressions for the first-law efficiency.

Additionally, one other kind of definition of energy efficiency is sometimes used, based on how much the actual process deviates from the thermodynamic limit. According to this definition, a perfect process would have a value of infinity for either its first-law or second-law efficiency; that is, the efficiency would be the tons of iron produced per amount of energy or free energy beyond the thermodynamic limit. Hence, for a perfect process, the denominators in these measures would be zero. No real process achieves the thermodynamic limit, of course, and so no real process has an infinite efficiency according to this last kind of definition.

It is also possible to use a more realistic counterpart of the (preceding) definition based on the comparison with the thermodynamic limit—namely, a comparison based on the most efficient possible process *subject to a chosen time or rate constraint*. This approach enables the user to compare, for example, the relative advantages and disadvantages (in energy efficiency terms) of higher-capacity but slow processes and lower-capacity but faster processes.

In practice, one very rarely encounters an explicitly stated definition of *energy efficiency*. Most commonly, people tend to use the very first definition, the amount of a task or product (the heating of a building for a specified time, the miles driven by a car, the tons of ore smelted, and so on) per *direct* unit of energy required for that task. When a different definition is being used, the user generally specifies *which* definition is being used. In this report, because the data have been taken from a very wide variety of sources, virtually none of which specified a definition, the panel assumed that the first and simplest definition was intended. This is not to imply that if the panel itself were to derive the efficiencies from primary data that it would use that same definition. The pragmatic course was taken here to allow the analysis to be carried out.

E Estimating the Net Costs and Benefits of Energy Savings

As described in this appendix, the question of whether an energy-efficient option will result in net cost or savings to the consumer depends on various considerations and can differ from consumer to consumer. For a product with a first cost higher than its less efficient equivalent, such as a compact fluorescent lamp (CFL) compared with an incandescent lamp, the savings would result from lower energy expenditures over the economic life of the product. However, those savings depend on the price of energy and the intensity of usage. For example, when replacing an incandescent lamp with a CFL in a fixture that is used for only a few hours per year, it would take many years for the higher initial cost of the CFL to be recouped, whereas a lamp used continuously would pay off the increased cost within a month. Similarly, high-priced energy increases the payoff of the energy-efficient product.

The answer to the question of whether there is a net cost or benefit also depends on the interest rate that the consumer must pay to finance the higher-cost, more efficient product. In some cases the customer has no ability to borrow the additional purchase price at any interest rate. More generally, customers can use their credit cards, borrowing the money at an interest rate of about 20 percent per year. In still other cases, the additional purchase price is paid from money in a checking account that earns no interest. An energy-saving product might be out of the question for the customer who cannot afford the additional cost, might be of marginal benefit for the customer using a credit card, and might be of large net benefit to the customer paying a zero interest rate.

Analysts attempt to measure energy savings to consumers by means of various measures. They often use the net present value (NPV) of the energy savings, computed using a particular interest rate, minus the initial purchase price. A variation on this method is to calculate the return on investment, or ROI. This method calculates the interest rate implicit in equating the higher initial cost to the stream of energy savings over time. For example, if an efficient air conditioner costs $300 more than a less efficient model and saves $30 per year in electricity, the ROI is approximately 10 percent. For the more efficient air conditioner, there is no net savings for someone who would have to pay a credit card interest rate, but a large savings for someone taking the money from a checking account. The ROI and NPV depend both on the annual savings and on how long the air conditioner provides these savings—for example, if the building in which the air conditioner is installed will be torn down and the air conditioner destroyed in 3 years, that is the relevant period over which to calculate the return.

F Equivalences and Conversion Factors

Energy savings are normally measured in megawatt-hours of electricity (MWh) or in million British thermal units (Btu). These energy savings can be converted directly into avoided million tons of emitted carbon (C) or carbon dioxide (CO_2). But most people have little feel for these strange units, and so news media, when reporting on energy topics, tend to convert energy and emissions savings to familiar equivalents: namely, avoided cars, homes, or power plants.

The tables in the section below show how to perform these conversions, but first it is useful to define the typical car, typical home, and typical power plant—things easily visualized—as used in the conversion tables.

- *A typical car* is defined here as one that has an average fuel economy of 24 miles per gallon (mpg) and is driven 12,000 miles per year, for a gasoline use of 500 gallons per year. Such a vehicle would be a passenger car (rather than something larger or heavier such as a van or sport-utility vehicle).
- *A typical home* is defined here as one having an average annual electricity use of 12,000 kilowatt-hours (kWh), corresponding to primary energy use at the power plant of 125 million Btu, plus an average annual 75 million Btu of fuel for heat (typically natural gas), for a total of 200 million Btu. Note that, for the discussion below, electricity accounts for about 2/3 of this 200 million Btu.

TABLE F.1 Metric Prefixes

Unit Multiple	Metric Prefix	Symbol	Value
10^3	kilo	k	Thousand
10^6	mega	M	Million
10^9	giga	G	Billion
10^{12}	tera	T	Trillion
10^{15}	peta	P	Quadrillion

- *A typical power plant* is defined here as one with a generating capacity of 500 megawatts (MW) that operates for slightly less than 5000 hours per year[1] and is thus selling 2.5 billion kWh per year (or 2.5 terawatt-hours [TWh]; tera = 10^{12} = trillion; see Table F.1). Although a typical 20-year-old power plant has a generating capacity of about 1000 MW, or 1 gigawatt (GW; giga = 10^9 = billion), the typical power plant as defined here is smaller because newly constructed power plants tend to have a capacity of about 500 MW.

The typical uses of energy and electricity given in the definitions above are shown in Column A of Tables F.2–F.4. Table F.1, "Metric Prefixes," is the basis of all notations used in Tables F.2–F.4.

WHICH TABLE TO USE

The conversion by energy (Table F.2), electricity (Table F.3), and C or CO_2 (Table F.4) differs by up to 50 percent. The choice of which table to use depends on one's "model." Those most interested in saving money, primary energy, and air pollution will prefer Table F.2, but those focusing on electricity trade-offs would use Table F.3 (which does not include cars), and those addressing CO_2 trade-offs would choose Table F.4.

[1]A more accurate number is 4850 hours per year (3300 billion kWh/680 GW from Table 7.1 in *Monthly Energy Review*, February 2001, and Table 35 in *1999 Electric Power Annual, Volume 2*, Energy Information Administration, Department of Energy, Washington, D.C., October 2000, respectively).

Table F.2 converts the energy use of cars, homes, and power plants to "primary energy" (also referred to as source energy). Thus, the primary energy associated with electricity production includes the energy burned at the power plant, not just the 30 percent delivered to the home.

Of the three conversion tables, Table F.2 would suffice for most purposes. As explained below, Tables F.3 and F.4 give slightly different conversions. Both cost to the customer and air pollution (nitrogen oxide [NO_x] and CO_2 emissions from combustion, as well as sulfur oxide [SO_x] emissions from coal combustion)

TABLE F.2 Energy Used Annually by a Typical Car and Home and Generated by a Typical 500 Megawatt Power Plant

	A Typical Annual Use (rounded)	B Conversion to Btu[a]	C Annual Energy Use (Btu)[b]	D Energy Use in Units of 1 Million Cars
Passenger cars, vans, sport utility vehicles, light trucks—U.S. stock (private and commercial): 247 million[c]				
1 typical car	500 gal[d]	1 gal = 125,000 Btu	62.5 million	—
1 million typical cars	500 million gal	1 gal = 125,000 Btu	62.5 trillion	1
Homes—U.S. stock: 111 million[e]				
1 typical home (electricity + gas/oil)	200 million Btu	—	200 million	—
1 million typical homes	200 trillion Btu	—	200 trillion	3.2
Power plants—U.S. stock: 3,300 TWh[f] *≡ 1,320 plants (½ GW)*				
Typical power plant (½ GW × 5,000 hours per year)	2.5 TWh	1 kWh = 10,500 Btu[g]	26.2 trillion	0.4

[a]See http://bioenergy.ornl.gov/papers/misc/energy_conv.html.

[b]For metric units (e.g., kWh) the metric prefix is used; for Btu, the "value" multiplier is used, as shown in Table F.1, "Metric Prefixes."

[c]Data from Table 1-11: Number of U.S. Aircraft, Vehicles, Vessels, and Other Conveyances, available at http://www.bts.gov/publications/national_transportation_statistics/html/table_01_11.html.

[d]Data from Table MF21 (for motor fuel use) and Table MV1 and MV9 (for private and commercial auto stock) in *Highway Statistics 1998*, available at http://www.fhwa.dot.gov/ohim/ohimstat.html. Table 1.10 for average annual miles in *Monthly Energy Review*, DOE/EIA-0035(2000-04), April 2000.

[e]Data from Residential Energy Consumption Survey, available at http://www.eia.doe.gov/emeu/.

[f]Data from Table 7.5 in *Monthly Energy Review*, DOE/EIA-0035(2000-04), April 2000.

[g]See Tables 2.6 and 7.5 in *Monthly Energy Review*, DOE/EIA-0035(2000-04), April 2000. In 1999, the U.S. electric grid consumed 34.5 quads of source energy to generate and sell 3,300 TWh of electricity. This yields a "heat rate" of 10,500 Btu/kWh.

TABLE F.3 Electricity Used Annually by a Typical Home and Generated by a 500 Megawatt Power Plant

	A Typical Annual Use	D Electricity Use in Units of 1 Million Homes
1 typical home	12,000 kWh	—
1 million typical homes	12 TWh	1
Typical power plant (½ GW × 5,000 hours per year)	2.5 TWh	0.2

are roughly proportional to primary energy, although the cost per Btu would vary among fuels.

Using Table F.2 (Primary Energy)

With respect to the use of Table F.2, suppose one learns that low-energy (low-E) windows are saving 1 quad per year (1 quad = 10^{15} Btu, or 1 quadrillion Btu), which is about 1 percent of total U.S. energy use. One can use Column C of Table F.2 to divide by 1 million cars.

Similarly one could calculate 5 million equivalent homes or 38 power plants avoided.

Using Table F.3 (Electricity)

Suppose one learns, however, that the 2001 refrigerator standard will save 30 billion kWh, or 30 TWh, annually. In this case, Table F.3 indicates that a typical power plant sells 2.5 billion kWh per year (or 2.5 TWh), so it can be seen that as a result of the standard 12 power plants are avoided. Likewise, according to Table F.3, 1 million homes use 12 TWh, and so the standard has freed up electricity to supply 2.5 million homes.

But as is noted above, for every 100 Btu of electric energy, homes use another 50 Btu of fuel, so there has not been enough energy and pollution saved to offset 2.5 million homes, but only about 1.7 million.[2]

[2]This 1.7 million home offset can be checked by converting 30 TWh to trillion Btu (using the grid's heat rate of 10,500 Btu/kWh) and then using Table F.2.

Using Table F.4 (CO_2)

Finally, suppose that manufacturers want to get CO_2 credit for the same refrigerator standards, so they use Table F.4 to convert the 12 power plants avoided into 18 million tons of CO_2 per year. As before, one can divide 18 million tons of CO_2 by the 1 million homes row (11 million tons of CO_2) to find about 1.6 million equivalent homes. CO_2 savings is often stated in million tons of carbon rather than CO_2, so it must be noted that 1 ton of carbon is equivalent to 44/12 = 3.67 tons of CO_2.

TABLE F.4 CO_2 Released by Cars, Homes, and Power Plants

	A Typical Annual Use (rounded)	B Conversion to CO_2[a]	C Annual CO_2 Release (metric tons)	D CO_2 Release in Units of 1 Million Cars
Passenger cars, vans, sport utility vehicles, light trucks—U.S. stock (private and commercial): 200 million[b]				
1 typical car	500 gal[c]	1 gal = 8.8 kg[d]	4.4	
1 million typical cars	500 million gal		4.4 million	1
Homes—U.S. stock: 100 million[e]				
1 typical home (electricity + gas/oil)	200 million Btu	1 million Btu = 55 kg[f]	11	
1 million typical homes	200 trillion Btu		11 million	2.5
Power plants—U.S. stock: 3,300 TWh[g] *≡ 1,320 plants (½ GW)*				
Typical power plant (½ GW × 5,000 hours per year)	2.5 TWh	1 TWh = 0.6 million tons[h]	1.5 million	0.34

[a]1 million tons/quadrillion Btu = 1 kg/million Btu; 1 ton of C corresponds to 3.67 tons of CO_2.

[b]Data from Tables MV1 and MV9 in Highway Statistics 1998, available at http://www.fhwa.dot.gov/ohim/ohimstat.html.

[c]Data from Table MF21 (for motor fuel use) and Tables MV1 and MV9 (for private and commercial auto stock) in *Highway Statistics 1998*, available at http://www.fhwa.dot.gov/ohim/ohimstat.html. Table 1.10 for average annual miles in *Monthly Energy Review*, DOE/EIA-0035(2000-04), April 2000.

[d]See Table B1, p. 104, *EIA Emissions of Greenhouse Gases in the US*, DOE/EIA-0573(98).

[e]Data from Figure 2.1, *A Look at Residential Energy Consumption*, DOE/EIA-0632(97).

[f]See Table A19 for million tons of carbon, p. 133, and Table A2 for primary quads, p. 118, *Annual Energy Outlook*, DOE/EIA-0383(2000).

[g]Data from Table 7.5, *Monthly Energy Review*, DOE/EIA-0035(2000-04).

[h]See Table A19 for million tons of carbon, p. 133, *Annual Energy Outlook*, DOE/EIA-0383(2000).

A comparison of Column D in Tables F.2 and F.4 shows a slight difference in their equivalence of cars and homes. Table F.2 shows that 1 million homes use as much energy as do 3.2 million cars, but Table F.4 shows that the same 1 million homes produce only as much CO_2 as 2.5 million cars. This is because, per Btu, gasoline produces 4/3 as much CO_2 as electricity or natural gas.

CONVERTING POWER PLANTS (OR PEAK SHAVING) TO "HOMES"

In the analysis above, energy (kWh), not power (kW or MW), is discussed. National newspapers often use "1 kW = 1 home" as the relevant conversion factor. This is nearly, but not quite, correct. A more realistic conversion is roughly 1.6 kW for an average California home, and roughly 2.4 kW for an average U.S. home. This is based on the assumption that an average California home uses approximately 8,000 kWh per year, whereas an average U.S. home uses 12,000 kWh. However, owing to fluctuations in power demand, a typical power plant runs for only about 5,000 hours per year rather than 8,760 hours per year.

G Acronyms and Abbreviations

ACEEE	American Council for an Energy-Efficient Economy
AEF	America's Energy Future
AEO	Annual Energy Outlook
AISI	American Iron and Steel Institute
APU	auxiliary power unit
ARRA	American Recovery and Reinvestment Act of 2009
ASHRAE	American Society of Heating, Refrigerating, and Air-Conditioning Engineers
BAU	business-as-usual
bbl/d	barrel(s) per day
BEV	battery-electric vehicle
BOF	basic oxygen furnace
Btu	British thermal unit
CAFE	corporate average fuel economy (standard)
CCE	cost of conserved energy
CEC	California Energy Commission
CEF	*Scenarios for a Clean Energy Future* study
CFL	compact fluorescent lamp
CHP	combined heat and power
CI	compression-ignition
CO_2	carbon dioxide
CPUC	California Public Utilities Commission
CVT	continuously variable transmission

DOE	Department of Energy, U.S.
DPF	diesel particulate filter
DSM	demand-side management
EAF	electric arc furnace
EIA	Energy Information Administration
EISA	Energy Independence Security Act of 2007
EJ	exajoule
EPA	Environmental Protection Agency, U.S.
ERFC	emphasis on reducing fuel consumption
ERI	Energy Recover, Inc.
ESCO	energy services company
ESP	electrostatic precipitator
FCV	fuel-cell vehicle
FEMP	Federal Energy Management Program
FHWA	Federal Highway Administration
FY	fiscal year
GDP	gross domestic product
GVW	gross vehicle weight
GW	gigawatt
HCCI	homogeneous-charge compression ignition
HDV	heavy-duty vehicle
HEV	hybrid-electric vehicle
HFCV	hydrogen fuel-cell vehicle
HVAC	heating, ventilation, and air-conditioning
IAC	industrial assessment center
ICE	internal-combustion engine
IDEC	indirect-direct evaporative cooling
IEA	International Energy Agency
IECC	International Energy Conservation Code
IEEE	Institute of Electrical and Electronics Engineers
IPCC	Intergovernmental Panel on Climate Change
IRP	integrated resource planning

IRR	internal rate of return
ITP	Industrial Technologies Program
ITS	intelligent transportation system
IWG	Interlaboratory Working Group on Energy-Efficient and Clean Energy Technologies
kw	kilowatt
kWh	kilowatt-hour
LBNL	Lawrence Berkeley National Laboratory
LDV	light-duty vehicle
LED	light-emitting diode
LEED	Leadership in Energy and Environmental Design
LIPA	Long Island Power Authority
low-E	low-emissivity
LPG	liquefied petroleum gas
MECS	*Manufacturing Energy Consumption Survey*
MEF	modified energy factor
MHDV	medium- and heavy-duty vehicle
mpg	miles per gallon
MW	megawatt
MWh	megawatt-hour
NAE	National Academy of Engineering
NAFTA	North American Free Trade Agreement
NAS	National Academy of Sciences
NEMS	National Energy Modeling System
NGL	natural gas liquid
NIACS	North American Industry Classification System
NO_x	nitrogen oxide
NPV	net present value
NRC	National Research Council
NSPS	New Source Performance Standards
NSR	New Source Review
NYPA	New York Power Authority
NYSEO	New York State Energy Office

NYSERDA	New York State Energy Research and Development Authority
O&M	operations and maintenance
OECD	Organisation for Economic Cooperation and Development
OEM	original equipment manufacturer
OLED	organic light-emitting diode
ORNL	Oak Ridge National Laboratory
PHEV	plug-in hybrid-electric vehicle
PSC	Public Service Commission
PTFE	polytetrafluoroethylene
PUHCA	Public Utilities Holding Company Act
PURPA	Public Utilities Regulatory Policies Act
PV	photovoltaic
R&D	research and development
RD&D	research, development, and demonstration
RHF	rotary hearth furnace
ROI	return on investment
SAF	submerged arc furnace
SBC	system benefits charge
SECP	State Energy Conservation Program
SEP	State Energy Program
SHGC	solar heat gain coefficient
SI	spark-ignition
SIC	Standard Industrial Classification
SO_2	sulfur dioxide
SO_x	sulfur oxide
STC	solar thermal cooling
SUV	sport utility vehicle
SWEEP	Southwest Energy Efficiency Project
TRB	Transporation Research Board
TWh	terawatt-hour

USABC	United States Advanced Battery Consortium
VMT	vehicle-miles traveled
WAP	Weatherization Assistance Program
ZEH	zero-energy home